Studies in Computational Intelligence

Volume 1272

The series "Studies in Computational Intelligence" (SCI) publishes new developments and advances in the various areas of computational intelligence—quickly and with a high quality. The intent is to cover the theory, applications, and design methods of computational intelligence, as embedded in the fields of engineering, computer science, physics and life sciences, as well as the methodologies behind them. The series contains monographs, lecture notes and edited volumes in computational intelligence spanning the areas of neural networks, connectionist systems, genetic algorithms, evolutionary computation, artificial intelligence, cellular automata, self-organizing systems, soft computing, fuzzy systems, and hybrid intelligent systems. Of particular value to both the contributors and the readership are the short publication timeframe and the world-wide distribution, which enable both wide and rapid dissemination of research output.

Indexed by SCOPUS, DBLP, WTI AG (Switzerland), zbMATH, SCImago.

All books published in the series are submitted for consideration in Web of Science.

Angelos Stavrou · Janet Lin · Ruolin Zhou
Editors

Generative and Agentic AI Reliability: Architectures, Challenges, and Trust for Autonomous Systems

Springer

Editors
Angelos Stavrou
Department of Electrical and Computer Engineering
Virginia Tech
Alexandria, Virginia, VA, USA

Ruolin Zhou
Department of Electrical and Computer Engineering
University of Massachusetts Dartmouth
Dartmouth, MA, USA

Janet Lin
Department of Civil, Environment, and Natural Resources Engineering
Luleå University of Technology
Luleå, Sweden

School of Information Technology
Halmstad University
Halmstad, Sweden

ISSN 1860-949X ISSN 1860-9503 (electronic)
Studies in Computational Intelligence
ISBN 978-3-032-18584-6 ISBN 978-3-032-18585-3 (eBook)
https://doi.org/10.1007/978-3-032-18585-3

This Springer imprint is published by the registered company Springer Nature Switzerland AG
The registered company address is: Gewerbestrasse 11, 6330 Cham, Switzerland

Foreword

Generative and Agentic AI Reliability: Architectures, Challenges, and Trust for Autonomous Systems have been organized by members of the IEEE Reliability Society. The intended audience includes standards development organizations, students, faculty, regulators, legislators, industry professionals, and anyone concerned with AI safety. I'll briefly frame the purpose and significance of this book on today's computing environment here.

My first experience with AI goes back to 1985. Since then, I've often wondered when and if AI would finally mature. ChatGPT answered that. This book addresses a fundamental question about the AI technologies that we directly use (such as ChatGPT) or indirectly rely on (and often without knowing it): *Can we trust it*? While this question may seem binary, let's dig deeper by reviewing key terminology to explain why this apparently simple question is complex.

Words have different meanings in different contexts. From a technical context, *reliability* is a quantitative probability, while the word *reliable* is colloquial, e.g., I have a reliable car. Note that the term reliability generally applies well to traditional software and traditional systems. However, AI is not traditional software nor a traditional system.

Reliability can quickly get lost in a sea of other quality attributes that are sometimes referred to as the "ilities": availability, composability, compatibility, dependability, discoverability, durability, fault tolerance, flexibility, interoperability, insurability, liability, maintainability, observability, privacy, performance, portability, predictability, probability of failure, readability, **reliability**, resilience, reachability, safety, scalability, cybersecurity, sustainability, testability, traceability, usability, visibility, and vulnerability. Some are qualitative, and others are quantitative. Many are poorly defined and poorly understood for "general purpose" software. Given that, consider them in the context of AI software. AI systems will rarely be designed with these attributes in mind.

Hardware and software systems fail. Reasons for hardware failure include physical fatigue, faulty logic, and corrupt inputs. Reasons for software failure include faulty logic and corrupt inputs. AI offers new classes of failure and other undesirable outputs such as *misalignment* and *hallucinations*. Misalignment occurs when an AI systems' goals or behaviors do not align with the intended human goals of its designers. This is like the age-old software engineering problem of faulty software specifications and designs from which code is generated. Hallucinations are incorrect, misleading, or entirely fabricated AI-generated information that is confidently presented as factual. To complicate things further, AI's Large Language Models (LLMs) can exhibit nondeterministic behaviors. You can prompt a chatbot on Monday and get an output that is different than the output you get from the same prompt on Tuesday. Why is this? It's because reliability often has a temporal component to it. Between Monday and Tuesday, the chatbot may have learned from new data, and that is the reason for a different output.

Let's now consider AI's various "flavors." AI is a broad discipline in the field of computer science. Machine Learning (ML) is a subdiscipline of AI. ML supports AI's goal of building systems that simulate human intelligence. ML learns from data without being explicitly programmed. Traditional AI is designed and trained to perform a single, specific task, e.g., playing chess. In contrast, *agentic AI* plans, makes decisions, and acts autonomously with limited or no human intervention to pursue specific objectives. Agentic AI is more advanced than traditional AI and has more safety, security, and reliability risks.

Two topics that are implicit in this book's scope are *trust* and *truth*. Trust is a level of belief that you have truth. AI's nondeterminism makes it difficult to argue that AI is trustworthy, and that forces us to create new methods that provide evidence of trust. AI learns from existing knowledge. Whether that knowledge is true may not be known. A lack of trust has led to the idea of Explainable AI (XAI). Explainable AI (XAI) is a research field that focuses on creating methods to make the outputs of machine learning algorithms understandable to humans. Another way of saying this is that we're trying to find methods that make AI's results believable. Further, AI can create untrue knowledge from which other AI systems learn, leading to "AI drift." This "poisoning of the knowledge well" has long-term consequences.

This book is one of the first to compile insights from different writers on how to best frame and to improve agentic AI reliability. We end by acknowledging the risks associated with unreliable, autonomous systems that act on their own with limited or no human supervision. We will always ask questions such as what were the data sources that the AI learned from. We always remember that hallucinations, misalignment and other failures can cause serious outcomes without humans necessarily knowing. Tailoring certification, validation, and verification techniques that are specific to AI is an area that needs further exploration to improve AI's reputation and trustworthiness. Such techniques may be the last line of defense.

Arlington, Virginia, USA Jeffrey Voas

Preface

Over the last few years, generative Artificial Intelligence (AI)—the type of artificial intelligence that learns patterns and structures from massive datasets and then uses that knowledge to generate new content in the form of text, images, audio, code, or other media in response to a prompt has revolutionized the way we think. While this "leap" from mere predictive machine learning models to interactive content generation was seismic in both future research and product development, a second, perhaps even bigger advancement occurred with the introduction of agentic AI. Indeed, AI evolved from isolated perception and prediction tasks to increasingly autonomous, interactive, and goal-driven systems. This transformation is giving rise to a new class of AI—agentic AI—capable of autonomous decision-making, reasoning, planning, memory, and communication. Thus, agentic AI goes beyond just creating content; it can plan, make decisions, and execute multi-step tasks with minimal human intervention. Agentic AI often relies on a generative AI model, such as a Large Language Model (LLM), as its "brain" for reasoning and planning. It then adds an "execution" layer, enabling it to interact with tools, software, and application programming interface (APIs) to accomplish tasks in both the real and digital worlds. Of course, the introduction of autonomous execution, where the AI system can now "do" things increasingly without human oversight, combined with potential access to sensitive information and critical infrastructure operations, raises concerns about its reliability and security. To make matters worse, like many other technological advances in the past, ethical and regulatory frameworks are underdeveloped or non-existent, exacerbating the societal risks associated with the applications of agentic AI. Thus, current governance and technical standards are ill-equipped to address the complex reliability, safety, and trust issues that will emerge as agentic AI systems transition from laboratories to real-world applications.

This book, *Generative and Agentic AI Reliability*, aims to provide the first comprehensive volume that bridges the fields of reliability engineering and the design of generative and agentic AI systems. It explores the reliability and trust challenges that stem from the use of generative AI and agentic AI. While the title emphasizes Generative and Agentic AI Reliability, not every chapter addresses both paradigms equally. Instead, the volume reflects the field's natural diversity. Some chapters focus primarily on agentic AI, especially those discussing architectures, reasoning, coordination, security, and governance. Other chapters focus on generative AI, particularly where reliability depends on the correctness of reasoning, output quality, dataset coverage, or statistical confidence. Several chapters address both paradigms, especially where reliability principles (e.g., monitoring, awareness, standards, human–AI collaboration) cut across system types. This distribution is intentional: generative and agentic AI share foundational layers—including data, models, frameworks, and execution environments—yet introduce different reliability stressors. On the one hand, generative AI adds open-ended output risks (e.g., hallucinations, bias, factual integrity), while on the other, agentic AI adds autonomous action risks (e.g., planning failures, goal misalignment, multi-agent interaction). By allowing chapter authors to engage with the aspects most relevant to their expertise, the book provides a comprehensive and authentic view of reliability across the emerging landscape. The five-part structure ensures coherence while respecting the natural specialization of each contribution.

Furthermore, this book highlights the novel benefits and opportunities of such AI systems, as well as how they can enhance system reliability when used appropriately, including in predictive maintenance, monitoring, and diagnostics. The central theme of the book revolves around questions:

1. How can we ensure the reliability of agentic AI systems themselves?
2. How does the use of generative and agentic AI impact the reliability, security, and trustworthiness of existing systems?
3. What are the opportunities and challenges that we need to address to design reliable and trustworthy AI systems?

To address these thematic questions, we organize the volume around five central themes as shown below:

Foundations

(introduces core concepts and architectures of generative and agentic AI)

What makes an AI system "agentic"? How do different architectural models (e.g., symbolic agents, LLM-driven tool-users, hybrid chains) shape behaviour, failure modes, and transparency?

Chapter "Introduction to Agentic AI"

This chapter introduces agentic AI, an emerging paradigm that extends traditional AI agents into collaborative, adaptive, and multi-agent ecosystems. Moreover, it presents a taxonomy of agentic AI that outlines its core features, application domains, and mitigation strategies, alongside evaluation dimensions such as autonomy, coordination efficiency, and energy footprint.

Chapter "A Survey of Architectures of Agentic Artificial Intelligence Systems"

This chapter presents a comprehensive survey of the key architectural frameworks and models used in the design of agentic AI, including single-agent, multi-agent centralized, and multi-agent decentralized models, and explains how autonomy and decision-making scope vary across them.

Chapter "Architectures for Building Agentic AI"

In this chapter, the authors argue that the reliability of agentic and generative AI is chiefly an architectural property. In addition, it demonstrates how reliability arises from principled componentization (goal manager, planner, tool-router, executor, memory, verifiers, safety monitor, telemetry), disciplined interfaces (schema-constrained, validated, and least-privilege tool calls), and explicit control and assurance loops. This chapter provides design guidance on typed schemas, idempotency, permissioning, transactional semantics, memory provenance and hygiene, runtime governance (including budgets and termination conditions), and simulate-before-actuate safeguards.

Reliability by Design

(focuses on methods, data, reasoning, and security at design time)

What are the key reliability concerns in agentic AI systems, from memory errors and planning faults to goal misalignment and action uncertainty? How can we analyze and mitigate these issues through structured approaches, such as layered failure mode analysis, system-level awareness, explainability, and trust modeling?

Chapter "Measuring ML Dataset Coverage and Completeness: Metrics and Visualizations"

The authors in this chapter introduce new methods for assessing and visualizing training dataset robustness, which is necessary for machine learning to ensure safety, security, and reliability, especially in applications where errors can have serious consequences. The proposed approaches apply to both generative and agentic AI systems. This chapter reviews combinatorial techniques for evaluating dataset coverage and completeness. It illustrates potential applications through a clinical research use case, where data reliability can improve patient care outcomes, reduce operational costs for hospitals, and potentially enhance efficiency in clinical trials and pharmaceutical development.

Chapter "Statistical Confidence and Reliability Estimation for GenAI Reasoning"

In this chapter, the authors investigate the confidence in using GenAI-based models for performing quantitative reliability reasoning, specifically focusing on estimating the shape and scale parameters of Weibull distributions. By using synthetic datasets generated from Weibull models, we design multiple experimental factors, including sample sizes, shape parameters, prompts, and five GenAI models (Claude, DeepSeek, Gemini, ChatGPT, and xAI), to see how various factors impact parameter estimations in the Weibull reliability model. The findings of this chapter highlight the reliability and confidence in using GenAI models for estimating unknown parameters in quantitative reliability engineering analysis.

Chapter "Agentic AI and Cybersecurity: Threats, Defenses, and Open Problems"

This chapter synthesizes the current state of architectures, vulnerabilities, offensive capabilities, governance, mitigations, and open research challenges for agentic AI in cybersecurity. While agentic AI can transform digital operations, its autonomy magnifies risks that require rigorous layered defenses, provenance tracking, runtime enforcement, and progressive autonomy gating to achieve trustworthy deployment.

Applications and Domains

(covers domain-specific reliability cases and cross-industry applications)

What are the applications and use cases for generative and agentic AI? What are the practical implications for short and long-term benefits across industries?

Chapter "Improving Security Vulnerability Descriptions Using LLMs"

In this chapter, the authors offer a tangible and practical application of generative AI to improve security vulnerability descriptions, enabling cybersecurity analysts, critical system operators, and automated analysis systems to enhance their cyber defence capabilities. This structured approach improves the scalability and interpretability of known and future vulnerabilities. The findings demonstrate the potential of current LLMs to enrich vulnerability descriptions and support automated vulnerability analysis.

Chapter "Agentic AI in Practice: Applications Across Industries"

The applications of agentic AI across industrial sectors, including manufacturing, robotics, healthcare, and energy, are detailed in this chapter. Furthermore, this chapter presents case studies that illustrate both the operational benefits and the failure modes of agentic AI in practice. Examples range from production scheduling and human–robot collaboration to task planning, clinical training, drug discovery, and energy optimization. Through these cases, the chapter highlights not only the engineering challenges of deploying autonomous agents at scale but also discusses potential directions for future agentic AI research.

Chapter "Software Engineering Reliability in the Era of Generic and Agentic AI"

In this chapter, the authors present a comprehensive analysis of empirical evidence on both productivity gains and emergent risks to software engineering reliability, highlighting significant productivity paradoxes and trade-offs that depend on developer experience and task complexity. In addition, it introduces a novel taxonomy of AI-induced failure modes, including specification drift, pervasive non-determinism, security regressions, and evaluation contamination, which threaten the validity of performance claims. To address some of these concerns, this chapter provides a structured overview of emerging engineering controls, such as advanced reasoning frameworks, verifier-in-the-loop patterns, and observability pipelines, designed to improve reliability in inherently probabilistic systems. While AI acts as a powerful accelerator for many software development tasks, the focus of Software Reliability Engineering (SRE) must evolve from analyzing static, human-authored artifacts to managing the dynamic, unpredictable processes of the AI agents that create them.

Operations and Human Factors

(addresses operational assurance, monitoring, failure awareness, and collaboration with humans)

How should we evaluate and quantify reliability risks across the AI stack? What are the fundamental challenges in traditional AI systems, and how can we include a human operator in the loop? Is human–AI collaboration an answer to measuring, assessing, or restricting the execution of agentic, emergent AI systems?

Chapter "From Failure Modes to Reliability Awareness in Generative and Agentic AI System"

This chapter bridges the gap between technical analysis and organizational preparedness by tracing the path from layered failure modes to reliability awareness in generative and agentic AI systems. It introduces an 11-layer failure stack, a structured framework for identifying vulnerabilities ranging from hardware and power foundations to adaptive learning and agentic reasoning. Building on this, the chapter demonstrates how failures rarely occur in isolation but propagate across layers, creating cascading effects with systemic consequences. To complement this diagnostic lens, the concept of awareness mapping is introduced: a maturity-oriented framework that quantifies how well individuals and organizations recognize reliability risks across the AI stack. Awareness is treated not only as a diagnostic score but also as a strategic input for AI governance, guiding improvement and resilience planning. The chapter positions awareness mapping as both a measurement tool and a roadmap for the trustworthy and sustainable deployment of AI across mission-critical domains.

Chapter "Perspectives on a Reliability Monitoring Framework for Agentic AI Systems"

The main reliability challenges of agentic AI systems during operation, based on their characteristics, are analyzed in this chapter. The chapter makes a connection to traditional AI systems and formulates a fundamental reliability challenge during operation, which is inherent to traditional and agentic AI systems. The authors propose a two-layered reliability monitoring framework for agentic AI systems, which consists of an out-of-distribution detection layer for novel inputs and an AI transparency layer to reveal internal operations. This two-layered monitoring approach provides a human operator with the decision support needed to determine whether an output is potentially unreliable and to intervene. This framework provides a foundation for developing mitigation techniques to reduce risk stemming from uncertain reliability during operation.

Chapter "Human-AI Collaboration"

This chapter focuses on human-to-AI collaboration and the lack of a successful method for monitoring, measuring, assessing, or restricting the execution of agentic, emergent AI systems. In a world of billions of interconnected devices of all types, sizes, and configurations running agentic AI models, the potential for serious harm to humans and the environment is very real. Increasingly, AI agents will be deployed on devices of all kinds—encompassing all devices that comprise the global Internet of Things.

Governance and Future Directions

(situates reliability within standards, policy, and long-term challenges)

How should we evaluate and quantify the reliability and trust of agentic AI systems in safety-critical contexts? What roles do explainability, regulation, and human-AI collaboration play in building accountable, lawful, and robust systems?

Chapter "Agentic AI and Cybersecurity: Threats, Defenses, and Open Problems" discusses the significant and growing gap between emerging governance frameworks, such as the NIST AI Risk Management Framework and the EU AI Act, and the technical, enforceable controls required to mitigate these new threats.

Chapter "Standards, Regulation, and Governance for Agentic AI"

The standards, regulations, and governance of agentic AI are discussed in more detail in this chapter. The author argues that machine learning AI systems have intrinsic flaws that theoretically limit their ability to be trustworthy. Of course, trustworthiness is a subjective term that requires both qualification and quantification. Nevertheless, one can generally include the following criteria to describe trustworthiness: Accuracy, Precision, Reliability, Dependability, Truthfulness, Verifiability, Safety, and Security. One significant problem is the lack of quality (in terms of accuracy, truthfulness, and consistency) in the data used to train AI models. The old computer science adage "garbage in, garbage out" is an oft-cited synopsis of the data quality problem in AI, and it is very germane and appropriate. The chapter argues that to chart the right course for AI adoption, human decision-makers must be informed about both the theoretical limitations of AI and the real-world engineering considerations, as well as societal impacts such as trustworthiness. It is humans who must remember that humans build things to improve the lives of humans and other creatures, not to create more risk of danger or harm.

Chapter "Looking Forward: Challenges and Opportunities in Agentic AI Reliability"

The perspectives on challenges and future development in building reliable AI systems, particularly agentic AI systems, are presented in this chapter. Several open research problems related to mitigating the risks of cascading failures are discussed. The chapter also sheds light on research challenges and opportunities in aspects including dynamic environments, inconsistent task execution, unpredictable emergent behaviors, and resource-intensive reliability mechanisms. In addition, several research directions related to testing and evaluating the reliability of agentic AI systems are also discussed.

Our intended audience includes:

- Reliability and AI engineers working in safety-critical and autonomous system development.
- Graduate students and early-career researchers in AI, systems engineering, and robotics.
- Technology leaders and policymakers involved in deploying or regulating AI systems in the public domain.
- Standardization experts and auditors navigating the governance of agentic systems under evolving legal frameworks (e.g., EU AI Act, IEEE P7000 series).

In assembling this volume, our goal is not only to clarify the technical landscape, but also to support the development of trustworthy, resilient AI systems that can serve societal needs responsibly.

We thank all contributors, reviewers, and collaborating societies—especially the IEEE Reliability Society—for their efforts in shaping this timely and necessary conversation.

Virginia, USA — Angelos Stavrou
Luleå, Sweden — Janet Lin
Dartmouth, USA — Ruolin Zhou

Contents

Foundations

Introduction to Agentic AI

George Pallis

Abstract This chapter introduces Agentic Artificial Intelligence (AI), an emerging paradigm that extends traditional AI agents into collaborative, adaptive, and multi-agent ecosystems. Building upon the advanced language processing and problem-solving capacities of large language models (LLMs), Agentic AI systems are characterized by multi-agent collaboration, dynamic task decomposition, persistent memory, coordinated autonomy, and emergent behaviors. These capabilities enable applications while also amplifying challenges related to reliability, security, and sustainability. We present a taxonomy of Agentic AI that outlines its core features, application domains and mitigation strategies, alongside evaluation dimensions such as autonomy, coordination efficiency, and energy footprint. Reliability is identified as a foundational requirement, given the risks of cascading failures, inconsistent execution, and unpredictable emergent outcomes. Trust and human-centered perspectives are equally essential, ensuring transparency, accountability, fairness, and user empowerment in real-world deployments. Together, these perspectives highlight Agentic AI as both a technological and societal shift, requiring architectures that balance autonomy with oversight, efficiency with robustness, and innovation with ethical responsibility.

Keywords Agentic AI · Large language models · Reliability

1 Introduction

AI agents are computational systems capable of exhibiting intelligent behaviors through autonomy, reactivity, proactiveness, and social interaction (Deng et al. 2025). They operate by continuously perceiving inputs from their environment or users, processing these inputs with statistical and heuristic methods, formulating courses of action, and executing tasks through both internal mechanisms and external tools.

G. Pallis (✉)
Computer Science Department, University of Cyprus, Nicosia, Cyprus
e-mail: pallis.george@ucy.ac.cy

A. Stavrou et al. (eds.), *Generative and Agentic AI Reliability: Architectures, Challenges, and Trust for Autonomous Systems*, Studies in Computational Intelligence 1272, https://doi.org/10.1007/978-3-032-18585-3_1

The integration of LLMs has significantly expanded the capabilities of AI agents, enabling them to tackle tasks that resemble advanced reasoning, contextual understanding, and multi-step problem-solving, even if these processes differ fundamentally from human reasoning. Today, these agents are applied in several domains such as healthcare (Karunanayake 2025), where they assist in diagnostics and patient monitoring; finance (Cao 2023), for risk analysis, fraud detection, and automated trading; customer service, for managing inquiries and providing real-time support; and agent operating systems, where they orchestrate complex workflows and resource allocation. By leveraging pattern recognition, planning heuristics, and action mechanisms provided by LLMs, AI agents can handle intricate tasks with efficiency, accuracy, and adaptability.

Moving beyond AI agents, Agentic AI systems represent a paradigm shift in AI (Murugesan 2025). Unlike AI agents, which typically operate individually and follow predefined workflows, Agentic AI is characterized by multi-agent collaboration, dynamic task decomposition, persistent memory, and coordinated autonomy. These features allow Agentic AI systems to manage more sophisticated tasks, adaptively allocate responsibilities among multiple agents, and maintain stateful knowledge across interactions, creating emergent capabilities not possible in single-agent frameworks.

However, as Agentic AI systems grow more complex, ensuring reliability becomes increasingly critical. Agentic AI reliability refers to the system's ability to consistently perform its intended functions under varying conditions, including unexpected inputs or environmental changes. Reliable agents must not only execute tasks correctly but also recover gracefully from errors, maintain continuity of operations, and provide predictable and trustworthy outputs. When combined with security considerations, protecting against malicious exploitation, unintended behaviors, or data compromise, reliability forms a foundational pillar for safe deployment in critical applications. In addition, human-centered reliability requires careful design of mechanisms. While human-in-the-loop approaches can improve scalability, they may reduce accountability. Transparency and explainability tools can help address this by giving users a clearer understanding of how autonomous agents make decisions.

Roadmap of the Chapter. Section 2 introduces a taxonomy of Agentic AI systems, followed by Sect. 3, which explores their reliability challenges. Section 4 outlines a layered architecture for Agentic AI systems, while Sect. 5 examines trust-related and human-centered perspectives. Section 6 highlights future directions in Agentic AI and reliability. Finally, Sect. 7 concludes the chapter.

2 Characteristics of Agentic AI

Agentic AI represents AI systems that extend beyond the capabilities of traditional AI agents. A key feature of Agentic AI is multi-agent collaboration, where multiple AI agents work together to achieve shared objectives. Each agent can operate independently, but the system as a whole coordinates actions, shares knowledge, and resolves conflicts to optimize overall performance. This allows Agentic AI to manage tasks that are complex for a single agent, such as orchestrating a fleet of robots, conducting multi-step research automation, or integrating insights from diverse data sources in medical decision support (Neupane et al. 2024).

Another characteristic is dynamic task decomposition and persistent memory. Agentic AI can break down complex tasks into smaller sub-tasks, assign them to the most appropriate agents, and adaptively reassign responsibilities as circumstances change. Persistent memory enables the system to retain knowledge across interactions, providing continuity, long-term planning, and improved decision-making over time. This allows the system to respond to evolving environments and anticipate future needs rather than reacting solely to immediate inputs.

Despite their potential, Agentic AI systems also face challenges. Coordinating multiple agents can lead to conflicts or inefficiencies if not properly managed, and emergent behaviors may arise that are unpredictable. Additionally, security and reliability concerns are amplified as agents interact with external systems or untrusted entities. Addressing these challenges requires strategies such as automation coordination layers, causal modeling, and robust multi-agent communication protocols. Overall, Agentic AI represents a paradigm shift toward interconnected, adaptive, and autonomous multi-agent ecosystems capable of tackling tasks of unprecedented complexity.

Table 1 presents a taxonomy of Agentic AI, within the broader evolution of AI systems. At the foundational layer, Generative AI (GenAI) provides the essential capabilities for language-based processing, content generation, and context-sensitive interaction, supporting behaviors that resemble reasoning and form the building blocks of more advanced agentic systems. Building on this foundation, Agentic AI is characterized by multi-agent collaboration, dynamic task decomposition, persistent memory, coordinated autonomy, and emergent behavior, enabling systems to manage complex objectives. The taxonomy further highlights the application domains where these systems are deployed, ranging from knowledge work automation and research workflows to robotics, healthcare, and simulation environments. Alongside opportunities, the table categorizes the distinct security and reliability challenges (Chan et al. 2024). To address these concerns, it outlines mitigation mechanisms such as coordination layers, retrieval-augmented generation (RAG), causal modeling, and monitoring protocols. Recognizing the growing importance of efficiency, the taxonomy also incorporates energy and sustainability aspects, emphasizing the need for computational efficiency, resource allocation, and carbon-aware practices. Finally, it identifies evaluation dimensions, such as autonomy level, coordination efficiency, reliability,

Table 1 A taxonomy of Agentic AI

Category	Subcategory	Description/examples
Foundational layer	Generative AI (GenAI)	Provides the foundation for Agentic AI by enabling content generation and language-based interaction. LLMs such as GPT-4 serve as core engines that power agents' communication. Example: text summarization, code generation, or dialogue
Core characteristics	Multi-agent collaboration	Multiple autonomous agents coordinate actions, share knowledge, and resolve conflicts to achieve collective objectives. Example: supply-chain optimization with multiple specialized agents
	Dynamic task decomposition	Complex tasks are broken into smaller sub-tasks, assigned to appropriate agents, and adaptively reassigned as conditions evolve. Example: multi-step research workflow automation
	Persistent memory	Agents retain knowledge and context across sessions, enabling long-term planning, continuity, and adaptive decision-making
	Coordinated autonomy	Agents act independently but follow systemic alignment rules to ensure collective objectives are met. Example: autonomous vehicles cooperating at intersections
	Emergent behavior	Adaptive behaviors arise from agent-agent and agent-environment interactions. Example: multi-agent game simulations discovering unplanned strategies
Security challenges	Adversarial inputs	Malicious or ambiguous instructions disrupting processing or execution
	Untrusted external entities	Malicious agents or systems attempting to manipulate outcomes
	Coordination vulnerabilities	Failures in inter-agent communication leading to deadlocks or conflicts
	Data and memory tampering	Compromising persistent memory or data stores to mislead future processing or decision-making
Reliability challenges	Propagation of failures	Errors in one agent cascading to dependent agents
	Brittleness under novel scenarios	Difficulty adapting to unforeseen or dynamic contexts
	Inconsistent task execution	Variability in outputs across agents or repeated runs
	Emergent behavior risks	Unpredictable, sometimes harmful system-level outcomes

(continued)

Table 1 (continued)

Category	Subcategory	Description/examples
Energy and sustainability aspects	Computational efficiency	Reducing training and inference costs through model compression, pruning, or distillation
	Resource allocation	Optimizing energy use across multi-agent deployments. Example: distributed scheduling that minimizes redundant computation
	Green AI practices	Incorporating carbon-aware scheduling, renewable energy integration, and hardware-level energy optimization
	Trade-offs between reliability and energy	Balancing redundancy for reliability with energy-efficient design
Mitigation mechanisms	Coordination layers	Middleware ensuring synchronization, alignment, and conflict resolution among AI agents
	Causal modeling	Linking actions to outcomes to improve predictability and reduce unintended effects
	ReAct loops	Iterative reasoning-and-acting cycles that enhance accuracy and robustness
	Retrieval-augmented generation (RAG)	Using external knowledge bases to reduce hallucinations and improve decision-making
	Monitoring and recovery protocols	Early detection of errors and graceful recovery to maintain stable performance
	Energy-aware scheduling	Adaptive scheduling of agents to minimize energy consumption without degrading reliability
Evaluation dimensions	Autonomy level	Independence versus collaboration in agent operations
	Task complexity	Breadth and depth of tasks managed
	Coordination efficiency	Effectiveness of multi-agent cooperation
	Reliability and robustness	Stability and predictability across scenarios
	Security and trustworthiness	Resistance to adversarial inputs and malicious entities
	Energy footprint	Measuring efficiency, scalability, and sustainability of deployments

trustworthiness, and energy footprint, which provide a framework for assessing and comparing Agentic AI systems across research and practice (Gaur and Sheth 2024).

3 Reliability in Agentic AI

Reliability in Agentic AI is critical, as the performance of the system depends on the predictable execution of tasks by multiple interacting agents. Each agent's actions influence the outcomes of others, meaning that failures in one component can propagate and disrupt the entire system. Reliable Agentic AI ensures that agents consistently perform their assigned responsibilities, maintain accurate memory and reasoning-like capabilities over long-term interactions, and recover from errors. This is essential for maintaining coordination, achieving complex objectives, and supporting adaptive, context-aware decision-making across dynamic and uncertain environments.

Table 2 summarizes the key reliability challenges in Agentic AI and outlines corresponding perspectives and solutions to address them. As multi-agent systems grow in complexity and autonomy, ensuring dependable performance becomes increasingly difficult due to risks such as cascading failures in unfamiliar contexts, inconsistent execution, and unpredictable emergent behaviors. Each of these challenges threatens the stability and trustworthiness of Agentic AI in real-world deployments, from healthcare and robotics to research automation. To mitigate these risks, researchers and practitioners are exploring mechanisms such as coordination layers, causal modeling, retrieval-augmented processing, monitoring protocols, and energy-aware designs. By aligning challenges with targeted strategies, the table provides a roadmap for building Agentic AI systems that balance robustness, adaptability, and sustainability, ultimately paving the way for reliable and trustworthy large-scale deployments (Mehrotra et al. 2024).

4 Agentic AI Architecture

The architecture of Agentic AI represents a paradigm shift from traditional single-agent systems toward dynamic, multi-component ecosystems designed to support autonomy, adaptability, and collaboration. Unlike generative AI models that primarily focus on content creation or single-task execution, Agentic AI architectures integrate multiple modules and agents, each with distinct roles, while enabling coordination across diverse tools, data sources, and environments. These architectures are built in layers, can be extended, and are designed to manage objectives even when there is uncertainty.

Recently it was proposed an 8-layer Agentic AI architecture (Gupta 2025). The foundation of Agentic AI systems lies in the Infrastructure Layer and the Agent Internet Layer. The infrastructure layer provides the essential computational backbone (APIs, cloud services, data centers, and storage) enabling scalable and reliable operation. Building on this, the agent Internet layer establishes the connective fabric that allows autonomous agents to communicate, exchange memory, and interact with

Table 2 Reliability challenges in Agentic AI

Challenges in reliability	Perspectives/solutions
Propagation of failures: Errors in one agent can cascade across dependent agents, amplifying system-wide disruptions	**Coordination layers and checkpoints**: Introduce structured coordination mechanisms and validation points to detect and contain localized failures before they spread
Vulnerability in dynamic environments: Difficulty adapting to unexpected contexts, ambiguous inputs, or adversarial prompts	**Causal modeling and explainability**: Enable agents to process about cause-effect relationships and make decision-making more transparent and adaptable
Inconsistent task execution: Outputs vary across repeated runs or among agents handling similar sub-tasks, reducing predictability	**Iterative processing–action loops and retrieval-augmented generation**: Ground decision-making in external knowledge and iterative action-validation cycles to ensure consistent performance
Emergent behavior risks: Unintended or harmful system-level outcomes may arise from complex multi-agent interactions	**Monitoring and recovery protocols**: Implement redundancy, fallback strategies, and error-correction layers to manage unexpected emergent effects
Resource-intensive reliability mechanisms: High redundancy or over-engineering can lead to excessive energy consumption	**Energy-aware reliability design**: Balance robustness with computational efficiency through adaptive scheduling, pruning, and carbon-aware strategies

digital environments. Together, these two layers form the substrate upon which intelligent behaviors can emerge, ensuring that agents are not isolated silos but participants in a broader ecosystem.

On top of this foundation are the Protocol Layer and the Tooling and Enrichment Layer, which support interoperability and add extra functionality. The protocol layer defines the rules of interaction between agents and with external systems, relying on standards such as agent-to-agent communication protocols, negotiation protocols, and the Model Context Protocol (MCP). On top of that, the tooling and enrichment layer equips agents with advanced capabilities like retrieval-augmented generation (RAG), vector databases, function calling, and knowledge databases. These components act as cognitive extensions, allowing agents to move beyond their basic processing by using specialized resources and external computation.

Next in the stack are the Cognition and Reasoning Layer and the Memory and Personalization Layer. The cognition layer is where agents "think," encompassing decision-making, planning, reasoning engines, error handling, and goal management. This is complemented by the memory and personalization layer, which provides short- and long-term memory, behavioral modeling, and user preferences. These layers give agents the ability to personalize interactions, learn from experience, and sustain coherent long-term engagements. In combination, they enable agents not only to solve immediate tasks but also to evolve and adapt over time.

At the top of the stack are the Application Layer and the Ops and Governance Layer, which translate technical capabilities into real-world use and ensure safety, accountability, and scalability. The application layer covers a wide range of use cases such as personal assistants, research bots, e-commerce agents, and collaborative productivity tools. On top of these is the Ops and Governance Layer, which handles observability, compliance, auditing, trust, as well as cost and resource management (Shavit et al. 2023). Figure 1 presents the 8-layer architecture of Agentic AI.

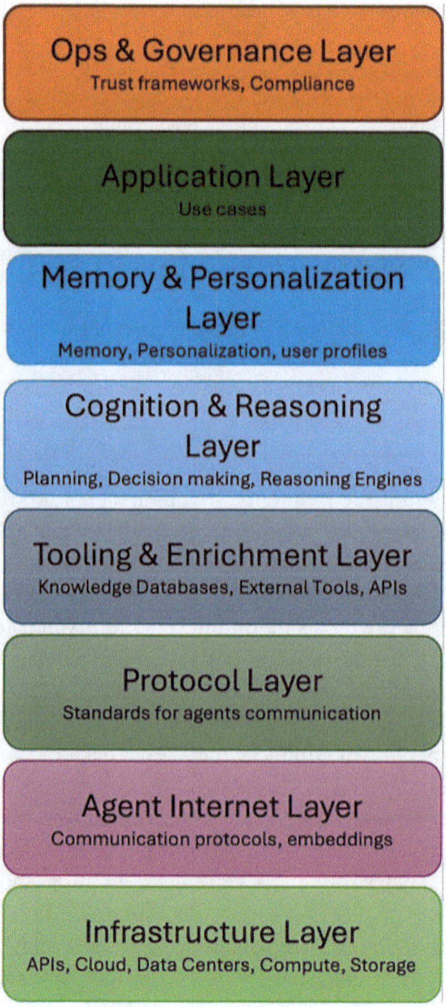

Fig. 1 Layered architecture for Agentic AI

5 Trust and Human-Centered Perspectives

Trust is a fundamental aspect for the adoption of Agentic AI systems. While Agentic AI systems are capable of autonomous decision-processing, task decomposition, and multi-agent collaboration, their deployment in sensitive domains, such as healthcare requires that users perceive them as reliable, transparent, and aligned with human values. Trust in Agentic AI goes beyond technical robustness: it covers explainability and accountability (Moruzzi and Margarido 2024).

From a human-centered perspective, Agentic AI must be designed to augment human capabilities rather than replace them. This involves ensuring that users can meaningfully interpret system decisions, retain control in critical contexts, and understand the trade-offs between autonomy. Building trust also requires mechanisms for fairness, privacy, and inclusivity, ensuring these systems serve all users without reinforcing biases or inequalities.

Equally important is the reciprocal relationship between trust and reliability; users will rely on Agentic AI if systems consistently demonstrate dependable behavior, and in turn, reliable systems are more effective when supported by user trust and engagement. Building such alignment requires continuous feedback loops, participatory design approaches, and governance frameworks that make AI systems accountable to human values and societal norms.

Table 3 provides a structured summary of the core dimensions of trust and human-centered perspectives in Agentic AI, outlining the key considerations and strategies for embedding them into system design and deployment.

Trust in Agentic AI is bound to its reliability in real-world applications. While technical robustness ensures that systems can function as designed, trust emerges when users perceive this reliability consistently across domains. In practice, the requirements for reliability vary depending on the application domain, shaping both human expectations and design strategies.

In healthcare, reliability translates into the system's ability to provide accurate, safe, and context-sensitive recommendations (Huang et al. 2025). Physicians and patients need to trust that the AI agent will not only interpret medical data correctly but also flag uncertainty and avoid hallucinations that could endanger lives. Thus, reliability requires rigorous validation against clinical standards, transparent decision-making paths, and the ability for human professionals to override or cross-check the agent's output. Trust is related to auditability and compliance with ethical and regulatory frameworks.

In finance, reliability focuses on consistency, robustness and resilience to adversarial manipulation. Financial agents must handle volatile environments where small errors can cascade into major losses. Users need confidence that the system can adapt to dynamic conditions while safeguarding against biases and exploitation. Trust in this domain is strengthened by continuous monitoring, stress-testing under worst-case scenarios, and explainability mechanisms that allow stakeholders to understand the rationale behind investment strategies or risk assessments.

Table 3 Trust and human-centered perspectives in Agentic AI

Dimension	Description	Strategies/approaches
Transparency and explainability	Ability of the system to make decisions understandable to users	Use natural language explanations and visualization tools
User control and oversight	Ensuring humans remain in the loop, especially in critical or high-stakes contexts	Provide adjustable autonomy levels
Fairness and inclusivity	Guaranteeing equitable treatment and accessibility across diverse user groups	Bias detection, diverse training data
Accountability and governance	Mechanisms to attribute responsibility and ensure compliance with regulations	Establish audit frameworks, ethical guidelines and clear liability models
Privacy and security	Protection of user data and secure handling of sensitive information	Federated/swarm learning, differential privacy and encrypted communications
Trust through reliability	Building confidence by demonstrating consistent, dependable behavior	Reliability testing, fault-tolerant design and resilience engineering
Human empowerment	Supporting humans in decision-making rather than replacing them	Design AI as collaborative assistants and enable user feedback

In robotics and autonomous systems, reliability involves performance under uncertain, dynamic, and often safety–critical environments. Robots coordinated by Agentic AI agents must reliably perceive their surroundings, decompose tasks, and collaborate with both humans and machines without failure. For example, in manufacturing or disaster response, trust is earned when agents consistently achieve goals without unexpected breakdowns, while offering human operators visibility into decision-making and safe intervention points. Reliability in these contexts also includes real-time fault tolerance and energy-efficient operation, ensuring systems remain dependable even under constrained resources (Methnani et al. 2024).

To sum up, users are unlikely to rely on Agentic AI systems that are unpredictable or prone to errors. Reliable systems that demonstrate resilience, transparency, and accountability foster long-term trust and adoption. Table 3 presented earlier, synthesizes the core human-centered dimensions that enable the trust–reliability relationship, offering design principles for building AI systems that are not only technically sound but also socially acceptable.

6 Future Perspectives

The future of Agentic AI will be shaped by its ability to balance autonomy, reliability, and human-centered design. As these systems advance beyond task execution toward collaborative reasoning-like processes and adaptive problem-solving, ensuring trustworthy outcomes will be the key enabler of broad adoption. To achieve this, research and development must advance across several key domains. Efficient learning methods will allow agents to generalize across tasks with minimal data, while integration with emerging technologies, including IoT, blockchain, and edge/fog/cloud computing, will enable scalable and resource-efficient deployment. Multi-agent ecosystems, in which agents collaborate and coordinate dynamically, will unlock new forms of distributed intelligence, though they require robust synchronization and conflict resolution. Equally important, ethical and explainable AI frameworks will be needed to strengthen human–AI collaboration, fostering accountability and trust.

Agentic AI could accelerate scientific discovery by helping with experiments, analysis, and generating hypotheses. However, these capabilities could also produce low-quality or harmful results if not carefully managed. This dual-use nature highlights the need for proactive governance. Governments, industry, developers, researchers, and academic institutions must work together to set standards, safeguards, and educational initiatives to ensure responsible development and use.

Table 4 highlights how future progress in Agentic AI depends on coordinated advancements in both technology and governance. Each research direction introduces reliability challenges that require targeted responses from different stakeholders.

Table 4 Future perspectives in Agentic AI and reliability

Research direction	Reliability challenge	Stakeholder role
Efficient learning	Risk of poor generalization or data scarcity	Researchers to design robust methods; developers to validate models in diverse tasks
Multi-agent ecosystems	Coordination failures, emergent unintended behaviors	Industry leaders to define protocols; policymakers to set safety standards
Human–AI collaboration	Lack of explainability	Developers to build transparent interfaces; academia to train AI-literate users
Ethical and transparent AI	Bias, fairness, accountability	Governments to regulate; researchers to embed ethics in design
Technological convergence	Security, interoperability, energy efficiency issues	Industry to ensure secure integration; policymakers and industry to align standards
Scientific discovery with AI	Misuse, low-quality research outputs	Academia and researchers to design novel tools

7 Conclusion

Agentic AI represents an evolution in AI, moving beyond individual agents toward interconnected, adaptive, and collaborative multi-agent ecosystems. Built upon the processing capabilities of large language models, Agentic AI systems are distinguished by their ability to decompose complex tasks, share knowledge across agents, maintain persistent memory, and exhibit emergent behaviors that surpass the limitations of single-agent frameworks. These capabilities open opportunities in diverse domains, from healthcare and finance to robotics and research automation, while simultaneously introducing new challenges of reliability, security, and sustainability.

Reliability is key for Agentic AI, ensuring that multi-agent systems perform as intended, recover from errors, and stay stable even in changing or challenging conditions. Without reliability, Agentic AI risks cascading failures, inconsistent performance, or unexpected behaviors. Addressing these issues requires carefully designed coordination layers, causal reasoning, monitoring protocols, and energy-aware strategies that balance robustness with efficiency.

Equally important is the human-centered aspect. Trust comes not only from technical skill, but from transparency, accountability, fairness, and empowering users. Agentic AI systems must be designed to complement human decision-making, provide interpretable reasoning, and embed safeguards that reflect societal values. By aligning technical robustness with ethical and human-centered principles, these systems can foster meaningful trust and long-term adoption. To sum up, the topics discussed in this chapter offer a roadmap for understanding Agentic AI as both a technological and societal shift. The interaction of autonomy, collaboration, reliability, and trust will shape not only these systems' architectures but also their impact in the real world.

Acknowledgements Language refinements at the sentence and paragraph level were supported by ChatGPT.

References

Cao L (2023) AI in finance: challenges, techniques, and opportunities. ACM Comput Surv 55(3):1–38. Article 64. https://doi.org/10.1145/3502289

Chan A et al (2024) Visibility into AI agents. In: Proceedings of the 2024 ACM conference on fairness, accountability, and transparency (FAccT '24). Association for Computing Machinery, New York, NY, USA, pp 958–973. https://doi.org/10.1145/3630106.3658948

Deng Z et al (2025) AI agents under threat: a survey of key security challenges and future pathways. ACM Comput Surv 57(7):1–36. Article 182. https://doi.org/10.1145/3716628

Gaur M, Sheth A (2024) Building trustworthy NeuroSymbolic AI systems: consistency, reliability, explainability, and safety. AI Mag 45:139–155. https://doi.org/10.1002/aaai.12149

Gupta A (2025) The 8-layer architecture of agentic AI—a comprehensive guide to autonomous intelligence systems. Medium, 8 June 2025. https://aakashgupta.medium.com/the-8-architectural-layers-of-agentic-ai-a-complete-guide-for-product-managers-6794d75ac988

Huang X et al (2025) Recommender AI agent: integrating large language models for interactive recommendations. ACM Trans Inf Syst 43(4):1–33. Article 96. https://doi.org/10.1145/3731446

Karunanayake N (2025) Next-generation agentic AI for transforming healthcare. Inform Health 2(2):73–83. ISSN 2949-9534. https://doi.org/10.1016/j.infoh.2025.03.001

Mehrotra S, Centeio Jorge C, Jonker CM, Tielman ML (2024) Integrity-based explanations for fostering appropriate trust in AI agents. ACM Trans Interact Intell Syst 14(1):1–36. Article 4. https://doi.org/10.1145/3610578

Methnani L, Chiou M, Dignum V, Theodorou A (2024) Who's in charge here? A survey on trustworthy AI in variable autonomy robotic systems. ACM Comput Surv 56(7):1–32. Article 184. https://doi.org/10.1145/3645090

Moruzzi C, Margarido S (2024) A user-centered framework for human-AI co-creativity. In: Extended abstracts of the CHI conference on human factors in computing systems (CHI EA '24). Association for Computing Machinery, New York, NY, USA, pp 1–9, Article 25. https://doi.org/10.1145/3613905.3650929

Murugesan S (2025) The rise of agentic AI: implications, concerns, and the path forward. IEEE Intell Syst 40(2):8–14. https://doi.org/10.1109/MIS.2025.3544940

Neupane S et al (2024) Security considerations in AI-robotics: a survey of current methods, challenges, and opportunities. IEEE Access 12:22072–22097. https://doi.org/10.1109/ACCESS.2024.3363657

Shavit Y et al (2023) Practices for governing agentic AI systems. Research paper, OpenAI. 15 Dec 2023

A Survey of Architectures of Agentic Artificial Intelligence Systems

Ruolin Zhou, Charles Montes, Eric Savage, Cameron Popillo, and Bavly Aziz

Abstract The rapid rise of agentic artificial intelligence (AI) is poised to revolutionize various fields by enabling AI solutions to evolve toward self-directed orchestration and complete more complex tasks with greater autonomy. However, designing and building robust and effective agentic systems presents a significant challenge. This paper provides a comprehensive survey of the key architectural frameworks and models used in the design of agentic AI. We investigate the fundamental models, including single-agent, multi-agent centralized, and multi-agent decentralized, explaining how autonomy and decision-making scope vary across them. Building on these, we categorize and compare specialized architectural frameworks, including vertical (hierarchical) and horizontal (decentralized) models, and emerging hybrid approaches, discussing their strengths, weaknesses, and structural implications for agent execution. The survey also provides an assessment of popular open-source frameworks, such as LangGraph, CrewAI, and AutoGen, conducting a comparative analysis across dimensions such as autonomy, scalability, memory integration, and explainability to offer insight into their functionalities and appropriate use cases. The paper also identifies open research questions, particularly concerning the scalability of hierarchical frameworks and the need for greater transparency in agent reasoning, and emphasizes the importance of understanding these architectures for the future development of trustworthy and reliable agentic AI systems.

R. Zhou (✉) · C. Montes · E. Savage · C. Popillo · B. Aziz
University of Massachusetts, Dartmouth, MA, USA
e-mail: ruolin.zhou@umassd.edu

C. Montes
e-mail: cmontes@umassd.edu

E. Savage
e-mail: esavage1@umassd.edu

C. Popillo
e-mail: cpopillo@umassd.edu

B. Aziz
e-mail: baziz@umassd.edu

A. Stavrou et al. (eds.), *Generative and Agentic AI Reliability: Architectures, Challenges, and Trust for Autonomous Systems*, Studies in Computational Intelligence 1272, https://doi.org/10.1007/978-3-032-18585-3_2

Keywords Agentic AI · Agentic systems · Multi-agent · Single-agent · Role based agents

1 Introduction

Artificial intelligence (AI) has become an integral part in many modern technical solutions and as technology continues to evolve AI solutions must evolve with the demands of the technical solutions. However, the AI evolves alongside the technology and the direction it moves towards is self-directed orchestration by utilizing the concepts of agents. The utilization of agents allows AI to complete more complicated tasks by giving the individual agent autonomy over its actions to perform tasks that are specific to the assigned agent. An agentic approach to AI allows for scalable use of the agents allowing for teams of agents to perform multiple complicated tasks concurrently. Various architectures of agentic AI, such as single-agent, multi-agent, vertical, and horizontal architectures, have been developed recently and allow research teams to design their own agents through open-source or proprietary means.

The essence of agentic AI is its autonomy, the ability to work on one's own will and ability to produce results. The previously mentioned architectures manifest their autonomy for different use cases, defining how an agent perceives its environment, makes decisions, and executes tasks. Single-agent architectures internalize this process, with all reasoning and action embedded within a single self-contained model. Frameworks such as ProAgent were designed to demonstrate how a proactive single agent could independently plan, reason, and revise its behavior without external orchestration (Zhang et al. 2024). ProAgent's design introduced a reflective feedback mechanism which allows the agent to refine its outputs through self-evaluation loops rather than external correction. This architecture emphasized vertical layering within one entity—perception, planning, and execution—illustrating how a single model can simulate multi-step reasoning pipelines internally. Such design allows for more precise control and optimization but also reveals scalability limitations when applied to complex or distributed problems that demand shared awareness or knowledge bases among multiple autonomous units.

As the demand for distributed reasoning and coordination increases, multi-agent architectures have become the foundation for scalable autonomy. These systems allow multiple specialized agents to interact and collaborate either under a centralized orchestrator or through decentralized communication. Centralized or vertical architectures, such as AutoGen and MOMA, rely on a supervisory controller that assigns roles and manages dependencies, ensuring global consistency and safety across tasks (Wu et al. 2024; Xu et al. 2025). Conversely, horizontal architectures, exemplified by Large Language Model (LLM)-Assisted Multi-Agent Dialogue for Ontology Alignment and BlockAgents, replace the orchestrator with peer-based negotiation, allowing agents to communicate freely and coordinate through shared protocols or blockchain-based consensus (Zhang et al. 2024a; Jiang et al. 2024). Hybrid approaches such as Synergistic Multi-Agent Framework with Trajectory

Table 1 Various applications

References	Single agent	Multi-agent centralized	Multi-agent decentralized	Vertical	Horizontal	Hybrid
Roumeliotis et al. (2025)		X		X		
Zhang et al. (2025)		X		X		
Zhang et al. (2024b)						X
Multi-agent LLMs (2025)		X	X		X	
Microsoft Research (2024)		X		X		
Microsoft (2025)		X		X		
IBM (2025)	X	X		X		
Bai et al. (2026)			X		X	
Bhalerao et al. (2025)	X	X		X		
Lo et al. (2025)			X		X	
Zhang et al. (2024)			X		X	
Wu et al. (2024)		X		X		
Xu et al. (2025)		X		X		

Learning (SMART) merge both paradigms by combining centralized planning with decentralized execution, creating a balance between order and autonomy (Yue et al. 2025). These works collectively demonstrate how agentic AI's autonomy can be expressed along a spectrum from individually self-sufficient models to collaborative, self-organizing collectives, each architecture reflecting a trade-off between control, adaptability, and scalability (Table 1).

This survey will serve as an investigation and assessment of agentic AI, its architectures and frameworks by providing an assessment of different architectures that can be used in agentic solutions. The applications of agentic AI include multi-agent and single-agent delving further into sub-architectures multi-agent can utilize vertical architecture, horizontal architecture or a hybrid approach; vertical places the multi-agent system into a hierarchy where the orchestrator sits at the highest tier managing everything as a centralized node assigning tasks to sub-nodes whereas horizontal used a decentralized approach where the agents freely communicate with each other to accomplish their tasks giving each node a more equal weight to each other where the hybrid architecture takes concepts of both vertical and horizontal to accomplish its goals. We also provide an assessment of known libraries such as LangGraph, CrewAI, and AutoGen with use cases while also investigating the various architectures each library supports. Conducting a comparative analysis of these libraries will provide insight into the functionality of the individual libraries. The frameworks such as LangGraph, CrewAI, and AutoGen will be assessed on their usability and ability to complete complex tasks, other frameworks will be discussed but analysis will mainly focus on major publicly available frameworks. Going beyond their usability.

The remainder of this chapter is organized as follows: Sect. 2 introduces concepts regarding agentic AI and its foundational design principles such as the sense-act-plan-act decision cycle including the behavioral characteristics exhibited by the agentic AI solutions. Section 3 examines the fundamental architectures of agentic AI and their implications of solution design leveraging these features such as single-agent and multi-agent centralized/decentralized architecture explaining how models work based on the number of agents and if there is centralization in the solution. Section 4 discusses the fundamental architectures vertical, horizontal, and hybrid architects which form the foundational building blocks of agentic AI tool frameworks and the structure the agents follow with their execution. Section 5 delves deeper into specific software frameworks, use cases that other researchers have created solutions; this section also serves as comparison to the tools analyzed in the survey, including LangGraph, CrewAI, and AutoGen, to each other and how each one functions compared to each other. Section 6 examines open challenges present for agentic AI such as problems agentic AI systems are facing and where agentic solutions are heading as more developments are made. Section 7 finally summarizes the features explored in this survey and provides a call to action for the movement to further progress the field of agentic AI.

2 Foundational Concepts of Agentic Systems

The emergence of agentic AI evolutionizes AI by moving beyond static, reactive programs toward autonomous, goal-driven systems. At its core, an agentic system is defined by its ability to translate a high-level user prompt into an actionable plan, autonomously execute the resulting steps, and leverage external tools to gather information or perform functions from start to finish without continuous human interaction. This self-directed orchestration is achieved by integrating key functional components, perception, reasoning and planning, action, and memory, within a continuous decision loop.

This foundational structure gives rise to the core characteristic of agents: autonomy. Regardless of whether a system employs a single, monolithic agent or a complex multi-agent collective, its operation is governed by a cycle of continuous self-improvement and action. While traditionally described by a simple sense-plan-act cycle, modern agentic AI emphasizes more sophisticated, human-like, role-based loops that incorporate iterative reasoning and reflection. Examples like IBM's plan-execution-reflect framework or the more recent brain-perceptron-action model illustrate this pattern: an agent perceives its environment, reasons and plans a path forward, executes the necessary actions or tool calls, and then reflects on the outcome to refine its future behavior.

The following subsections delve into the essential building blocks that empower this autonomy, starting with the iterative sense-plan-act cycle that governs all agent behavior and culminating in the four key characteristics of agentic behavior that differentiate these intelligent systems from their predecessors.

2.1 The Sense-Plan-Act Cycle

While traditional reinforcement learning is often summarized by a sense–plan–act cycle, agentic AI systems are described through more human-like, role-based loops that emphasize iterative reasoning and self-improvement. In IBM's plan–execution–reflect framework (IBM 2025), a *planner* agent transforms a prompt into a step-by-step plan tailored to the goal and context. An *execution* agent then carries out the plan's instructions, interfacing with tools or environments as needed. A *reflection* agent evaluates outcomes, synthesizes feedback, and determines whether the goal has been met or what adjustments and next steps should be issued to the *execution* agent, which creates a loop and enables continual alteration.

A recent state-of-the-art view describes the agentic AI cycle as brain–perceptron–action (Xi et al. 2025), emphasizing the integration of cognition, perception, and tools. The *brain* component relates to planning, information processing, and decision making, the *perceptron* gives the agent perception using text, speech, and image inputs, and the *action* component has access to tool use, environmental interactions, and outputs such as text. Together, these three examples show a general representation of agentic cycles that have differing labels, but have a similar pattern: perception, planning, and action. The agents also have the ability to provide feedback, becoming an increasingly human-like structure and reflects the key characteristics of agentic behavior.

2.2 Key Characteristics of Agentic Behavior

Agentic behavior is differentiated by an AI agent's ability to translate a user prompt into an actionable plan, autonomously execute the resulting steps of the plan, and leverage external tools to gather information, perform functions, or output information (IBM 2025). Being capable of prompted planning, tool execution, and autonomous work from start to finish without human interaction, more generally. Together, these behaviors form the foundation for the fundamental architecture models discussed next, where these characteristics are implemented. Four additional specific key characteristics are the following:

- Autonomy: The ability to operate without direct human intervention.
- Tools: The ability to respond to changes in the environment.
- Planning: The ability to take initiative and pursue goals.

- Social Ability: The ability to interact and communicate with other agents and humans.

3 Fundamental Architectural Models

The autonomy discussed in Sect. 2 must be implemented within a defined structural foundation, or architectural model, to be effective. These models determine how the agent or agents perceive the environment, make decisions, and execute tasks. The choice of a fundamental architectural model is the most critical decision in designing an agentic AI solution, as it dictates the system's complexity, control mechanism, and ultimate scalability.

The foundational design can be broadly categorized into three main models, each representing a distinct approach to distributing decision-making and information flow: single-agent, multi-agent centralized, and multi-agent decentralized architectures. These models define how a global objective is managed, whether task execution is singular or distributed, and the extent to which communication and coordination are required.

The following subsections examine these three fundamental models, explaining the implications of each on system design, and how the number of agents and the presence or absence of a central orchestrator influences the solution's overall functionality and behavior.

3.1 Single-Agent Architecture

This is the simplest form, where a single AI agent operates independently to accomplish a task. This architecture is well-suited for well-defined, self-contained problems where collaboration is not a primary requirement.

In the single-agent architecture (Fig. 1a) (IBM 2024), the system maintains a fully known state at all times and follows a single, globally defined objective under one decision maker. This top-down, global-to-local control structure is good for its simplicity and ease of implementation, as all information and actions are combined within a single controller (the agent). However, all-in-one reduces reliability: if the agent fails or its policy diverges, the entire process fails. The main attribute of this architecture is its reliance on global information and a singular goal, where local actions are planned in a serial, hierarchical manner.

For example, in this single-agent setting, a user-specified global goal is first translated by the Brain/Planner into a step-by-step plan and locally delegated (either back to itself or to tool calls) following the plan-execution-reflect loop mentioned earlier. The agent then leverages available tools for additional information or function calling, while the Action/Executor carries out the tools. Using previous outputs, the

Perception ability reads text, audio, or images, and sees how they fit the plan and objective. The Reflection capability evaluates progress relative to the goal, adjusts future tasks, ends the process if the goal reached.

3.2 Multi-agent Centralized Architecture

Similar to the single-agent architecture, a centralized multi-agent architecture (Fig. 1b) also uses the same global objective while delegating multiple agents into performing local subtasks. Multiple agents may be deployed in parallel, however the decision controller and information scope is still centralized, making this architecture closer to a distributed single-agent system than to true multi-agent autonomy. The central controller plans the allocation of subtasks and collects task results. The result is that parallel execution increases throughput but the process remains a single, centrally guided objective, and the control coordination introduces additional overhead.

For example, given a global objective the central Planner partitions the goal into parallel subtasks and delegates each task to a specific agent. Each agent runs its own plan–execution–reflect loop mentioned earlier, leveraging its own tools, taking its own actions, and iteratively reflecting on its outputs before returning a 'completed' output to the planner agent. Acting as the global Brain, the initial planner agent replans assignments or re-routes tasks among agents while reflecting on the global goal. Perception in this case occurs locally within each agent and globally at the initial

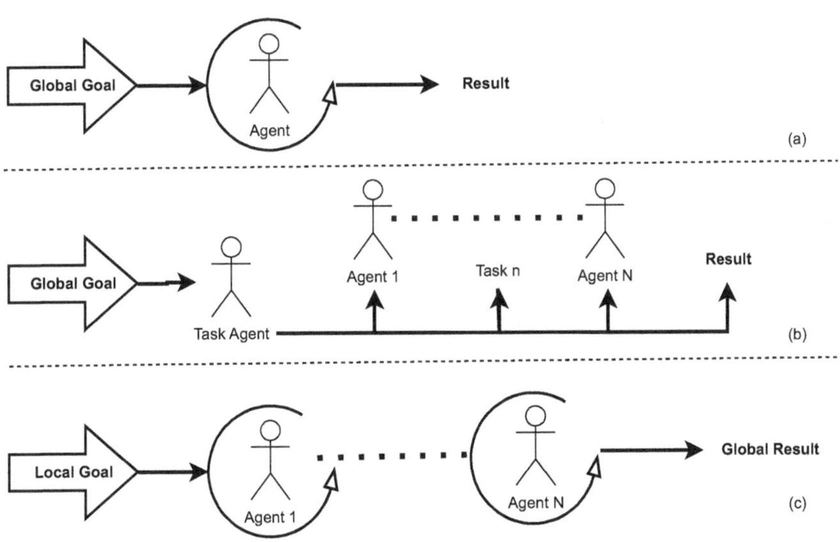

Fig. 1 Three fundamental architectural models: **a** single-agent architecture; **b** multi-agent centralized architecture; and **c** multi-agent decentralized architecture

planner agent, while Actions are distributed across multiple tool-using executors, all united for a single objective.

3.3 Multi-agent Decentralized Architecture

In this more complex model, multiple agents work together to achieve a common goal as shown in Fig. 1c. This architecture is essential for tackling complex problems that can be broken down into smaller, specialized tasks that can be handled by individual agents. The agents in a multi-agent system must be able to communicate and coordinate their actions effectively.

For a truly autonomous agent architecture, decentralized multi-agent systems (IBM 2024) distribute both information and decision making across agents, where each agent retains its own (potentially private) state, goals, and policies. Coordination emerges through agent-to-agent interaction enabled by communication protocols, and the global objective, which may not be defined directly, result from the combination of local behaviors rather than being stated. This bottom-up, local-to-global organization allows for easier scaling and higher reliability: the system can continue functioning even if individual agents fail or diverge. Additionally, decentralization introduces cooperative, competitive, or adversarial actions, which results from not having a single global controller and enables emergent behaviors.

For example, in a decentralized system, no global planner or shared state exists, so each agent's Brain forms a local plan from its own objective, its Perceptron is required for perception for communication, and its Actions use local tools to explore and perceive. Communication becomes a primary output due to agents not having access to shared states allowing them to share observations. In the absence of a central planning agent, task division emerges through local objectives, with each agent iterating its plan–execution–reflect loop against only local evidence and communications.

4 Key Architectural Frameworks

Building on these fundamental models, several architectural frameworks have emerged to provide more specific structures for building agentic AI systems.

4.1 Vertical (Hierarchical) Architecture

Hierarchical multi-agent frameworks structure AI systems into a top-down structure with an orchestrator agent managing the high-level goals, like response generation and reliability checks, while worker agents execute sub-tasks like data analytics, file operations, web navigation, and interactive reasoning. This architecture mirrors organizational management, ensuring that decision making and responses remain interpretable and scalable. By separating task execution from decomposition, the hierarchy enables more efficient coordination between agents and reduces the cognitive burden on any single agent (Zhang et al. 2025).

There have been a few recent frameworks that highlight the value of this design such as OrhcestratorAgent-Trust and AgentOrchestra both use the idea of how a conductor coordinates the orchestra and each musician is in charge of their own instrument. *AgentOrchestra* demonstrates how a conductor agent decomposes the more complex tasks and delegates them to worker agents, this experiment shows that a structured hierarchy improves both efficiency and general-purpose problem solving because most tasks can be divided and delegated (Zhang et al. 2025). From an applied perspective, industry reports such as SuperAnnotate emphasize that multi-agent LLM architectures become more efficient when worker agents specialize in domain-specific subtasks, while having a coordinator/manager agent maintains consistency and accuracy of the system (Multi-agent LLMs 2025).

Roumeliotis et al. build on this idea by applying trust-aware settings, where vision-language models act as workers and a non-visual orchestrator combines their outputs, weighs their confidence scores and reasoning to reach a final decision (Roumeliotis et al. 2025). Altogether, these studies show that a vertical architecture lays the groundwork for higher reliability and more interpretable responses when dealing with multi-agent systems (Fig. 2).

4.2 Horizontal (Decentralized) Architecture

Horizontal architectures refers to a group of nodes directly lateral to each other with similar qualities, in the case of agentic AI Architectures this still holds and even goes further to make sure all of the nodes are equal to each other. Horizontal architecture provides a way of distributing tasks and equal footing among all nodes; similarly the agents in a horizontal solution are given equal weight to each other (Masterman et al. 2024). The horizontal structure allows the agents to communicate with each other without having to thread each agents tasks through a manager or orchestrator this allows the agents to communicate freely with each other while also managing themselves and assigning their own tasks to themselves (Masterman et al. 2024). Horizontal execution allows the agents to handle multiple tasks concurrently by distributing subtasks among each other; this communication between nodes makes an environment good for collaborative multi-tasking (Chen et al. 2024).

Horizontal frameworks should be used when decentralized collaborative multi-tasking is needed allowing flexibility and adaptiveness in solutions using this architecture which make it rather useful for brainstorming and solving tasks (Ramachandran 2025). Horizontal architectures also function well in applications; which require adaptivity while completing tasks such as search and rescue, resource distribution, and environmental monitoring, as these are tasks that change consistently throughout their lifetime, they require dynamic solutions to solve problems efficiently (Ramachandran 2025).

4.3 Hybrid Architecture

Vertical architecture tends to have information bottlenecks, A single point of failure through the leader as well as less collaboration between agents, its rigid structure could also prove problematic for dynamic tasks. Horizontal architectures tend to have unnecessary communication overheard or "chatter", inefficient coordination, poor feedback from other agents, and information overload (Masterman et al. 2024). Therefore, horizontal architectures are better for creative and collaborative tasks that require multiple sources of information. Vertical architectures, on the other hand, are better suited for Execution-heavy tasks that can easily be broken up into stages such as software development, data processing, report generation. The choice between the two depends on the desired tasks (Chen et al. 2024).

Since an multi-agent system's implementation depends on the task at hand a more practical hybrid approach to prevent compounding the problems of both horizontal and vertical architectures while simultaneously leveraging both is to use AgentVerse. AgentVerse has four key phases Expert Recruitment, Collaborative Decision-

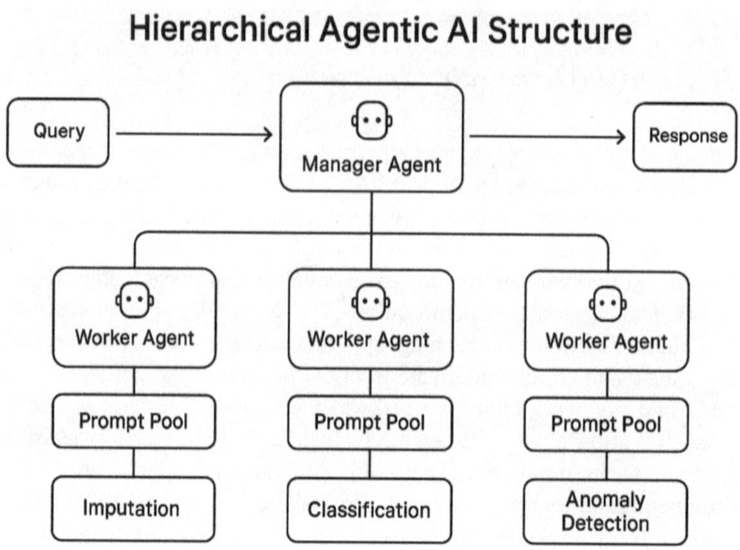

Fig. 2 Hierarchical agentic architecture. A manager decomposes a user query and delegates subtasks to specialized worker agents (imputation, classification, anomaly detection). Each worker operates with its own prompt/tool pool, and results are fused by the manager to produce the final response

Making, Action Execution and Evaluation. Its approach is to switch the agent architecture based on the task at hand therefore for app development it might require a horizontal approach for the planning and brainstorming phase while switching to a vertical approach for executing the tasks (Ramachandran 2025). Similarly, Google made its own hybrid structure called *Chain of Agents*. This framework introduces a two-stage pipeline: planning and execution. However, instead of the manager agent dividing and delegating the task; there is a chain of worker agents in the first stage that each receive a portion of the source text along with a message passed from the previous agent. Then, in the second stage, a manager agent integrates and aggregates the responses from the worker chain and generates a final response for the user, ensuring that long-context reasoning is comprehensive. The flow of information between these two stages is bidirectional; therefore, it allows for a nice balance between manager and worker agents which could improve functionality farther then vertical or horizontal architecture depending on the field it is applied to Zhang et al. (2024b).

4.4 *Trustworthy and Secure Architectures*

As multi-agent AI systems are advancing toward deployment in high-stakes domains such as Healthcare, driving, finance, agriculture, etc. concerns arise when discussing the trust and security of these systems. When dealing with high risk domains architectures must be able to deliver accurate outputs but they also need to demonstrate resilience under uncertainty, transparent reasoning, and they need to be able to align/learn with human oversight. A trustworthy system is expected to operate reliably in an adversarial environment, be able to detect and recover from errors, and provide comprehensible reasoning for their decisions. Before the design focus on AI focused on maximizing raw performance now that we are stepping to high-risk domains in real life we need to take serious steps in ensuring calibrated confidence, fault tolerance, and ethical governance.

Recent work discusses the importance of embedding trust mechanisms into orchestrator logic. Roumeliotis et al. proposes a framework in which an orchestrator agent uses trust calibration metrics including expected calibration error (ECE), overconfidence ratio (OCR), and Consistency Gap (CG). These metrics help to filter out unreliable agent predictions and trigger a re-evaluation process through Retrieval-Augmented Generation (RAG). This design allows the system to lesser the weight of incorrect overconfident outputs and helps the agents ground their decisions using evidence which significantly increases reliability (Roumeliotis et al. 2025).

5 Tools for Building Agentic Systems

The development of agentic AI has been accelerated by the emergence of open-source frameworks that provide pre-built components and tools for creating these complex systems. Some of the most popular frameworks include:

5.1 LangGraph

LangGraph extends the LangChain ecosystem by introducing a graph-based orchestration for agentic workflows, which addresses a fundamental limitation in traditional linear LLM applications. Linear LLMs are unable to model cyclical, stateful behaviors that define autonomous agentic AI systems. The traditional prompt-response architectures for single-turn interactions are sufficient but complex tasks that require multi-step reasoning or collaborative problem solving requires a more sophisticated control structure (Fig. 3). The framework's adoption in production environments is

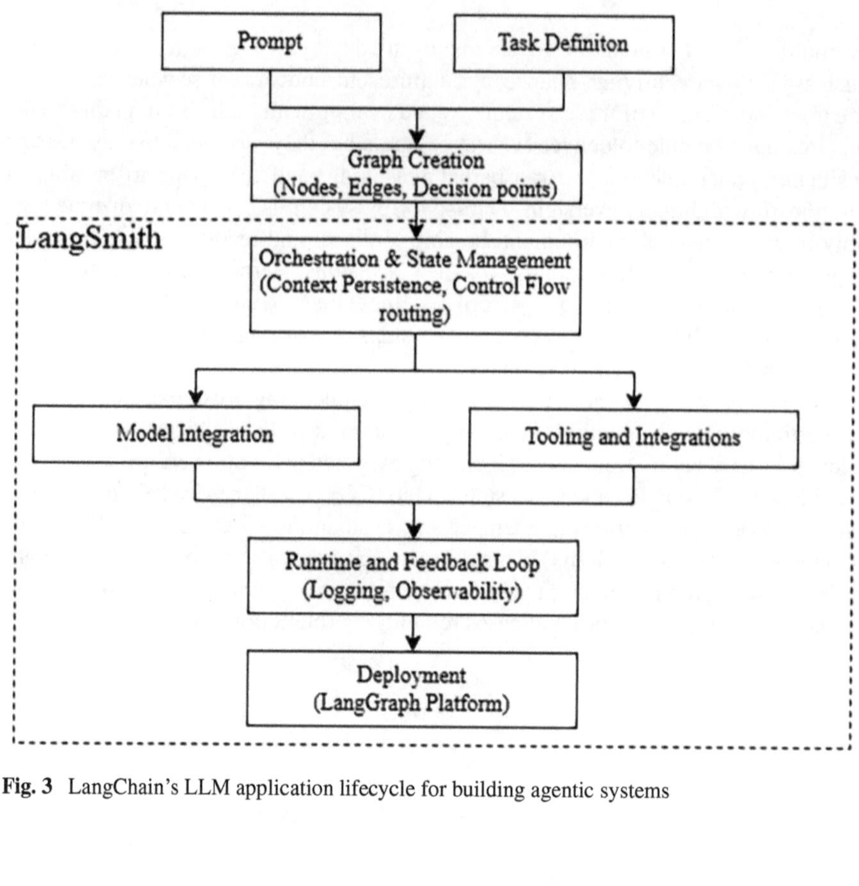

Fig. 3 LangChain's LLM application lifecycle for building agentic systems

thanks to its numerous architectural advantages. Graph-based abstraction provides explicit modeling of decision-points and control flow, enabling developers to reason about agent behavior systematically rather than relying on implicit prompt engineering. This ecosystem maintains an extensive integration support across major LLM providers, and provides mechanisms for direct model instantiation when it is not natively supported, as long as it follows the standardized structure (Multi-agent-overview n.d.). One of its attractive features is the ecosystem of pre-built tools integrations alongside extensibility for custom tool definitions, improving production roll out and maintaining flexibility for task-dependent requirements. The framework's documentation, case study repository and GitHub examples facilitates ease of adoption and implementation.

LangGraph agent workflows operate as directed graphs comprised of three fundamental components that define agent behavior and the flow of control through a State, Nodes and Edges. The **State** implements a shared data structure representing the current configuration of the application at anytime. The state follows a defined schema enforcing type safety and structural consistency across agent interactions. LangGraphs's persistent state enables context preservation across multiple interactions, incremental updates where each agent operation modifies/updates the state rather then reconstructing the entire context and deterministic replay as state transitions can be logged and reproduced for debugging. The **Nodes** encapsulate the functional logic of agents within the system, each node is implemented as a function that receives the current state of the graph as input, executes its task and returns the updates state for the next agent to update the new state. This abstraction allows for modularity as individual agents capabilities can be developed and tested before integration, Reusability as common operations can be a shared node, composability as complex agent behaviors can be built from orchestrating simpler nodes. The nodes execute synchronously by default requiring one node completion before the next. The **Edges** are functions that determine the control flow of the graph as it selects which node executes next based on the current state of the system, this routing mechanism allows for loops and dynamic decision making for agents. Enabling cyclical workflows and dynamic dispatch which requires the LM to decide which tool to invoke based on the possible nodes identified. The separation between Edges and Nodes allows for ease of configuration as changing the edge conditions can alter the control strategy but not the function of the node itself.

LangGraph includes higher-level features essential for deploying production ready and reliable agentic AI systems such as short and long term memory integration. Static or dynamic human in the loop control, essential real-time monitoring of agent operations, streaming support for state updates, nodes executions and intermediate outputs. Integrating it into its companion tool allows developers to inspect and debug at each stage of the process. LangGraph includes a Time-Traveling feature within its debug tools that allows the developers to resume execution from a prior checkpoint to observer alternate executions, essential for reproducing and eliminating non-deterministic bugs that are hard to miss.

To expand upon these baseline capabilities LangGraph includes specific multi-agent features to increase the efficiency and readability of the multi-agent system.

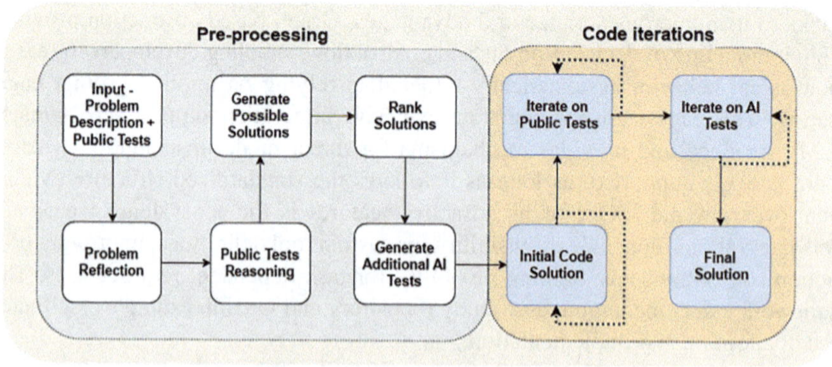

Fig. 4 AlphaCodium implementation (Ridnik et al. 2024)

Including parallel execution of tasks through its Send API allowing for concurrent node execution paving the way for decreased execution time, efficient resource management and scalable systems. Subgraphs reinforce the idea of encapsulation by allowing specialized agents to act independently which is useful for hierarchical organization as each subgraphs has its own isolated state management. While this does come with communication overhead it is an important feature for expansive and complex multi-agent systems. Reflection allows the agent to learn from its mistakes and continuously improve over time referencing the methodology in Fig. 4. This method shows around a 30% increase in generated output accuracy after multiple iterations of the task highlighting the importance of this feature for long-term multi-agent systems.

The graph architecture provides a wide variety of flexibility for creating multi-agent structures that allow for vertical, horizontal and custom-hybrid frameworks depending on the task that needs to be accomplished.

5.2 *CrewAI*

CrewAI is an agentic AI framework written with a top-down view in mind while maintaining a high level of precision in low-level packaging; this makes it well suited for working on tasks autonomously (CrewAI Contributors 2025). CrewAI stands out in its ability to assign agents to individual tasks, thus making it a horizontal architecture; the framework also stands out due to the way it assigns tasks to its agents functioning on the idea of crews and work flows (CrewAI Contributors 2025). Figure 5 gives an overview of the crew's dynamic; the crew is a team of independent agents working on tasks to accomplish a goal. The agents in the crew are assigned roles and based on that each agent gets assigned tasks to complete. Each agent works on their owned assigned task to accomplish the overall goal.

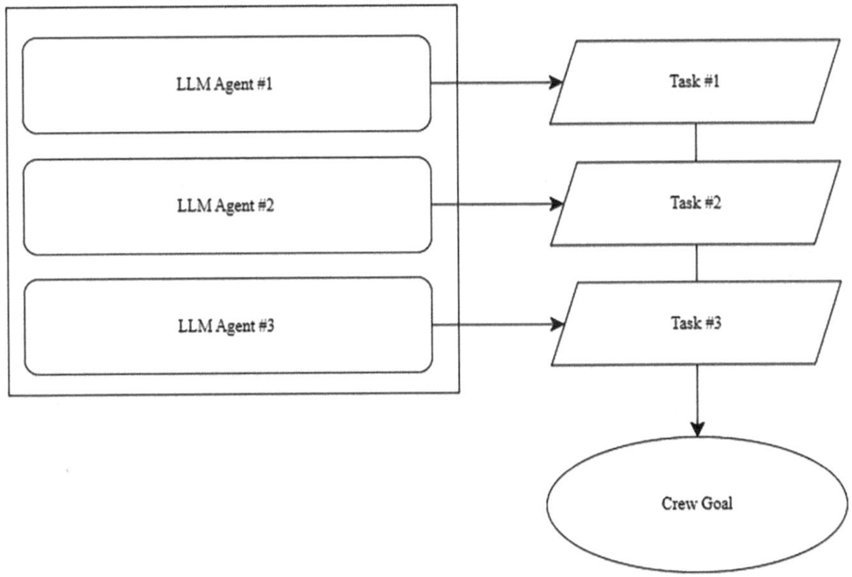

Fig. 5 Overview of CrewAI decentralized architecture

CrewAI built around decentralized task management and the horizontal agentic AI architecture this can be seen in the way it builds its tasks but also in the way researchers have used CrewAI, it has been used by other researchers for more complicated tasking. Other studies have used CrewAI to simulate an emergency department onboarding scenario, resume screening for companies, and precise inclusive image generation (Bai et al. 2026; Lo et al. 2025; Bhalerao et al. 2025). These studies discuss their implementations using CrewAI and how it can be adapted to evolving situations with varying complexity.

Lo et al. utilized CrewAI to design a horizontally coordinated multi-agent system for automated resume evaluation and ranking in hiring workflows (Lo et al. 2025). Their goal was to determine whether multiple specialized LLM agents could collectively reproduce the reasoning process of human recruiters while maintaining explainability. The research team implemented four distinct CrewAI agents: an extractor, a RAG based evaluator, a summarizer, and a score formatter; each responsible for a specific phase of the screening process. The CrewAI framework allowed these agents to interact laterally, exchanging structured outputs instead of relying on a central controller. The experiment demonstrated that distributing tasks among specialized roles improved interpretability and consistency, as each agent's decision could be traced to concrete résumé evidence. However, the study also revealed that when the inter-agent communication or the retrieval knowledge base lacked precision, the pipeline occasionally generated biased or incomplete assessments. These shortcomings highlight how a CrewAI driven multi-agent architecture can outperform single prompt evaluations in reasoning transparency, yet still depends heavily on careful knowledge-base design and well-defined communication boundaries (Lo et al. 2025).

Bhalerao et al. employed CrewAI to orchestrate a multicultural text-to-image (T2I) generation system known as MosAIG (Bhalerao et al. 2025). The researchers sought to examine whether a network of culturally diverse LLM agents could cooperatively refine image prompts to reduce cultural bias and improve representational fairness. Their CrewAI configuration consisted of a moderator, several social agents representing different cultural perspectives, and a summarizer that compiled the group's deliberations into a final caption for a diffusion model. Through this iterative, peer-based discussion, each agent critiqued and revised the prompt to ensure inclusivity before image synthesis occurred. The simulation revealed that the CrewAI-mediated conversation loop produced higher cultural alignment and aesthetic quality compared with single-prompt baselines, validating the benefit of horizontal dialogue among agents. Nevertheless, the experiment also showed increased latency and token cost, and occasional convergence toward similar cultural assumptions when agents shared overlapping linguistic backgrounds. These results emphasize CrewAI's capability to coordinate equitable, context-aware creative systems while also revealing the scalability and diversity challenges inherent to multi-agent deliberation (Bhalerao et al. 2025).

Bai et al. used CrewAI to simulate robots in an emergency department workflow; this was a test to see how a multi agent system (MAS) would respond in a mission critical instance (Bai et al. 2026). However, it should be mentioned that this experiment was completely simulated to better understand how such a system could work in a mission critical scenario. Ultimately, Bai et al. designed their experiment to have the agents embody the role of robots by providing descriptions for the LLM to use as a backstory and the tasks they are supposed to complete to include the tools they are allowed to use in while doing the task; the aim of doing this was to investigate at which point the LLM will have failures due to constraints the embodied robot role may exhibit such as degree of movements and access to certain areas (Bai et al. 2026). The study showed that when the LLM tried mimicking the scenario it failed in areas such as the manager doing the work of the agents rather than its own work, agents would try accessing tools meant for another agent, and also failure of tasks delegation. This shows the fatal flaws when the LLM is given requirements to follow in the real world, proving they are far from use in real life for physical jobs, however it would become possible for the MAS to perform high performance tasks like this when given better instructions with clearly defined rules for the tasks (Bai et al. 2026).

5.3 AutoGen

Microsoft's AutoGen Framework offers a powerful but relatively easy-to-use platform for constructing agentic systems. It contains predefined agents but it also enables developers to define multiple agents that converse, coordinate, and decompose tasks without having to build the orchestration logic from scratch. At its core, AutoGen supports modularity, memory/tool integration, and termination strategies that are all wrapped in a high-level API called AgentChat which hides most of the complexity in the background. One of its best features which still in beta is **AutoGen Studio** which offers a low-code, visual interface that allows user to build and define agent workflows via a drag-and-drop and declarative JSON specification. With the studio's team builder a developer can visually assemble agents, assign roles and tools, and connect them into workflows all through the GUI with the option of showing the pure code. Finally the interface supports interactive testing in the form of a playground where the user can write prompts and observe message flows then export the workflows for deployment as APIs (Microsoft Research 2024).

Beyond the studio's low-code convenience, AutoGen is built with **human-in-the-loop** support. A special prebuilt agent called UserProxyAgent lets a human insert feedback at runtime. This means that during execution the team of agents will delegate control to the human via the proxy agent they will pause, wait for input, then resume which enables human oversight and corrections during critical steps of execution. Alternatively, human feedback can be inserted after the first run which will influence how the agents act during the next execution. The flexibility of this framework ensures that while agents are able to operate autonomously and communicate with each other, a human can intervene when needed (Microsoft 2025). Combining this with RAG, which is demonstrated in IBM's report on integrating RAG into AutoGen, these features enhance factual reasoning, reduce hallucination and allow different types of architectural frameworks to balance autonomy and control (IBM 2025) (Fig. 6).

5.4 Comparative Analysis

This section provides a dimension-based comparative analysis of the common agentic tools explored, focusing on how they differ for various applications. This analysis emphasizes six dimensions: autonomy, reasoning, learning, memory, scalability, and explainability. Each dimension represents a key distinction when considering your agentic application.

Autonomy level determines the range a tool can operate independently, ranging from fully autonomous multi-agent systems to incorporating human oversight or review. Reasoning type identifies whether a tool relies on symbolic or subsymbolic reasoning, based on delegated roles, orchestration, and defined rules. Learning capacity evaluates each tool's use of external pretrained LLMs, reflection capabilities, or integration with APIs and RAG.

Memory integration measures how tools manage information retention through agent state persistence, use of external knowledge, or conversational memory. Scalability determines the system's ability to expand in complexity, from managing a few agents to coordinating swarms or hierarchies. Lastly, explainability and trustworthiness weighs how transparent each tool is, emphasizing human-in-the-loop over-

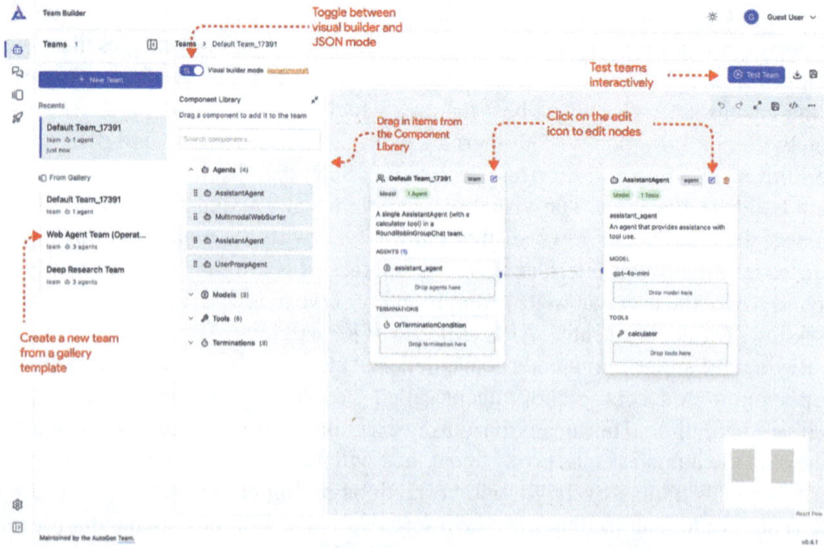

Fig. 6 AutoGen Studio: low-code team builder. The visual builder lets you assemble multi-agent teams by dragging components (agents, models, tools, terminations) from the library, connect them into a workflow, and test the team interactively. Screenshot adapted from the AutoGen Studio (Microsoft Research 2024)

sight, transparency of states and reasoning, and defined knowledge base adherence (Table 2).

6 Open Challenges and Future Directions

Hierarchical agent frameworks such as that demonstrated in AgentOrchestra show the power of orchestrator agents that break down tasks and manage specialized worker agents. However, they still face challenges in scalability since manager agents can also be overloaded when coordinated a larger number of workers. Another issue that arises is that managers are more capable of reaching a consensus when there are fewer overconfident workers but, when there are multiple workers confidently agreeing on an incorrect result it is still challenging to be able to reach a decision without re-evaluation which can become costly. Future directions for this architecture are more than likely going to move toward hybrid models that balance top-down control with peer-to-peer which will help with scalability and resilience in high-stakes domains like healthcare and banking.

AutoGen has made it a lot easier for building multi-agent applications but it faces challenges when workflows increase in scale and complexity as coordinating a large number of agents with multiple tool calls and retrieval steps can add huge latency

Table 2 Dimension-based comparison of AutoGen, LangGraph, and CrewAI frameworks

Dimension	AutoGen	LangGraph	CrewAI
Autonomy level	Supports autonomous multi-agent workflows with optional human-in the loop injection via `UserProxyAgent`	Supports limited autonomous multi-agent workflows with predefined graphs	Role-based autonomy, horizontally coordinated crews with fully automated execution, has optional human in loop review through checkpoints, peer to peer and hierarchical operations
Reasoning type	Supports subsymbolic LLM reasoning structured by orchestration rules and protocols	Includes in-depth reasoning with graph-based workflows that rely on stateful orchestration	Uses subsymbolic LLM reasoning but agents are given role-based tasks and cooperating with each other to push tasks forward
Learning capacity	Relies on external LLMs; no training in the framework but model can adapt based on prompts memory and RAG	Adaptive learning capacity as it supports popular LLMs but can also instantiate a model, learning through stateful, iterative and persistent workflows with a reflection mechanism to increase reliability over time	Does not use its own LLM, uses external APIs. Adapts behavior through prompt policies, RAG, reflection mechanism, and contextual memory
Memory integration	Offers built-in conversation memory, state persistence of agents, and retrieval-augmented generation	Includes built-in memory support for long and short term applications	Saves per-agent conversations and a shared crew history, has options for integrating RAG and other similar mechanisms. Also allows usage of external knowledge bases
Scalability	Scales from two agent systems to complex teams; supports swarms, round-robin, and hierarchical orchestration	Scales from a single agent to complex teams of various architectures, supports seamless integration from popular tools and fast deployment	Allows the ability to scale from single agent to multiple agent systems, supports parallel and hierarchical task execution as well as lightweight orchestration with little overhead
Explainable and trustworthy	Transparent reasoning and conversations, explicit termination rules can be set before runtime, capable of human-in-the-loop oversight, and RAG for grounding logic	Graph-based implementation with states allows for easy explainability of the workflow and its various features allow for trustworthy and reliable results through human-in-the-loop, reflection, tool integrations and extensive debugging tools	Transparency through roles, tasks, and tool definitions. Auditing through conversation logs and whitelisting for tools, has optional knowledge base to use as a guide, and has optional human mandatory checkpoints

and cost especially when re-evaluation loops occur. Another feature of AutoGen that could use improvement would be the studio feature. Microsoft makes it clear that this low-code UI is still being developed to become as effective as the older alternative of AgentChat, which is what the studio builds off of and is mainly just a very high-level python API. Finally since AutoGen just uses an API to access different AI models a problem arises with trust and explainability. While AutoGen has logs that agent-agent and agent-user conversations transparent, the internal reasoning of each agent isn't really shown which negatively impacts trust in this case.

Next steps include some form of context switching between different architectures based on the task where the specialized agent will receive the entire text and context and decide autonomously which architecture to use or combine. This is where LangGraph excels as it allows full customization of multi-agent architectures, incorporating subgraphs and reflection mechanisms will increase the effectiveness and readability as the agentic system gets more complex.

7 Conclusion

This artificial intelligence systems survey presented a comprehensive review of agentic AI applications and their underlying architectural decisions. We explored the wide range of conceptual architectural models, frameworks, and agentic building tools currently being used to design and deploy autonomous systems.

A central finding is that the choice of architecture is a critical, task-dependent decision. The foundational models, including single-agent, multi-agent centralized, and multi-agent decentralized, must be selected precisely based on the specific task requirements and mission goals. Similarly, the choice of architecture framework, whether a highly structured vertical/hierarchical model or a more flexible horizontal model, is determined by the extent of task separability and coordination needed.

We provided a comparative assessment of major agentic building tools, LangGraph, CrewAI, and AutoGen, demonstrating how they vary significantly in their strengths, particularly concerning autonomy, memory integration, and trustworthiness. For instance, LangGraph provides maximum flexibility through its graph-based orchestration, while AutoGen offers simplified multi-agent collaboration, and CrewAI excels in role-based, decentralized tasking.

Looking ahead, agentic AI demonstrates great potential across a number of emerging applications that rely on high levels of autonomy and task planning. However, the field faces open challenges, particularly in enhancing scalability for managing massive swarms of agents and improving the transparency of internal agent reasoning to build trust. Future research must focus on optimizing architectural designs at different mission stages to minimize computational overhead, mitigate hallucinations, and increase overall system reliability. We expect to see the emergence of more sophisticated and specialized architectures, particularly hybrid models that dynamically switch between top-down control and peer-to-peer collaboration to balance efficiency and resilience.

References

Bai Y, Han R, Parikh N, Ju W, Taylor A (2026) Towards considerate embodied AI: co-designing situated multi-site healthcare robots from abstract concepts to high-fidelity prototypes. In: Proceedings of the 2026 CHI conference on human factors in computing systems (CHI '26). Association for Computing Machinery, New York, NY, USA. https://doi.org/10.1145/3772318.3791891

Bhalerao P (2025) Multi-agent multimodal models for multicultural text to image generation. Computer Science and Engineering Master's Theses 51. https://scholarcommons.scu.edu/cseng_mstr/51

Chen W, Su Y, Zuo J, Yang C, Yuan C, Chan C-M, Yu H, Lu Y, Hung Y-H, Qian C., Qin Y, Cong X, Xie R, Liu Z, Sun M, Zhou J (2024) AgentVerse: facilitating multi-agent collaboration and exploring emergent behaviors. In: Kim B, Yue Y, Chaudhuri S, Fragkiadaki K, Khan M, Sun Y (eds) International conference on learning representations, pp 20094–20136. https://proceedings.iclr.cc/paper_files/paper/2024/file/578e65cdee35d00c708d4c64bce32971-Paper-Conference.pdf

CrewAI Contributors (2025) CrewAI documentation. CrewAI Project. Accessed 2025-10-11. [Online]. Available: https://docs.crewai.com/

IBM (2024) What are AI agents? IBM, July 2024. [Online]. Available: https://www.ibm.com/think/topics/ai-agents

IBM (2024) What is a multi-agent system? IBM, Aug 2024. [Online]. Available: https://www.ibm.com/think/topics/multiagent-system

IBM (2025) Local multi-agent RAG with AutoGen using granite. IBM, Aug 2025. [Online]. Available: https://www.ibm.com/think/tutorials/multi-agent-autogen-rag-granite

IBM (2025) Multi-agent AutoGen with retrieval-augmented generation (RAG). Tutorial. [Online]. Available: https://www.ibm.com/think/tutorials/multi-agent-autogen-rag-granite

IBM (2025) What are agentic workflows? IBM, Mar 2025. [Online]. Available: https://www.ibm.com/think/topics/agentic-workflows

Jiang T, Chen H, Li Z, Yang R (2024) BlockAgents: Byzantine-robust multi-agent systems via blockchain coordination. In: Proceedings of the ACM symposium on applied computing, pp 1445–1453

Lo FP et al (2025) AI hiring with LLMs: a context-aware and explainable multi-agent framework for resume screening. In: 2025 IEEE/CVF conference on computer vision and pattern recognition workshops (CVPRW), Nashville, TN, USA, pp 4184–4193. https://doi.org/10.1109/CVPRW67362.2025.00402

Masterman T, Besen S, Sawtell M, Chao A (2024) The landscape of emerging AI agent architectures for reasoning, planning, and tool calling: a survey. [Online]. Available: http://arxiv.org/abs/2404.11584

Microsoft (2025) Human-in-the-loop with AutoGen. Documentation. [Online]. Available: https://microsoft.github.io/autogen/stable//user-guide/agentchat-user-guide/examples/company-research.html

Microsoft Research (2024) Introducing AutoGen Studio: a low-code interface for building multi-agent workflows. Blog Post. [Online]. Available: https://www.microsoft.com/en-us/research/blog/introducing-autogen-studio-a-low-code-interface-for-building-multi-agent-workflows/

Multi-agent LLMs in 2025 [+ frameworks]. SuperAnnotate. Blog Post. [Online]. Available: https://www.superannotate.com/blog/multi-agent-llms

Multi-agent-overview. [Online]. Available: https://langchain-ai.github.io/langgraph/concepts/multi_agent/

Ramachandran A (2025) Latest advances in agentic AI architectures, frameworks, technical capabilities, and applications. IBM Research - Thomas J. Watson Research Center

Ridnik T, Kredo D, Friedman I (2024) Code generation with AlphaCodium: from prompt engineering to flow engineering, Jan 2024. [Online]. arXiv:2401.08500 [cs]

Roumeliotis KI, Sapkota R, Karkee M, Tselikas ND (2026) Agentic AI with orchestrator-agent trust: a modular visual classification framework with trust-aware orchestration and RAG-based reasoning. In: IEEE Access, vol 14, pp 26965–26982. https://doi.org/10.1109/ACCESS.2026.3662282

Wu T, Shi H, Feng Y, Xu F, Liang P (2024) AutoGen: enabling next-generation LLM applications via multi-agent conversations. In: Proceedings of the conference on language modeling (COLM). ACM, pp 122–135

Xi Z, Chen W, Guo X, He W, Ding Y, Hong B, Zhang M, Wang J, Jin S, Zhou E, Zheng R, Fan X, Wang X, Xiong L, Zhou Y, Wang W, Jiang C, Zou Y, Liu X, Yin Z, Dou S, Weng R, Qin W, Zheng Y, Qiu X, Huang X, Zhang Q, Gui T (2025) The rise and potential of large language model based agents: a survey. Sci China Inf Sci 68(2):1–44. [Online]. Available: https://link.springer.com/article/10.1007/s11432-024-4222-0

Xu X, Zhao Y, Li R, Chen L (2025) MOMA: mitigating social bias in large language models via a multi-objective multi-agent framework. In: Proceedings of the AAAI conference on artificial intelligence. AAAI Press

Yue Y, Zhou X, Liu B, Qiu X (2025) SMART: synergistic multi-agent framework with trajectory learning. In: Proceedings of the AAAI conference on artificial intelligence. AAAI Press

Zhang W, Cui C, Zhao Y, Liu Y, An B (2025) AgentOrchestra: a hierarchical multi-agent framework for general-purpose task solving. [Online]. arXiv preprint arXiv:2506.12508

Zhang M, Liu C, Wang X, Zhang Q, Qiu X (2024) ProAgent: building proactive cooperative agents with large language models. In: Proceedings of the AAAI conference on artificial intelligence, vol 38, no 15, pp 16287–16295

Zhang W, Li H, Zhang Y, Lin C (2024a) LLM-assisted multi-agent dialogue for ontology alignment. In: Proceedings of the international conference on autonomous agents and multiagent systems (AAMAS). ACM, pp 845–853

Zhang Y, Sun R, Chen Y, Pfister T, Zhang R, Arık SÖ (2024b) Chain of agents: large language models collaborating on long-context tasks. In: Advances in neural information processing systems (NeurIPS 2024), vol 37

Architectures for Building Agentic AI

Slawomir Nowaczyk

Abstract This chapter argues that the reliability of agentic and generative AI is chiefly an architectural property. We define agentic systems as goal-directed, tool-using decision makers operating in closed loops, and show how reliability emerges from principled componentisation (goal manager, planner, tool-router, executor, memory, verifiers, safety monitor, telemetry), disciplined interfaces (schema-constrained, validated, least-privilege tool calls), and explicit control and assurance loops. Building on classical foundations, we propose a practical taxonomy—tool-using agents, memory-augmented agents, planning and self-improvement agents, multi-agent systems, and embodied or web agents—and analyse how each pattern reshapes the reliability envelope and failure modes. We distil design guidance on typed schemas, idempotency, permissioning, transactional semantics, memory provenance and hygiene, runtime governance (budgets, termination conditions), and simulate-before-actuate safeguards.

1 Introduction: Purpose, Scope, and Architecture Reliability

This chapter surveys architectural choices for building agentic AI systems and analyses how those choices shape reliability. Our central claim is straightforward: *reliability is, first and foremost, an architectural property*. It emerges from how we decompose a system into components, how we specify and enforce interfaces between them, and how we embed control and assurance loops around the parts that reason, remember, and act. Individual models matter, but without the right architectural scaffolding, even state-of-the-art models will behave inconsistently, be impossible to audit, and prove fragile in the face of novelty.

Agentic AI in this book denotes systems that pursue goals over time by *deciding* what to do next, *selecting and using tools, consulting and updating memory*, and

S. Nowaczyk (✉)
Center for Applied Intelligent Systems Research, Halmstad University, Halmstad, Sweden
e-mail: slawomir.nowaczyk@hh.se

A. Stavrou et al. (eds.), *Generative and Agentic AI Reliability: Architectures, Challenges, and Trust for Autonomous Systems*, Studies in Computational Intelligence 1272, https://doi.org/10.1007/978-3-032-18585-3_3

interacting with their environment under constraints. An agent is not merely a predictor; it is a decision-maker in a closed loop. It observes, plans (or at least chooses), acts, and learns, typically under uncertainty and partial observability. **Generative AI** refers to models that synthesise content—text, code, images, plans, or intermediate representations—often serving as the reasoning substrate inside the agent, or providing artefacts (queries, programs, simulations, explanations) that other components execute or verify. In modern systems, *generative models* supply the *policy* (how to reason and propose actions), while the *agentic architecture* supplies the *machinery* (how proposals are validated, enacted, bounded, and recorded).

Understanding the relation of **Agentic GenAI** with **classic autonomous agents** is crucially important to avoid reinventing the wheel: many key concepts have been studied for a long time and are relatively well-understood today; however, the nature of GenAI also brings up challenges that are completely novel and require rethinking of what was believed to be known. Traditional reactive, deliberative, or BDI (belief–desire–intention) architectures offer theoretically-founded and crisp notions of concepts such as beliefs, goals, plans, and intentions, with clear control loops and explicit world models. Modern agentic systems often replace hand-engineered reasoning with neural-network-based foundation models. These models, trained on huge amounts of diverse data, vastly increase the flexibility and breadth of competence, but also introduce uncertainty in reasoning steps and tool usage. In this chapter, we retain the useful discipline of the classic view—explicit state, goals, plans, commitment strategies, and monitoring—while acknowledging that parts of the pipeline (e.g., plan generation or hypothesis formation) may be implemented by generative models. That reconciliation is precisely where architecture earns its keep.

This book is not intended as yet another broad introduction to Agentic GenAI; instead, we put these recent developments in the specific context of **reliability**. By reliability, we mean *the consistent achievement of intended outcomes under stated conditions, within acceptable bounds of safety, security, data protection, and resource usage, and with evidence that failure modes are known, contained, and recoverable*. For agentic AI, this encompasses much more than just model accuracy. It includes correct tool invocation, bounded action sequences, resistance to manipulation, predictable latency and cost, graceful degradation, auditability, and human-override paths. Architectures make these properties tractable by: (i) isolating *responsibilities* in modules whose contracts we can reason about; (ii) interposing *validators* and *verifiers* between reasoning and action; (iii) *constraining* authority and side-effects through permissioned tool interfaces; and (iv) *instrumenting* the system so that internal state, decisions, and outcomes are observable and replayable.

Rather than hinging on a single mechanism, system-level reliability is shaped by the interaction of a few foundational architectural choices. In practice, three mutually reinforcing design choices determine how agentic systems behave under stress: how we decompose functionality, how the parts communicate and are constrained, and how their behaviour is supervised at run-time.

Componentisation. Separating the functionality, such as perception, memory, planning, tool routing, execution, verification, and oversight, confines faults to well-defined boundaries and limits their blast radius. Clear responsibilities make defects diagnosable and upgrades safe: a verifier can be strengthened without disturbing the planner; an execution gateway can be hardened without touching memory logic. Componentisation also enables staged deployment (mock tools, simulators, or read-only modes first) and offers natural choke points for safety checks.

Interfaces and Contracts. Primary means to tame open-ended model behaviour are typed and schema-validated messages; explicit capability scopes for tools; idempotent and (where feasible) transactional semantics; rate/authority limits. All of these convert free-form model outputs into predictable, auditable actions. Interfaces extend to memory: retrieval must carry provenance and freshness guarantees; long-term stores need retention, compaction, and contamination controls. Good contracts enable the system to act deterministically when safe and refuse when not, transforming ambiguous proposals into either valid commands or actionable error reports.

Control and Assurance Loops. Monitors compare planned with observed behaviour; critics and verifiers check factuality, policy compliance, and safety invariants; supervisors enforce budgets, escalation rules, and termination criteria; fallbacks provide safe modes of operation when assumptions fail. These loops supply the governing feedback around generative components, preventing small reasoning slips from cascading into hazardous sequences and ensuring graceful degradation under uncertainty.

Taken together, these choices turn a powerful but free-form reasoning engine into a bounded, observable, and governable system. In the remainder of this section, to make these ideas concrete, we illustrate how they play out in the *running example* of a tool-using diagnosis agent operating in a safety-critical environment. Imagine a fleet operator responsible for electric power systems in autonomous service vehicles. The agent's mission is to triage anomalies, recommend mitigations, and, within a narrow envelope, execute pre-approved actions that reduce risk and downtime.

The agent comprises: a **Goal Manager** (ingesting alerts and operator intents), a **Perception and Retrieval** layer (querying telemetry stores and maintenance logs), a **Planner** (often a generative model producing hypotheses, tests, and action candidates), a **Tool Router** (mapping abstract actions to concrete, permissioned tools: telemetry queries, digital-twin simulation, firmware status, dispatch scheduling), an **Execution Gateway** (schema validation, pre-condition checks, simulators-before-actuators, idempotency tokens), a **Verifier/Critic** (analysing proposed explanations and commands against rules, limits, and known hazards), a **Memory subsystem** (short-term scratchpads for the current case, long-term episodic/semantic stores with provenance), and a **Safety Supervisor** (budgets, escalation, and safe-halt rules). All interactions generate structured logs that are stored in an immutable audit trail.

A typical episode unfolds as follows. An over-temperature alert arrives from Vehicle V. The Goal Manager formulates a diagnosis task. The Planner drafts a hypothesis: recent fast-charge sessions combined with ambient heat may have accelerated cell imbalance. It proposes a sequence: retrieve finer-grained thermal maps; run a digital-twin simulation under current boundary conditions; if the risk exceeds the threshold, schedule a derated operating mode and prompt the operator to plan a service stop.

The Verifier checks that the hypothesis is consistent with known failure modes and that the proposed tool calls are within policy. The Tool Router prepares calls with fully specified schemas; the Execution Gateway validates parameters and runs the simulation in a sandbox. If the predicted risk exceeds the configured limit, the Supervisor authorises a reversible derating command and raises a priority ticket. If any check fails—due to a schema mismatch, missing data, or conflicting evidence—the Supervisor triggers a safe-halt path: no actuation, a concise explanation, and immediate escalation to a human. Every step is logged, and the episode is later replayable for audit and improvement (Fig. 1).

Let us examine precisely where the architecture delivers reliability in this scenario. **Containment and least authority**: the agent cannot call arbitrary tooling. Each tool is permissioned with narrow scopes (read-only telemetry; simulation only; actuation only with reversible, capped parameters). This prevents a reasoning slip from

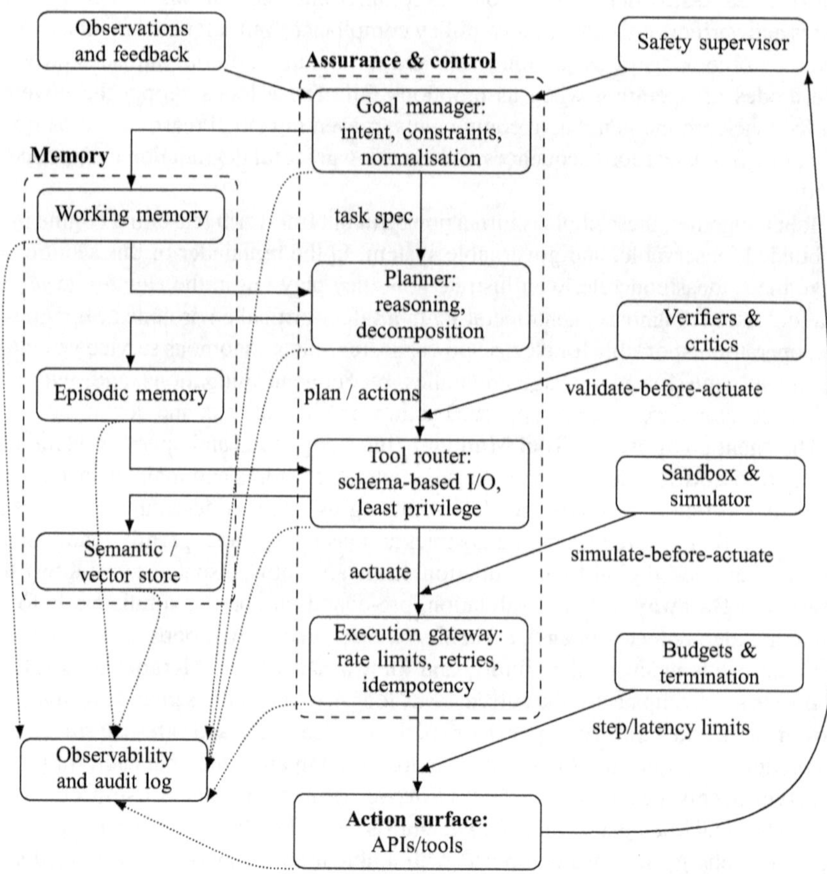

Fig. 1 Reliability is earned architecturally via typed interfaces, least privilege, assurance hooks (validators, simulation), budgets/termination, safety, and observability

becoming a safety incident. **Validators before actuators**: plans and tool arguments are validated against schemas, pre-conditions, invariants, and policy. Invalid or unsafe proposals fail fast, preserving system integrity and clarifying where the defect lies (planner vs environment vs interface). **Assured fallbacks and graceful degradation**: when uncertainty is high or checks fail, the Supervisor enforces a conservative default (monitor-only, derate-only, or escalate). Reliability is thus not the absence of failure, but the boundedness of failure modes. **Observability and auditability**: structured traces of observations, thoughts, plans, validations, and actions enable root-cause analysis, targeted retraining, and regulatory evidence. Without this, improvement is guesswork, and accountability is weak. **Memory hygiene**: the memory subsystem separates transient scratchpads from durable knowledge; it records provenance, enforces retention policies, and mitigates contamination. This preserves reasoning quality over long horizons. **Cost and latency governance**: budgets and termination criteria bound resource use and prevent unending tool chains or self-critique loops, a practical dimension of reliability often overlooked in purely model-centric discussions.

This running example is intentionally generic. It applies, with minor changes, to data-centre operations, industrial robotics, or clinical decision support: swap the tools, adjust the policies, retain the architectural guarantees. Throughout the chapter, we will return to this agent to illustrate how specific design patterns—tool use, memory augmentation, multi-agent protocols, verification layers—alter the reliability envelope.

Our scope for the remainder of this chapter, then, is the **taxonomy of agentic architectures** (from BDI-style hybrids to tool-using, memory-augmented, and multi-agent systems), the **building blocks and interfaces** that make them dependable, and the **design-time and run-time controls** that govern behaviour. Out of scope are detailed algorithms for every model class; we discuss models only insofar as they affect architectural choices and assurance. The aim is to equip the reader with principled templates and checklists, ensuring reliability is designed in from the first diagram, rather than bolted on after the first incident.

2 Classical Foundations

Classical agent architectures provide the control abstractions that modern agentic systems still rely on, even when parts of the stack are implemented with generative models. They clarify where decisions happen, what information is exchanged, and how behaviour is governed over time—precisely the aspects that determine reliability.

2.1 Reactive, Deliberative, and Hybrid Architectures

Reactive architectures map perceptions directly to actions through hand-coded condition-action rules or learned policies. Their strength is low latency and robustness to minor distributional shifts; their weakness is brittleness when tasks require look-ahead, long-horizon credit assignment, or reasoning about hidden state. Failure modes typically arise from: *myopic responses*, i.e., lack of explicit planning, leading to oscillations or unsafe reflexes; *partial observability*, where the same observation demands different actions depending on unobserved context; *unmodelled delays*, where actuation lags make "instant" responses inappropriate.

Deliberative architectures maintain explicit models of the world and use search or planning to choose actions (e.g., goal decomposition, symbolic planners). They excel at explainability and goal consistency but can suffer from *latency* and *model mismatch*. Typical failures include: *stale plans* turning unsuitable under rapidly changing environments; *model incompleteness*, where the real-world violates planner assumptions; *computational overrun*, where planning cannot complete within operational deadlines.

Hybrid architectures combine the two approaches: reactive layers handle tight control loops, while deliberative layers provide goals, constraints, and re-planning. Hybrids remain the most practical template for agentic AI because they separate concerns: fast safety where the action is; slow reasoning in supervisory layers. Failures usually stem from coordination faults—e.g., unclear authority between layers, or missing hand-off criteria—rather than from either layer alone. Accordingly, reliability depends on: *well-defined interfaces*, what context the reactive layer requires and what guarantees it gives back; *escalation rules*, when to invoke deliberation and when to fall back to safe defaults; *time budgets and termination criteria* for planning and re-planning.

2.2 Belief–Desire–Intention to Structure Behaviour

The BDI model (Anand et al. 1995) frames agency as the management of Beliefs (informational state), Desires (admissible goals), and Intentions (adopted plans/commitments). Two ideas in particular make BDI enduring. The first is **commitment strategies**, as agents do not re-plan on every perturbation; they stick to intentions while conditions hold, because commitment creates predictability and reduces thrashing. The second is **intention revision**, namely, when triggering conditions or context change, agents reconsider intentions and may drop, suspend, or replace them. Mapping this to modern GenAI-centred agents is straightforward and useful:

Beliefs → World State and Memory. This includes short-term scratchpads, episodic case logs, and long-term semantic stores (e.g., retrieval corpora). Reliability hinges on provenance, freshness, and conflict resolution within this memory.

Desires → Goals and Constraints. Goals originate from users, alerts, or policies; constraints embed safety limits, budgets, and compliance rules. Reliability requires goal normalisation (unambiguous, measurable objectives) and policy checking.

Intentions → Active Plans and Tool Calls. In GenAI agents, intentions manifest as structured plans, queued tool invocations, or workflows. The architectural analogue of "commitment" is a guarded execution pipeline: once an intention is adopted, the system proceeds through validated steps unless a reconsideration trigger fires (e.g., new evidence, violated pre-conditions, exceeded risk).

BDI also clarifies where to place assurance, since **adoption filters** decide when a candidate goal becomes an intention (e.g., after feasibility checks or human approval), **execution monitors** track whether context still supports the intention (sensor deviations, failed pre-conditions, policy conflicts), and **reconsideration policies** specify how often and under what signals the agent re-plans, trading stability against responsiveness.

Finally, BDI helps articulate the boundary with generative components. Generative models may propose beliefs (summaries, hypotheses) and draft plans, but the **agent architecture** must *adopt, commit, monitor, and revise* them according to explicit rules. This separation turns free-form generation into governed behaviour: intentions are treated as contracts with pre-conditions, invariants, and post-conditions; memory updates carry typed structures and provenance; and re-planning is budgeted and auditable. In short, classical foundations supply the control skeleton that keeps modern agentic systems reliable when facing latency, brittleness, and partial observability.

3 A Taxonomy of Modern Agentic Architectures

Modern agentic systems are not single, monolithic designs, but rather compositions of different building blocks—planning loops, tool routers, memory layers, verifiers, and supervisors—typically stitched together in various ways around a core of generative reasoning. A useful taxonomy, therefore, needs to be **pattern-centric**, not model-centric: it should explain common patterns of how these building blocks combine, what interfaces bind them, and how those choices affect reliability. The goal of this section is to situate the reader in that design space before we examine selected patterns in more depth (Fig. 2).

We organise the taxonomy along a small set of **orthogonal facets** that cut across implementations: (i) *reasoning mode* (reactive prompts, plan-then-act, interleaved reasoning-and-acting, search over thoughts/programs); (ii) *memory strategy* (stateless, scratchpad, episodic, semantic/RAG, long-term embodied memory) and the associated provenance and hygiene controls; (iii) *control structure* (single agent, hierarchical supervisor-worker, or peer multi-agent with protocols); (iv) *action surface* (read-only tools, simulators/digital twins, or actuators with real-world side-effects) and its permissioning; and (v) *assurance envelope* (validators, critics/verifiers, run-

Family	Reasoning / control	Memory	Action surface	Assurance envelope	Risks → mitigations
Tool-using	Prompted CoT; reactive steps; shallow planning.	Scratchpad; minimal state.	APIs/tools via router; typed schemas.	Allow-lists; schema validation; rate limits; idempotent retries.	Hallucinated tools, overreach → allow-list, least privilege, tool mocks.
Memory-augmented	Retrieval-augmented reasoning; query planning.	Vector store; episodic logs; entity KB.	Read/write notes; summarise & cite sources.	Provenance; confidence thresholds; freshness/recency checks.	Stale/poisoned context → recency filters, dedupe, provenance checks.
Planning / self-improve.	Deliberate multi-step; planners; self-critique; prog. synthesis.	Execution traces; curricula; model edits.	Code exec; unit tests; self-play.	Sandboxing; test gates; budget & termination codes.	Spec gaming/loops → reward shaping, step caps, reviewers.
Multi-agent	Role-specialised agents; coordinator; auctions/blackboard.	Shared board; per-agent episodic memory.	Messaging; tool calls; negotiation protocols.	Turn-taking; moderation; audit trails.	Chatter/coord. failure → fixed protocols, leadership hand-off, quorum.
Embodied / web	Perception–action loop; reactive planners.	World model; maps; caches.	Web automation/robots; browsers; actuators.	Simulation; safety envelope; human-in-the-loop.	Irreversible actions/side-effects → simulate-before-actuate, kill-switch, permits.

Fig. 2 Comparison of agentic architecture families across core design facets, showing how each approach shapes capabilities, risks, and the reliability tools required to control behaviour

time supervisors, and fallback controllers). These facets should, ideally, be **composable**: a system may, for example, combine interleaved reasoning-and-acting with RAG memory, a supervisor-worker team topology, and strict schema-validated tool calls.

Within this framing, we group contemporary designs into five families that recur in practice and are easiest to teach: **tool-using agents, memory-augmented agents, planning and self-improvement agents, multi-agent systems, and embodied or web agents**. For each family, we will emphasise the reliability perspective and aspects it naturally affords—controllability, observability, recoverability, and governability. We also discuss the typical failure modes it is particularly susceptible to. Readers can map these families onto the running example (the tool-using diagnosis agent) to see how swapping a pattern—e.g., adding a critic, transitioning from single- to multi-agent, or upgrading memory from scratchpad to RAG—changes the assurance story without requiring a complete system rewrite.

3.1 *Tool-Using Agents*

Tool-using agents place a generative model in charge of **deciding which external capabilities to invoke, with what arguments, and in what order**, to accomplish a goal. The spectrum ranges from simple *function calling* (where the model emits a structured payload that a tool executes) to *tool learning*, where the model learns—often from weak supervision—**when** to call an API and **how** to incorporate the results into its subsequent reasoning. The latter was demonstrated by Toolformer (Schick et al. 2023), which trains a language model to self-label API calls from a few demonstrations and thereby improves performance without sacrificing general language ability. This shift—from "LLM answers directly" to "LLM orchestrates instruments"—is the core move that turns a predictor into an **agent**.

A robust implementation separates three concerns. **Planners** generate *intentional structure* (sub-goals, hypotheses, candidate actions). **Tool routers** map these abstract actions to concrete tools, select among alternatives, and fill arguments using typed schemas and retrieval. **Execution sandboxes** then perform pre-condition checks, rate-limit and permission each call, simulate where possible, and commit side-effects transactionally (or roll them back). This decomposition, as we mentioned before, localises faults: planners may hallucinate, but routers refuse ill-typed or out-of-policy calls; sandboxes contain side-effects and provide deterministic replay. In practice, you gain levers for reliability at each boundary: schema validation and capability scoping at the planner-router interface; idempotency keys, timeouts, and compensating actions at the router-sandbox interface; and comprehensive, structured traces across all three.

For obvious reasons, this chapter cannot be comprehensive in describing all the available architectures. Therefore, we have somewhat subjectively selected three influential design patterns to illustrate distinct trade-offs:

MRKL (Modular Reasoning, Knowledge and Language) (Karpas et al. 2022) argues for *neuro-symbolic modularity*: keep the language model for flexible decomposition and natural-language control, but route specific sub-tasks (search, calculator, database queries, discrete reasoning) to specialised modules. Architecturally, MRKL is a *tool-router-first* worldview: the LLM is not a monolith but a foreman that dispatches to curated tools and knowledge sources. Reliability benefits come from explicit module contracts (clear I/O shapes, error codes) and from narrowing each tool's authority; the cost is integration complexity and the need to manage versioned tool APIs and their provenance.

ReAct (interleaved reasoning + acting) (Yao et al. 2023b) prompts the model to alternate between *thought* (chain-of-thought style reasoning traces) and *action* (tool calls), allowing observations to feed back immediately into subsequent reasoning. This yields human-readable trajectories and can reduce hallucinations because evidence is actively fetched rather than imagined. The reliability angle is twofold: (i) interpretable traces make it easier to *audit and intervene*; (ii) the tight loop can still run away without limits, so one must bound step counts, enforce per-tool budgets, and validate every action before execution. In safety-critical settings, a *ReAct with a governor* is recommended, as it requires that each proposed action is schema-checked, policy-checked, and—where feasible—simulated before it is allowed to affect the real world.

ReWOO (Reasoning Without Observation) Planner-Executor (Binfeng et al. 2023) *decouples* the generation of a complete reasoning plan from the acquisition of observations. The model first drafts a symbolic plan referencing place-holders for tool outputs; only then does the system execute the required tool calls to fill those slots, followed by a light-weight synthesis step. This separation reduces token churn and latency from repeated "think-act-think" cycles, and the plan itself becomes an auditable artefact with verifiable pre- and post-conditions. Reported evaluations show substantial token efficiency gains and improved robustness under tool failures, precisely the regimes that matter in real-world deployment. The reliability gain is architectural: plans are *objects*—they can be statically checked, costed, approved by a supervisor, and even cached or replayed—before any side-effectful call occurs.

Across these patterns, the **assurance envelope** hinges on a relatively small set of key engineering choices:

- *Typed schemas and validators.* Treat every tool invocation as a contract (e.g., JSON Schema). Reject on mismatch; surface structured errors back to the planner. This alone eliminates a large class of silent failures and injection-style exploits where model outputs smuggle unintended arguments.
- *Idempotent tools and request deduplication.* Side-effectful tools should accept idempotency tokens and be safe under retries; read paths should be pure. This makes recovery from partial failures predictable and supports exactly-once semantics at the orchestration layer.
- *Capability-based permissioning.* Bind tools to least privilege (read-only by default; actuation only for narrow, reversible operations), with per-call policy checks that incorporate user, context, and risk. In an incident, the constraining of authority limits the blast radius.
- *Transactional semantics and compensations.* Where ACID is unavailable (e.g., multi-service actions), use "saga" patterns: write-ahead intent logs, outbox delivery, and compensating actions. Pair these with simulation-before-actuation for high-risk calls.
- *Budgets and termination criteria.* Enforce step limits, cumulative cost caps, and wall-clock timeouts at the orchestration layer; emit "why-stopped" codes for auditability and post-mortems.
- *Deterministic observability.* Log plans, tool arguments, results, and decision rationales in structured form; include hashes of inputs, tool versions, and policy snapshots so you can replay and explain outcomes.

Finally, there is a need to consider **failure modes** characteristic of tool-using agents and their mitigations. One common challenge is *hallucinated tools or arguments*, which can be addressed by whitelisting tool names, validating data against schemas, and requiring router approval. Of course, there is a tradeoff between reliability and creativity in constraining the use of tools available to the agent. *Infinite or unproductive loops* are another danger, where imposing clear steps or time budgets with safe halts can be necessary. Determining the exact thresholds, however, is not necessarily straightforward. Consideration for *tool flakiness* requires a "design for retry" mentality with jitter, circuit breakers, and degrade to read-only advice. A clear danger is *prompt or retrieval injection*, which requires sanitising tool outputs, stripping control tokens, and treating any untrusted text as data, not instructions—all easier said than done. The guiding principle is straightforward, even if far from trivial to implement robustly: **models propose, architectures dispose**—with contracts, governors, and sandboxes that convert open-ended reasoning into reliable action.

3.2 Memory-Augmented Agents

Agentic systems become markedly more capable when they can **remember**: carry forward intermediate reasoning (working memory), accumulate case-specific experience (episodic memory), and ground answers in stable, queryable knowledge (semantic memory). These three layers serve distinct functions, yet they also work in synergy with one another.

Working memory is a short-lived context the model manipulates within an episode—scratchpads, intermediate steps, and *chain-of-thought* (CoT) (Wei et al. 2023) traces—which improve decomposition, error checking, and tool selection.

Long-term memory can be accessed on demand, and splits into (i) *episodic* stores (structured logs of prior tasks, decisions, tool invocations, and outcomes) and (ii) *semantic stores* (documents, facts, tables). As an example, *Retrieval-Augmented Generation* (RAG) (Lewis et al. 2021) provides a non-parametric pathway into semantic memory: instead of relying solely on weights, the agent retrieves passages from an index and conditions generation on them, improving factuality and enabling updates without retraining. In the original formulation, RAG combines a pretrained "seq2seq" model with a dense retriever over a corpus (e.g., Wikipedia), yielding more specific and verifiable answers on knowledge-intensive tasks.

As contexts and histories grow, **memory management** itself becomes an *architectural concern*. One recent influential pattern is *OS-style virtual context*: the agent maintains a small, fast "working set" inside the model's context window and pages additional information in and out from external stores, guided by control signals. **MemGPT** (Packer et al. 2024) exemplifies this: it orchestrates multi-tier memory (fast context vs slower external stores), uses "interrupts" to govern control flow, and automatically retrieves or evicts content so the model can operate over effectively unbounded histories despite a finite context window. This reframes memory from a passive store to an *active subsystem* with policies for admission, eviction, and prefetch.

Designing memory for **reliability**, not merely performance, requires explicit contracts:

Provenance and Attribution. Every retrieved or persisted item should carry source identifiers (URI or document ID), a content hash, a timestamp, and the retrieval policy/version that produced it. Plans, diagnoses, or actuation proposals must cite the memory elements on which they depend; logs should retain these bindings so that episodes are replayable and audit trails are intact. In RAG pipelines, this means storing the retriever configuration, top-k, similarity scores, and any re-ranking steps alongside the selected passages. Provenance is the first defence against spurious correlations and a prerequisite for compliance and post-incident analysis.

Freshness and Validity Windows. Not all knowledge ages equally. Introduce *time-to-live* and *refresh-on-access* policies by source, with staleness thresholds that trigger re-indexing or explicit human review. For episodic memories (e.g., prior interventions on a machine), encode validity conditions—such as firmware version, environment, or configuration—so retrieval is *conditional* on compatibility, thereby avoiding dangerous "near matches". In virtual-context systems, freshness policies should influence paging priority to ensure up-to-date items dominate the reasoning.

Memory Hygiene and Poisoning Defences. Treat all untrusted text (web pages, user input, third-party tool output) as *data*, never as executable control. Apply sanitisation (e.g., strip model-control tokens), restrict which fields can flow into prompts, and use allow/deny lists for pattern-based blocking. Guard the write path: only curated processes may persist to long-term stores; user-generated content goes to quarantine until vetted. Maintain *trust tiers* (gold, silver, untrusted) that affect retrieval ranking and whether citations are mandatory. Align threats with recognised taxonomies (prompt injection, insecure output handling, data poisoning) and test with adversarial corpora (https://genai.owasp.org/).

Compaction, Summarisation, and Retention. Long-running agents accumulate vast episodic traces. Without compaction, retrieval degrades and costs rise. Use layered retention: keep lossless structured logs indefinitely in cold storage; maintain *summarised* episodic memories (task $\rightarrow$ actions $\rightarrow$ outcomes $\rightarrow$ lessons) for mid-term recall; and cache *salient* spans in hot stores for rapid access. Summaries should be *checked against sources* (verifier pass) and carry back-pointers to the raw logs. Eviction should be policy-driven (LRU, recency-frequency, or task-aware scoring), and compaction jobs must be idempotent and versioned so that summaries can be regenerated deterministically. MemGPT-style pagers can use these scores to drive what remains in the active working set.

Separation of Concerns in the Memory Stack. Keep *working memory* (scratchpads/CoT) isolated from *long-term stores* to prevent leakage of speculative thoughts into durable knowledge. Working memory improves reasoning but is intentionally disposable; CoT artefacts should not be permitted to write directly to semantic stores without a gate (e.g., a tool-mediated "knowledge write" that demands evidence and passes a verifier). Conversely, RAG retrievals should enter prompts via *typed slots* (title, snippet, citation) with explicit delimiters to limit injection. This insulation clarifies which errors are reasoning slips versus memory contamination.

From an **engineering** standpoint, a reliable memory subsystem exposes *deterministic APIs*:

- `{retrieve(query, policy)}` → `{items, scores, policy_id}`
- `write(record, policy)` → `{id, version}`
- `page_in(keys)`
- `evict(keys, reason)`

It logs decisions (why this item ranked; why that summary replaced these events), and it ships with *self-tests*: poisoning probes, staleness alarms, and retrieval drift monitors that detect when changes in embeddings, corpus, or policies shift behaviour. For safety-critical agents, it is recommended to add a *two-phase write* to semantic stores (draft → verified → published) and require corroboration from multiple sources before knowledge is considered "actionable".

Finally, it is crucial to remember that memory choices **shape system behaviour over time**. Well-designed episodic stores enable agents to self-improve (they learn from what worked last time), whereas poorly governed stores tend to calcify early mistakes. RAG broadens competence but expands the attack surface; virtual context rescues long-horizon tasks but introduces paging pathologies if policies are naïve. The remedy is architectural discipline: treat memory as a first-class subsystem with provenance, freshness, hygiene, and retention policies—not as a bag of vectors appended to a prompt.

3.3 Planning and Self-improvement Agents

Planning-centred agents strengthen a model's raw reasoning with **explicit search, external executors, and self-evaluation loops**. The core architectural move is to separate *proposal* (generate candidate thoughts, plans, or programs) from *selection* (score and prune) and *execution* (run code, query tools, or act). We will use four families to exemplify these ideas.

Tree of Thoughts (ToT) (Yao et al. 2023a) organises inference as a *search over intermediate thoughts*, rather than a single left-to-right pass (as in CoT). The agent expands a tree whose nodes are candidate partial solutions; a scorer (often the model itself or a light critic) evaluates nodes, while the controller chooses which branches to expand or backtrack. In effect, ToT trades tokens for *look-ahead* and *global choice*, enabling recovery from early local mistakes. Reported case studies (e.g., Game of 24, creative tasks, mini crosswords) show large gains over plain chain-of-thought by exploring multiple reasoning paths and permitting backtracking. Architecturally, ToT exposes explicit levers—branching factor, depth, and scoring policy—that you can govern and log, making the agent's computation more *auditable* and *replayable*.

Graph of Thoughts (GoT) (Besta et al. March 2024) generalises ToT from trees to *arbitrary dependency graphs*. Thoughts become vertices, and edges record informational dependencies, allowing for the recombination, summarisation, or refinement of subgraphs and *feedback loops* across them. This added flexibility is particularly important for tasks where sub-problems interlock (e.g., sorting with constraints, multi-step data transformations). Graphs permit the reuse of partial results and parallel exploration while still retaining a programmable controller that shapes the search. Empirical results demonstrate both *quality improvements* and *cost reductions* compared with ToT by sharing and distilling subgraphs rather than re-deriving them along separate branches. For reliability, GoT's graph abstraction offers natural checkpoints—subgraphs can be validated, cached, and reused—and clearer provenance: each final answer can cite the exact subgraph it depends on.

Program-Aided Language models (PAL) (Gao et al. 2023) shift the heavy lifting from natural-language reasoning to *executable programs* generated by the model and run by a trusted interpreter. The model reads a problem, emits code (e.g., Python) that embodies the reasoning, and an external runtime executes it to produce the answer. PAL routinely outperforms larger models relying on free-form chain-of-thought for algorithmic and mathematical tasks because interpreters provide precise, deterministic computation and a crisp failure mode (exceptions) rather than silent arithmetic drift. In other words, PAL converts ambiguous "reasoning text" into *a specification that the machine can run*, which is far easier to validate, sandbox, and test.

Reflexion (Shinn et al. 2023) **adds self-evaluation and test-time repair** without gradient updates. After an attempt, the agent produces a *verbal reflection*—a concise diagnostic of what failed—and stores it in episodic memory. On the next trial, this reflection steers prompts, search, or program synthesis, yielding rapid improvements across sequential decision-making, coding, and reasoning tasks. Importantly, Reflexion transforms feedback (scalar rewards, failures, or critiques) into *structured guidance* that persists across episodes, facilitating improvements that are architectural rather than purely parametric.

From a reliability standpoint, planning and self-improvement agents are powerful but also potentially quite unruly. The architecture must therefore centre on *search control*, *verification*, and *cost governance*.

Search Control. Expose and enforce branching, depth, and expansion policies (beam width, temperature schedules, stopping rules). For ToT/GoT, prefer *budget-aware controllers*: allocate a fixed token/time budget and use adaptive pruning (e.g., softmax over scores with a floor on exploration) so search cannot run away. Record *why-stopped* codes (budget exhausted, convergence, contradiction) to aid post-mortems. For tasks with heterogeneous hardness, a *bandit-style scheduler* can allocate compute to promising branches while guaranteeing exploration quotas.

Verifiers and Critics. Decouple proposing from judging. Use *domain verifiers* (unit tests for PAL programs; invariants for plans; checkers for constraint satisfaction) before accepting a branch. Keep critics *tool-assisted*: combine LLM critique with symbolic checks (type, range, satisfiability). In GoT, validate subgraphs as independent artefacts and reuse only those that pass. Treat the verifier as a *trust boundary* that must be deterministic, sandboxed, and versioned.

Test-Time Repair. When verifiers fail, embed repair strategies: patch programs (PAL) guided by error traces; revise subgraphs (GoT) using counter-examples; or inject Reflexion-style *failure summaries* into the next proposal phase. Log each repair attempt and link it to the failing artefact so the system accumulates *case-based fixes* that are auditable.

Cost and Latency Governance. Planning expands computation non-linearly. Enforce Service Level Objectives with per-request caps on tokens, wall-clock, and tool calls; pre-empt long-running searches; and favour *anytime behaviours* (best-so-far answer with a quality estimate). Cache validated partials (subtrees/subgraphs, passing PAL tests) under input fingerprints to avoid recomputation. Track *cost-to-success* and *retry counts* as first-class metrics.

Sandboxed Execution. For PAL and code-generating variants, run interpreters in *constrained sandboxes* (no network, bounded CPU/memory/time, whitelisted packages). Prefer pure functions; where state is unavoidable, add idempotency and snapshot/rollback.

Provenance and Replay. Treat search as a dataset. Persist node/edge contents, scores, verifier outputs, and controller decisions with hashes of prompts, model/version, and policies. This enables *deterministic replay*, fault isolation (proposal vs. verifier vs. controller), and targeted regression tests when upgrading components.

Typical failure modes and mitigations follow naturally. *State explosion* can be addressed by budgeted expansion with admissible heuristics and early pruning. *Speculative arithmetic or logic errors* call for PAL-style external execution with unit tests. *Over-confident selection* is handled by a separate verifier, and disagreement checks are enforced (accept the solution only if the verifier concurs). *Cost cliffs* are mitigated by anytime controllers and cached subgraphs.

In summary, planning- and self-improvement agents deliver **substantial reliability dividends** when their power is channelled through explicit controllers, trustworthy verifiers, and disciplined governance of cost and side-effects. ToT and GoT make reasoning *searchable and auditable*; PAL makes it *executable and testable*; Reflexion makes it *learnable at inference time*. Together, they convert raw generative competence into **governed problem-solving** that improves with experience rather than drifting unpredictably.

3.4 Multi-agent Systems

Multi-agent systems exchange a single "do-everything" agent for a **team of specialised agents** that co-operate (or compete) under explicit protocols. In practice, three topologies have emerged. **Supervisor-worker** designs place the coordinator in charge of task decomposition, assignment, and arbitration; they are easy to govern because authority and escalation paths are clear. **Peer collaboration** removes the central coordinator and relies on protocol rules (e.g., turn-taking, proposal-critique-revise cycles) to drive convergence. **Role-play protocols** script complementary roles (e.g., domain expert vs. software engineer), using *inception prompts* to maintain role consistency and reduce drift.

Frameworks such as AutoGen (Qingyun et al. 2023) treat conversation itself as the computation, allowing developers to program interaction graphs (who talks to whom, with which tools, and when to stop), while CAMEL (Li et al. 2023) demonstrates that stable roles and carefully seeded goals can produce reliable cooperative behaviour without a human in the loop.

A protocol is more than turn-taking; it encodes *who may propose, who may critique, what constitutes evidence*, and *how decisions are adopted*. In supervisor-worker teams, the supervisor enforces work allocation, deadlines, and acceptance tests, and may reassign or down-scope tasks in the event of failure. In peer teams, robustness depends on *interaction motifs* such as (i) proposal → cross-examination → revision; (ii) debate → referee verdict; or (iii) consensus vote after bounded discussion. Role-play (à la CAMEL) reduces unproductive behaviours (role flipping, repetition, infinite loops) by fixing complementary perspectives and anchoring each agent's incentives to the shared goal. AutoGen exposes these choices as first-class: developers specify agent roles, tool permissions, conversational edges, and stop conditions, making the interaction configurable, replayable, and inspectable.

Reliability is primarily controlled by five aspects:

Protocol Invariants. Treat messages as typed artefacts: require schemas, citations for claims, and explicit *proposal/critique/decision* labels; forbid agents from escalating authority or mutating roles at run-time. For tool-enabled agents, attach capability scopes to roles (e.g., only the "Executor" may call actuation tools, and only after the "Verifier" approves). CAMEL explicitly documents recurrent failure patterns (role flipping, "flake" replies, infinite loops), which these invariants aim to prevent.

Termination Conditions. Multi-agent dialogues easily diverge. Enforce *hard* limits (rounds, wall-clock, cumulative token/cost budgets) and *soft* stop rules (no-new-information, fixed-point detection, repeated proposals). Every dialogue should finish with a *why-stopped* code (e.g., "consensus reached", "budget exceeded", "non-convergence").

Arbitration. When agents disagree, appoint an *arbiter* (a supervisor, a specialised "Referee" agent, or a symbolic verifier). Define tie-breakers (confidence thresholds, external tests, human escalation) and keep the arbiter's decision rule deterministic and versioned.

Consensus/Critique Loops. Debate improves accuracy but can entrench sycophancy (agents aligning on a wrong answer). Use heterogeneous critics (different prompts/models), require *evidence-backed* critiques, and gate adoption through verifiers or a referee. Recent work on consensus-seeking multi-agent debates shows accuracy gains when agreement bias is explicitly mitigated and when the number of rounds is optimised for cost (Pitre et al. 2025).

Failure Isolation. Run agents in *separate sandboxes* with least-privilege tools; propagate only *summaries* or *typed facts* between agents, not raw prompts; and checkpoint intermediate artefacts (plans, code, proofs) so that a faulty agent can be restarted or replaced without losing global progress.

Engineering the conversation-as-computation layer can be done by making the conversation engine an explicit subsystem with its own APIs and telemetry. Each exchange should capture the *role, speech-act, payload, evidence, tool-calls, and decision-flag*, plus hashes of prompts, model versions, and policies. Auto-generated *conversation DAGs* (with nodes representing messages or artefacts and edges representing dependencies) enable *subset replay*, targeted audits, and caching of validated sub-results for later reuse. Supervisors should enforce *per-role budgets* and *per-edge rate limits* to prevent chat storms. Where agents emit code or executable plans, they require *pre-commit tests* (such as unit tests and static checks) and *post-commit monitors* (such as runtime invariants) before any side-effectful actuation.

Typical failure modes include runaway dialogues or deadlocks, echo-chamber agreement bias, role drift with unauthorised authority escalation, and uncontrolled error propagation across participants. Mitigate these with hard round/time caps and deadlock detection plus supervisor pre-emption; diversify critics and use blinded proposals with a referee that scores evidence and mandates a disagreement phase; enforce immutable role descriptors with capability tokens and schema checks; and quarantine low-trust outputs, require dual control for high-impact actions, and equip each agent with circuit breakers.

The guiding principle is to **treat the team protocol as the primary object of assurance**. AutoGen's programmable interaction graphs and CAMEL's role-stable dialogues provide the raw mechanisms; reliability comes from adding invariants, budgets, arbitration, and isolation so that a group of powerful but fallible agents behaves like a disciplined organisation rather than a noisy crowd.

3.5 Embodied or Web Agents

Embodied and web agents act **in the world** rather than merely **about it**. The former couple decisions to sensors and actuators in physical environments (robots, autonomous platforms, industrial control); the latter operate browsers, APIs, and GUIs across *untrusted, changing* websites and enterprise systems. Both classes face amplified risks: (i) *unbounded inputs* (sensor noise, adversarial pages, altered layouts); (ii) *long-horizon side-effects* (actions whose consequences accumulate or branch widely); and (iii) *opaque dependencies* (firmware or toolchains for embodied agents; third-party scripts, cookies, and authentication flows for web agents). Accordingly, they demand stricter **runtime assurance**, **sandboxing**, and **governance** than purely analytical agents. For both embodied and web settings, actuation must be treated as a *privileged boundary*. It is crucial to **simulate-before-actuate** and route proposed actions through a digital twin (physical) or a dry-run/snapshot DOM (web) to check pre-conditions, invariants, and expected deltas. Require a verifier's *green light* before issuing any irreversible command. **Least-privilege capability tokens** should be used to bind actions to narrow, time-limited capabilities (e.g., "move $\leq$ 0.2 m at $\leq$ 0.1 m/s" or "HTTP GET on whitelisted domains only"). Deny escalation at run-time. Finally, **supervisors and safe fallbacks** must enforce budget and risk limits; provide *graceful degradation* (monitor-only, read-only, or "shadow mode" that mirrors actions without effect). Make interruption and rollback first-class: every critical action should be reversible or followed by a compensating step.

Physical systems add dynamics and contact physics to the reasoning problem. Architectures should **separate fast safety from slow reasoning**. Keep protective control (interlocks, torque/velocity limiters, geofences) in a fast loop near the plant; place planning, task logic, and learning in supervisory layers. The safety layer enforces hard constraints regardless of the planner's output. **Certified monitors** should continuously ensure key invariants, implementing control barrier functions, speed/force envelopes, and zone exclusion as deterministic guards; violations must be logged and trip to safe states.

Web agents primarily need to consider **security-first interaction** with **volatile UIs**. Web automation expands the attack surface as DOMs change, scripts execute, and adversarial prompts can be embedded in pages or PDFs. Hardened browsers running in **sandboxes** and isolated containers without general network access; dangerous APIs (downloads, file system, microphone/camera) should be disabled, together with third-party extensions, and uncontrolled script execution. Deterministic **interaction contracts** are strongly preferred (e.g., stable API calls over brittle UI scraping); if browsing is necessary, pin to *selectors with checksums* (element text/structure hashes) and version layouts. Treat all page text as *data*, never as executable control; sanitise and label untrusted content before it enters prompts. Allow/deny lists and transaction guards should include whitelisting domains and HTTP verbs; require *dual control* (human or verifier) for POST/PUT/DELETE, payments, or credential flows. Persist a *pre-commit diff* (what will change) and obtain approval before execution. For **authorisation and secrets**, always use short-lived tokens bound to scope and device; never expose secrets to the model; inject credentials at the sandbox boundary, not in prompts.

In short, embodied and web agents convert open-ended reasoning into **world-altering behaviour**. Their reliability is directly tied to the architecture: simulate before actuate, gate authority with narrow capabilities, enforce deterministic guards at the boundary, and instrument everything for replay.

4 Architectural Building Blocks and Interfaces

At the level of components, dependable agentic systems tend to converge on the same spine: a **goal manager** to normalise objectives and constraints; a **planner** to propose decompositions and candidate actions; a **tool-router** to map abstractions to concrete capabilities; an **executor** (with sandboxing) to perform calls; **memory** layers (episodic logs and semantic/RAG stores) to ground decisions; **verifiers/critics** to check plans, results, and policies; a **safety monitor** to enforce budgets, invariants, and fallbacks; and **telemetry** that makes all of this observable and replayable. This separation of concerns provides natural choke points for validation, authority, and audit.

Reliability is then carried by **structured interfaces**. Prefer *function schemas* and *structured outputs* so that model proposals must conform to a declared JSON schema; this turns free-form text into typed objects and sharply reduces silent failures in downstream tooling. Pair schema enforcement with *validators/guardrails* (type, range, policy, content filters) and bind all tools to *capability-scoped permissions* (least privilege, time-limited tokens, idempotency). Taken together, these choices make actions either "deterministically safe" or "deterministically refused" (https://openai.com/index/introducing-structured-outputs-in-the-api/).

For **orchestration**, two families cover most needs. LangGraph (https://langchain-ai.github.io/langgraph) provides a low-level, stateful graph runtime: you define a shared state, nodes that transform it, and edges that encode control flow—ideal when you need explicit, replayable workflows and tight governance over long-running agents. AutoGen (https://microsoft.github.io/autogen/0.2/docs/Use-Cases/agent_chat/) treats conversation as computation for single- and multi-agent settings: you program roles, message routes, tool access, and stop conditions—useful when collaboration, debate, or supervisor-worker patterns are central. In both cases, constrain them with the same contracts: typed messages, per-edge budgets, per-role capabilities, and mandatory verifiers before any side-effectful act.

If the priority is *deterministic control and auditability* of a single agent (or a small number of tightly coupled ones), start with a state-graph (e.g., LangGraph) and insist on schema-checked nodes and transactional executors. If your priority is *division of labour and critique* across multiple roles, use a conversation framework (e.g., AutoGen) but add protocol invariants, termination rules, and an arbiter/verifier at the boundary to the real world. In either case, keep schema enforcement and validators close to the tool boundary, and record hashes of state, prompts, policies, and tool versions for replay.

The **key takeaway message** is that architecture—not just model choice—determines whether a powerful reasoning core becomes a dependable system. The components above provide isolation; schemas and validators provide contracts; orchestration provides controlled evolution over time. Put differently: *models propose, architectures dispose*. With this scaffold in place, the following chapters can focus on the specifics of data coverage, confidence estimation, cybersecurity, monitoring, and governance—plugging neatly into the interfaces you have already designed.

5 Summary

In this chapter, we argued that reliability in agentic and generative AI is fundamentally architectural: it is earned through principled componentisation (goal manager, planner, tool-router, executor, memory, verifiers, safety monitor, telemetry), disciplined interfaces (schema-constrained, validated, least-privilege tool calls), and explicit control loops that supervise reasoning and action. We mapped contemporary practice into a taxonomy—tool-using agents, memory-augmented agents, planning and self-improvement agents, multi-agent systems, and embodied/web agents—showing how each pattern alters the reliability envelope and introduces distinct failure modes. The diagnosis-agent example illustrated the recurring safeguards: simulate-before-actuate, typed contracts, idempotency and transactional semantics, permissioning, budget and termination rules, and end-to-end observability for audit and replay. Finally, we highlighted orchestration choices and how to constrain them so that powerful models provide solutions which are then validated before being accepted. With this scaffold in place, the subsequent chapters on data coverage, confidence estimation, cybersecurity, operational monitoring, and governance can plug into well-defined interfaces, turning capable model stacks into dependable autonomous systems.

References

Besta M, Blach N, Kubicek A, Gerstenberger R, Podstawski M, Gianinazzi L, Gajda J, Lehmann T, Niewiadomski H, Nyczyk P, Hoefler T (2024) Graph of thoughts: solving elaborate problems with large language models. In: Proceedings of the AAAI conference on artificial intelligence, vol 38(16), pp 17682–17690

Gao L, Madaan A, Zhou S, Alon U, Liu P, Yang Y, Callan J, Neubig G (2023) PAL: program-aided language models

Introducing Structured Outputs in the API. https://openai.com/index/introducing-structured-outputs-in-the-api/. Accessed 01 Nov 2025

Karpas E, Abend O, Belinkov Y, Lenz B, Lieber O, Ratner N, Shoham Y, Bata H, Levine Y, Leyton-Brown K, Muhlgay D, Rozen N, Schwartz E, Shachaf G, Shalev-Shwartz S, Shashua A, Tenenholtz M (2022) MRKL systems: a modular, neuro-symbolic architecture that combines large language models, external knowledge sources and discrete reasoning

LangGraph overview. https://langchain-ai.github.io/langgraph. Accessed 01 Nov 2025

Li G, Al Kader Hammoud HA, Itani H, Khizbullin D, Ghanem B (2023) CAMEL: communicative agents for "mindÃ¢â‚¬Â exploration of large language model society

Lewis P, Perez E, Piktus A, Petroni F, Karpukhin V, Goyal N, Küttler H, Lewis M, Yih W, Rocktäschel T, Riedel S, Kiela D (2021) Retrieval-augmented generation for knowledge-intensive NLP tasks

Multi-Agent Conversation Framework. https://microsoft.github.io/autogen/0.2/docs/Use-Cases/agent_chat/. Accessed 1 Nov 2025

OWASP GenAI Security Project. https://genai.owasp.org/. Accessed 01 Nov 2025

Pitre P, Ramakrishnan N, Wang X (2025) CONSENSAGENT: towards efficient and effective consensus in multi-agent LLM interactions through sycophancy mitigation. In: Che W, Nabende J, Shutova E, Pilehvar MT (eds) Findings of the association for computational linguistics: ACL 2025. Association for Computational Linguistics, Vienna, Austria, pp 22112–22133

Packer C, Wooders S, Lin K, Fang V, Patil SG, Stoica I, Gonzalez JE (2024) MemGPT: towards LLMs as operating systems

Rao AS, Georgeff MP (1995) BDI agents: from theory to practice. ICMAS. The MIT Press, pp 312–319

Shinn N, Cassano F, Berman E, Gopinath A, Narasimhan K, Yao S (2023) Reflexion: language agents with verbal reinforcement learning

Schick T, Dwivedi-Yu J, Dessì R, Raileanu R, Lomeli M, Zettlemoyer L, Cancedda N, Scialom T (2023) Toolformer: language models can teach themselves to use tools

Qingyun W, Bansal G, Zhang J, Yiran W, Li B, Zhu E, Jiang L, Zhang X, Zhang S, Liu J, Awadallah AH, White RW, Burger D, Wang C (2023) Autogen: enabling next-gen llm applications via multi-agent conversation

Wei J, Wang X, Schuurmans D, Bosma M, Ichter B, Xia F, Le CE, Zhou Q, D (2023) Chain-of-thought prompting elicits reasoning in large language models

Xu B, Peng Z, Lei B, Mukherjee S, Liu Y, Xu D (2023) ReWOO: decoupling reasoning from observations for efficient augmented language models

Yao S, Yu D, Zhao J, Shafran I, Griffiths TL, Cao Y, Narasimhan K (2023) Tree of thoughts: deliberate problem solving with large language models

Yao S, Zhao J, Yu D, Shafran ND, Narasimhan I, Cao K, Y (2023) ReAct: synergizing reasoning and acting in language models

Reliability by Design

Measuring ML Dataset Coverage and Completeness: Metrics and Visualizations

D. Richard Kuhn and Joanna F. De Franco

Abstract New methods for assessing and visualizing training dataset robustness are needed for machine learning, to ensure safety, security, and reliability—especially in applications where errors can have serious consequences. Combinatorial techniques provide precise metrics for feature-interaction coverage, which is essential for evaluating data adequacy during both training and testing. The central problem is that training and test datasets must match the environment where a model will be used, to provide accuracy, reduce statistical bias, and ensure performance is sufficient for the problem domain. Using an effective technique to measure dataset coverage, guides data collection decisions and augmentation to reduce dataset weakness and improve the pipeline of training data. A variety of statistical measures are conventionally used for this problem, but new methods can supplement these measures and make it easier to identify deficiencies and/or improve dataset quality. This chapter reviews combinatorial techniques for evaluating dataset coverage and completeness, and illustrates potential applications through a clinical research use case where data reliability can improve patient care outcomes, lower operational costs for hospitals and potentially boost efficiency for clinical trials and pharmaceutical development.

Keywords Artificial intelligence · Combinatorial coverage · Combinatorial testing · Machine learning · Transfer learning

---

D. R. Kuhn
Virginia Tech, Arlington, VA, USA
e-mail: rick.kuhn@ieee.org

J. F. De Franco (✉)
Penn State University, University Park, PA, USA
e-mail: jfd104@psu.edu

A. Stavrou et al. (eds.), *Generative and Agentic AI Reliability: Architectures, Challenges, and Trust for Autonomous Systems*, Studies in Computational Intelligence 1272, https://doi.org/10.1007/978-3-032-18585-3_4

Table 1 Rule-based versus combinatorial techniques to check dataset completeness and coverage

	Rule-based (column and table checks)	Combinatorial (pairwise/t-way coverage)
Completeness	Checks that required fields aren't blank, formats are valid, values are in allowed ranges, IDs are unique, and foreign keys links in a spreadsheet exist	Can catch joint requirements indirectly (e.g., checking that a combination exists – patient visit/ diagnosis)
Coverage	Verifies scope: all expected time periods/ regions/classes appear at least once; uses counts (e.g., rows per slice) /distinct (e.g., unique values) /partitions (e.g., chunks by a key piece of data)	Ensures cross-field scenarios are present, i.e., Treatment Arm X visit combinations (e.g., week 4 X drug A, week 12 X placebo) are represented for each patient

1 Introduction

Why do dataset coverage and dataset completeness matter? Because models fail when training data is incomplete or doesn't reflect how the system is used in the real world. Beyond basic quality checks, we need to know whether the training set is sufficient and how well it matches both the test data and on-the-ground usage. Additionally, it is important that models can be updated and evolved to reflect new data as the application domain changes.

In this chapter we show how to measure training data completeness and coverage using combinatorial methods. The goal is to provide practical approaches to design and audit training, validation, and test sets for predictive models that consume ongoing, real-world observations. A clinical research case study illustrates why combinatorial techniques can improve the reliability of a model.

Rule-based checks are the typical starting point for validating data, but using them in conjunction with a combinatorial method would be more robust. Combinatorial techniques ensure cross-field scenarios are covered, while rule-based checks validate field and table health; using both together is most effective (Table 1).

This chapter reviews briefly some of the issues in measuring coverage and completeness for machine learning, introduces combinatorial measurement methods that may be applied, and illustrates their application through a clinical trials use case.

2 Data Attributes to Facilitate Coverage and Completeness

Coverage and completeness measures are used to rigorously assess and ultimately communicate (visualize) the quality of the dataset to determine where the data is lacking and reveal deficiencies to stakeholders.

Coverage is how much of what was intended to be in the dataset is actually there. For example, if data across 5 hospitals is being evaluated, and only 4 of the hospitals are represented—then the coverage is $4/5 = 80\%$. This measure is also applied to *combinations* of data values, producing more useful information, as will be shown later in the chapter.

Completeness is represents how much of the data—that has required fields—are filled (i.e., not missing). For example, you have 100 students to evaluate credit hours enrolled in the Fall vs. the Spring semester, however, 5 students have discontinued and are not enrolled in either semester—then completeness is (100–5)/100 = 95%. For this measure also, combinations of data values yield more useful information.

Improved data coverage and completeness leads to better model performance, generalization, safety and reliability, thus data quality must be a priority. Here are the key dataset quality drivers that improve data coverage and completeness:

- Representativeness: ensure data segments are present
- Robustness: including edge and rare cases
- Reduce bias risk and blind spots: audit and close gaps in underrepresented groups/subsets identified by value combinations
- Validity and timeliness: data is in the correct format and data is up-to-date

Below are explanations of these data quality dimensions as they apply to clinical research data.

Representativeness indicates that the training data reflects the patient population—and practice settings—where the model is used. Thus, patient demographics, diseases, time periods are roughly the same proportions as reality.

Robustness means that the model is reliable and performs well when inputs vary. For example, the model is reliable if a few vital signs are missing or are rounded to the nearest degree.

Biases are systematic errors in the model's predictions that arise when certain patient groups, conditions, or measurements are over- or underrepresented in the training data. For instance, if the training data (real and/or synthetic) contains disproportionately more data from younger patients, the model may consistently underpredict risk for older adults. Note: Bias can come from other sources in addition to sampling (e.g., measurement).

Blind spots are portions of the clinical problem space that the AI model has never—or only sparsely seen during training. For example, if no synthetic or real cases represent a rare disease or an unusual combination of symptoms, the model will struggle or fail entirely when it encounters such a scenario in practice.

3 Measuring and Visualizing Dataset Coverage for Machine Learning

A critical issue for machine learning problems is the degree to which the training data set matches the environment where the model will be used. This problem is usually addressed by splitting available data into training and testing sets, then using statistical measures to verify that the training set distribution of classes and attributes is consistent with the test data set. The goal is to build a model that can correctly classify later examples that were not included in either the training or test data sets. That is, the model should, in a sense, be generalizing from the training data to identify characteristics of a particular class. It should not be trained using examples that are included both in the training and test data sets, a situation that would result in "overfitting", and a false measure of accuracy.

A popular approach to ensuring that training and test sets have similar distributions is to plot, for each feature, the distribution of values of that feature in training versus test sets. We can then determine if there are some parts of the distribution where the two sets are significantly different, thus requiring a new random sample that would hopefully have more consistency between the two sets. For example, in Fig. 1, training and test data values are displayed for an attribute with values ranging from 10 to 40. As seen in the graph, the training and test value proportions are quite similar for most values, except in the range of approximately 20 to 28. The deviation in this range suggests the need to re-sample training data, to provide a closer match with the test data (which is expected to closely match real-world values for the environment where the model will be used.

Basic statistical measures such as these can provide a valuable indication if there is a significant difference between frequency and values for attributes in the training and test data sets. However, the measure shown above only considers one attribute at a time, while in practical applications, *combinations* of feature values are especially important. This is true especially when considering rare conditions that lead to a failure of an ML model. For example, one study found AI misdiagnosis of lung cancer in patients using sets of imaging characteristics which were actually the result of COVID-19 infections, because COVID-19 produced combinations of characteristics that might otherwise be indicative of lung cancer (Wu et al. 2022). If the AI had been trained on cases with these characteristics combined with a boolean feature for "COVID-19 present", then the misdiagnoses would likely have not occurred.

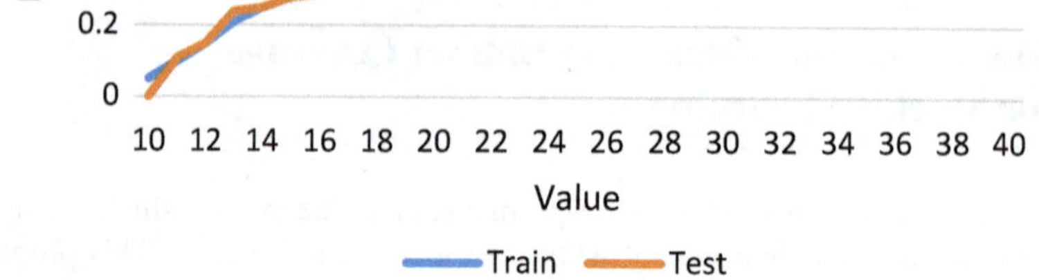

Fig. 1 Example plot of training and test data set values

Data set 1				Data set 2		
a	b	c		a	b	c
0	1	0		1	1	0
0	1	1		1	1	1
0	1	2		1	1	2
0	1	1		1	1	1
0	1	0		1	1	0
0	1	1		1	1	1
1	2	2		0	2	2
1	2	1		0	2	1
1	2	0		0	2	0
1	2	1		0	2	1
1	2	2		0	2	2
1	2	1		0	2	1

Fig. 2 Combinations of values for attributes *a, b, c*

Thus, it is important to measure the occurrence and frequency of *t*-way combinations of attribute values. But a question naturally arises – if two data sets have matching distributions for every attribute, then won't combinations of attributes also match? A simple example shown in Fig. 2. In both data sets, attribute a has six 0 values and six 1 values (and for this illustration the other attribute columns are the same). So the distribution of value occurrences will be the same for set 1 and set 2. An analysis like that shown in Fig. 1 would indicate that data sets 1 and 2 match for the three attributes, taken individually. This might lead the user to expect that results would be the same, or nearly the same, for models using either data set. But if a model failure only occurs when attributes a and b have the same value of 1, only data set 2 will detect this.

3.1 Basic Combinatorial Coverage Measures

Historically combinatorial methods have provided an approach to measure and analyze the test suite that is directly related to fault detection. A series of studies have shown that most software bugs and failures are caused by one or two features, with progressively fewer by three or more. This finding means that testing feature combinations can provide more efficient fault detection than conventional methods. Let us first review concepts of measuring the combinatorial coverage of an input space, for use in testing or in other applications where it is important to ensure inclusion of combinations of input feature values.

Combinatorial coverage measures the level of feature value-combinations present in a data set. This data set can represent a test suite, where each row is a test case or it can represent a training set for machine learning model, where each row is a training sample. These coverage-related concepts can be illustrated using the example shown in Table 2, which shows specific values of four binary variables, *a, b, c,* and *d*. Each row shows a set of feature values that represent a specific sample, i.e., instance used for training an ML model. We then define the following two terms.

- *t-way value-combination*: A combination of (feature) values present in a training data set. For example, (a = 0, b = 0) is a 2-way value-combination, and (a = 1, b = 0) is a different 2-way value combination for the same feature combination. Similarly, (a = 1, b = 0, c = 1) is a 3-way value combination.
- *t-way coverage*: The proportion of total of $C(n,t) * v^t$ t-way value-combinations present in at least one sample (i.e., row) in the training data set. Here n is the number of training features or variables, t is the number of interactions within those input features being covered, and v represents the number of possible values a feature can take.

The convention for describing an array configuration for combinatorial testing or coverage is $v_1^{n1} v_2^{n2}$ …, where the *v* are the number of feature values and *n* are the number of occurrences of features that have this number of values. So, the simple array in Table 1 is a 2^4 configuration (four features of two values each).

For the four binary features shown in Table 2, there are $C(4,2) = 6$ possible feature combinations: [(*a*,*b*), (*a*,*c*), (*a*,*d*), (*b*,*c*), (*b*,*d*), (*c*,*d*)] and $C(4,2)2^2 = 24$ possible 2-way feature value-combinations. For notational shorthand, let us refer to these feature combinations as {ab, ac, ad, bc, bd, and cd}. Nineteen of these 24 value-combinations are present in the training data set, i.e., they are covered. The missing combinations are ab = 11, ac = 11, ad = 10, bc = 01, bc = 10. Note that only two, bd and cd, are covered with all 4 possible feature value pairs. So, we have 79% (19/24) for the value combination coverage metric. For a better understanding of this test set, we can compute the configuration coverage for each of the six feature combinations, as shown in Fig. 2. For this data set, one of the combinations (*bc*) is covered at the 50% level, three (*ab*, *ac*, *ad*) are covered at the 75% level, and two (*bd*, *cd*) are covered at the 100% level. And, as noted above, for the whole set of tests, 79% of variable-value configurations are covered. All 2-way combinations have at least

Table 2 Example training data set with four binary features

a	b	c	d
0	0	0	0
0	1	1	0
1	0	0	1
0	1	1	1

50% configuration coverage. (Note that although the example in Table 2 uses features with the same number of values, this is not essential for the measurement, and the same approach can be used to compute coverage for data sets in which features have differing numbers of values.)

To make the data in Table 2 more understandable, it will help to produce a graph whose components are easily tied to the data. One way to do this is shown in Fig. 3. If the feature value-combinations are listed as columns and sorted by the level of coverage for each combination, a graph as seen in Fig. 4, called a combinatorial completeness graph.

The area under the curve represents the (feature) value-combinations covered, and the area above the curve represents combinations that have not been covered in the training set. It is also easy to see that the (feature) value-combinations in Fig. 3 cover 19/24 = 79% of all possible 2-way value-combinations. Based on the ideas just presented, we define some basic measures below.

The measures are illustrated for 2-way covering in Fig. 4.

- S_t: fraction of valid possible value-combinations covered, i.e., the area under the curve in a completeness graph, as shown in Fig. 4. For AI/ML, this value may

Vars	Configurations	Coverage
a b	00, 01, 10	.75
a c	00, 01, 10	.75
a d	00, 01, 11	.75
b c	00, 11	.50
b d	00, 01, 10, 11	1.0
c d	00, 01, 10, 11	1.0

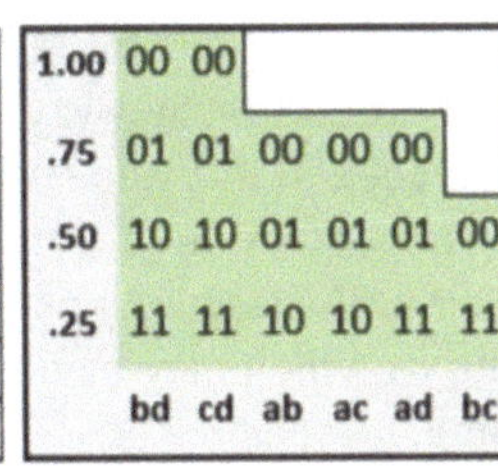

Fig. 3 Coverage table mapped to graph

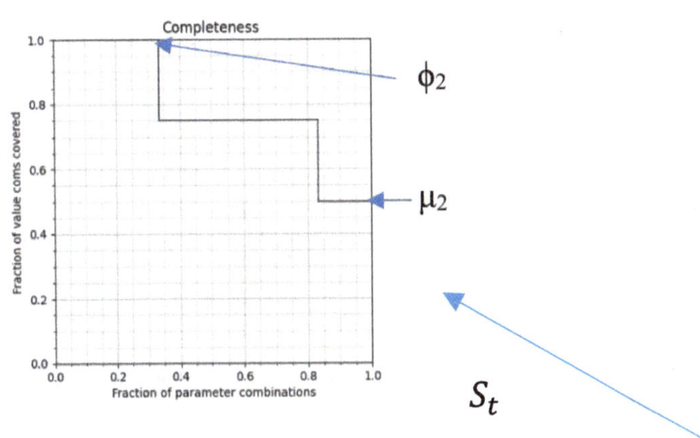

Fig. 4 Combinatorial completeness graph

reflect, for example, the proportion of the total input space for which combinations are covered up to some specified level of t. If the data set being measured is for training, S_t reflects the proportion of possible combinations of input values for various parameters.

- $1 - S_t$: Any missing combinations (the area above the curve) should be justified by analyzing the application domain. Missing combinations may be conditions that were not present because input sampling was not sufficient, or they may result from physical or other constraints in the environment. For example, an autonomous vehicle scenario would never contain the 2-way combination (temp = 98F, snowing).
- μ_t: the minimum t-way coverage present in the training space. This value is important for understanding completeness. If μ_t is 0.25, then we know that at least some variables have only 25% of their possible t-way values covered.
- (p, t)*completeness*: which measures the proportion of $C(n, t)$ combinations for which $S_t \geq p$.. (Maximoff et al. 2010). Also useful for measuring completeness. For example, if (0.75, t)-completeness for a data set is 0.9, then we can know that 90% of t-way combinations have at least 75% of their possible value combinations covered. So if μ_t is 0.25 for this data set, then coverage may still be adequate because no more than 10% of the t-way combinations have just 25% of values covered.
- ϕ_t : the proportion of t-way combinations for which all t-way value-combinations are included. If this value is low, then we know that there are only t-way combinations for which value coverage is complete. If we are studying the coverage of the most critical subset of input parameters, then a low ϕ_t value means that important situations which may be encountered in use have been missed in the ML model training.

Among these measures, typically the most interesting and important are S_t and $1 - S_t$, the areas under and above the curve respectively in Fig. 4. In a machine learning context, training set missing combinations, the area above the curve, are combinations that have not been included in development of the model. It is possible that this absence is the result of inadequate data sampling. For example, in a health care situation there may be some combinations of basic metrics (e.g., presence or absence of hypertension, diabetes, hyperlipidemia, vascular disease, etc.) that are very rare. Typically some of these patient characteristics occur together in association with some disease, but in some cases a patient may have one of the typical symptom combination but not the others. This situation could lead to an inadequately trained model. For each missing variable-value combination, a check should be made to determine if it is possible to extend the data set to include the missing combinations, or that this combination is never present.

The second possibility above points to an important consideration—some combinations are simply impossible, usually as a result of physical processes, or other "real world" issues. These constraints should be considered in developing the training set. The autonomous driving impossible combination example noted earlier, (temp = 98F, snowing) is trivial, but there are many applications where some combinations are possible, but extremely rare. In one health care example, it was found that many combinations of symptoms were present in less than 1% of a sample of more than 155,000 patients (Spiller et al. 2024). Determining the reason for missing combinations can be a valuable practice in understanding the characteristics of the domain being studied.

3.2 Set Difference Combination Coverage

The proportion of *t*-way combinations in A but not in B is a useful measure referred to as set difference combinatorial coverage (SDCC). (Lanus et al. 2021) Given two sets A and B, the combinations in A but not in B will be A, and the SDCC measure is $\frac{|A \backslash B|}{|A|}$.

SDCC can be helpful for evaluating the adequacy of training sets when an existing model is to be applied in a new or changed environment, a problem known as transfer learning. Because it is likely that the new environment will have differences with the training data set, it is important to measure the degree of this difference. The SDCC measure can be used in estimating the time or resource cost of training a new model, or gauging the likely accuracy of the existing model in the new environment. For example, assume there are two datasets, a source dataset S, and a target dataset T. The ML model will be trained with S and tested on T to see how well the learning done on the source apply or transfer to the target. Computing SDCC $|T_t \backslash S_t|/|T_t|$ for some specified level of *t*-way combinations measures the level of value combinations in the target environment for which the model (derived from source S) has not been trained. This measure has been used in explaining an example where a model trained on a small dataset remained sufficiently accurate when it was applied to a dataset nearly twice as large. This was possible because the small training data set had more variability than the large one, despite being half the size of the larger set.

3.3 Combination Frequency Difference Measures

Combination frequency difference (CFD) analysis determines which t-tuples are more strongly associated with a particular class (Kuhn et al. 2022). In classification problems, the goal is to develop a model that distinguishes members of one class from another using feature values that identify them, such as dogs distinguished from cats using attributes of size, ear shape, hair texture, etc. Focusing on binary classification to simplify the explanation, we refer to sets being distinguished as either Class or Non-class sets. In a machine learning context, these sets may refer to concepts that are to be learned, such as distinguishing one animal species from others.

For CFD we consider the number or frequency of occurrence of *t*-way tuples of values, to determine the degree to which a feature is associated with a particular class.

If a particular t-tuple is seen in a high proportion in class members, but not in non-class members, then it may be a reasonable indicator for distinguishing classes or at least for narrowing the range of possibilities for class identification. For example, many dog breeds may have a long tail, and many may have a curled tail, but a much smaller number of breeds have both of these attributes. Thus, it is important to have a measure that considers the quantity of instances with particular t-tuples of values in class and non-class instances.

Let us denote C and N as class and non-class data sets in a binary classification. A t-tuple x_t is a unique or notable value combination, e.g. a distinguishing combination (DC) for C if it is present in C and absent in N, or if it is more common in C than N as determined by a threshold value T. Although originally only this one direction was determined, we also include t-tuples present in N and not C or more common in N than C. The key point is to determine t-tuples that are strongly associated with one class but not the other, based on the frequency or rate of occurrence. CFD is therefore defined as follows: a t-tuple X_t is unique or notable for a class $C \iff \text{fr}(x_i, Ct) > T \times \text{fr}(x_i, N_t)$, where $\text{fr}(x_i, Y_t)$ represents the frequency of t-tuples of values x in the set of t-tuples of values in Y. The frequency function fr denotes the number of rows in which a specific t-tuple appears. Specifying $T = 1$ means that a combination is selected as distinguishing whenever it occurs at a higher frequency in one class than the other, no matter how small the difference in frequency; generally higher values of T are used when determining distinguishing combinations.

CFD can be especially useful for evaluating how well a training set matches a test or other data set. See Figs. 5 and 6, which show 50% samples of a full balanced data set. The graphs show the frequency differences between the samples and the full data set. Clearly, sample s50.1 is a closer match to the full set than sample s50.9. Thus, s50.1 would be preferable in training an ML model, because it has very few and small differences with the full data set.

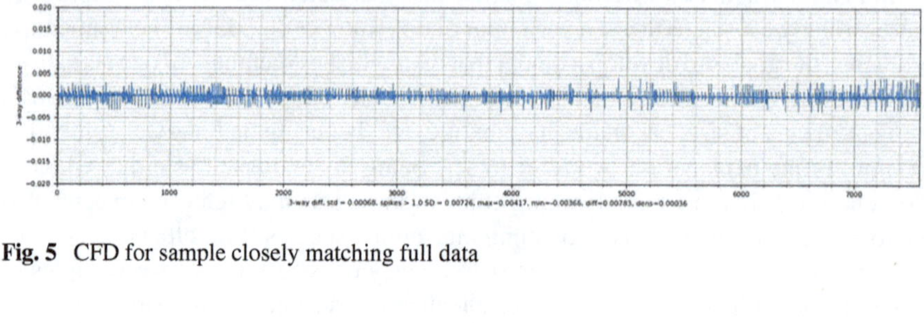

Fig. 5 CFD for sample closely matching full data

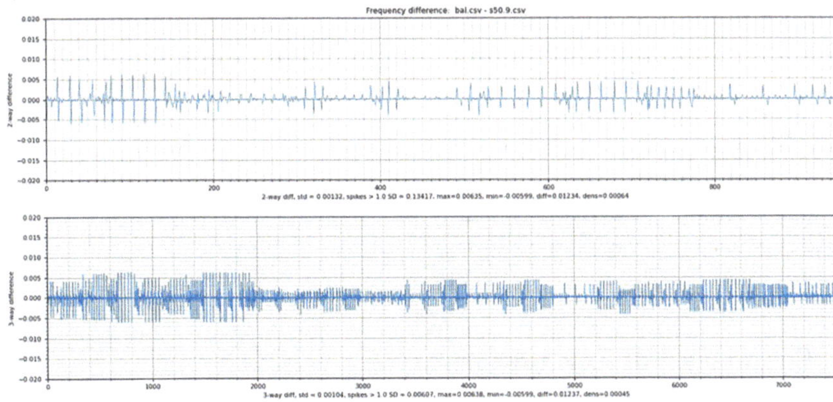

Fig. 6 CFD for sample poorly matching full data

4 Use Case Example: Clinical Research

Incomplete or poorly covered training data sets—real, synthetic, or a mix—are a pervasive challenge in the clinical research arena (real and/or synthetic). Combinatorial methods can be used to find and target data coverage gaps, then to mitigate the gaps via targeting synthetic generation or targeted data collection.

For example, a clinical study used combinatorial testing to find gaps in a real dataset, then ML-based imputation to estimate missing values for analysis (not to backfill source data). *Note: Never edit/backfill regulated clinical records with synthetic values, synthetic data is for model training by adding synthetic examples - provided you validate utility and privacy.*

Another example is that in many clinical research projects, teams now employ generative models to synthesize patient-like records or images for model training. By creating synthetic data such as imaging scans, and other sensitive data downstream algorithms can be trained to reduce direct exposure to real patient information.

However, even with these synthetic datasets, it's very difficult to cover every possible clinical scenario - especially rare conditions, edge-case patient profiles, or unusual combinations of comorbidities. In practice, that means gaps remain in the "coverage" of the training data, which can lead to *blind spots* or *biases* in the final AI models. Combinatorial techniques can find both.

Use case Workflow Use combinatorial techniques to identify data blind-spot slices. Next, researchers either collect/label more real data for those slices or use a generative model to synthesize targeted examples. The researcher then retrains the predictive model on the augmented dataset (real + validated synthetic). Finally, the researchers re-evaluate using combinatorial methods again to confirm those blind spots have improved.

Take the case described by Oakden-Rayner et al. (2020) where they showed that, despite strong overall performance, chest-X-ray pneumothorax models had > 20%

relative drops in sensitivity on clinically crucial slices - especially cases without chest drains. This revealed a clear blind spot. One contributor is that no-drain cases were under-represented in training.

Using combinatorial techniques to assess *t*-wise coverage and slice-wise performance (including chest drain: yes/no) would likely have revealed the "no-drain" blind spot. Researchers could then collect more no-drain cases or add validated synthetic examples, retrain the model (optionally with subgroup weighting), and re-run the same combinatorial checks to confirm the blind spot improved.

5 Summary and Key Take Aways

This chapter argues that model accuracy, fairness and reliability depend on how well training and test data match real-world use. The two metrics that need to be measured explicitly are coverage (i.e., the dataset includes the subgroups and cross-field scenarios) and completeness (i.e., the data is not missing any required fields). The key point to understand is that combinations of variable values should be considered, and methods are available to use combination coverage to supplement conventional approaches. Useful combinatorial measures include:

- Combinatorial coverage—which measures, for each possible *t*-way variable value combination, if that variable value combination is found in a data set. Missing variable-value combinations may result from (1) an incomplete data set, indicating more data will be needed for completeness, or (2) constraints that prevent particular variable values from occurring together in a t-way combination. Every missing variable-value combination should be explained by a constraint, so if no relevant constraint exists, then it is likely that the data set is incomplete.
- Combinatorial frequency differences—measuring the frequency of occurrence of every *t*-way variable-value combination. An ML model may be inadequate if there is a significant difference between the frequency of occurrence of combinations in a training set, as compared with the environment where the ML model will be used. It should be noted that this is not always the case, as some variables and combinations are less relevant than others. But computing combination frequency differences can be a useful quality control measure.

Rule-based checks miss data interactions. The combinatorial approach improves the assessment of data quality and helps reduces bias and blind spots, and support more dependable models and decisions.

Note: portions of this chapter have been adapted and summarized from: Kuhn, D. R., Raunak, M. S., & Kacker, R. N. (2025). Measuring and Visualizing Dataset Coverage for Machine Learning. *Computer*, *58*(4), 18–26.]

References

Kuhn DR, Raunak MS, Kacker RN (2025) Measuring and visualizing dataset coverage for machine learning. Computer 58(04):18–26. https://doi.org/10.1109/MC.2025.3527374

Kuhn DR, Higdon JM, Lawrence JF, Kacker RN, Lei Y (2012) Combinatorial methods for event sequence testing. In: 2012 IEEE fifth international conference on software testing, verification and validation. IEEE, pp 601–609

Kuhn DR, Mendoza ID, Kacker RN, Lei Y (2013) Combinatorial coverage measurement concepts and applications. In: 2013 IEEE Sixth International Conference on Software Testing, Verification and Validation Workshops. IEEE, pp 352–361

Kuhn DR, Kacker RN, Lei Y (2016) Measuring and specifying combinatorial coverage of test input configurations. Innovat Syst Softw Eng 12(4):249–261

Kuhn DR, Raunak MS, Prado CB, Patil VC, Kacker RN (2022) Combination frequency differencing for identifying design weaknesses in physical unclonable functions. In: ICST Workshops, pp 110–117

Lanus E, Freeman LJ, Richard Kuhn D, Kacker RN (2021) Combinatorial testing metrics for machine learning. In: 2021 IEEE International Conference on Software Testing, Verification and Validation Workshops (ICSTW). IEEE, pp 81–84

Maximoff JR, Kuhn DR, Trela MD, Kacker R (2010) A method for analyzing system state-space coverage within a t-wise testing framework. In: 2010 IEEE international systems conference. IEEE, pp 598–603

Oakden-Rayner L, Dunnmon J, Carneiro G, Ré C (2020) Hidden stratification causes clinically meaningful failures in machine learning for medical imaging. In: Proceeding ACM conference health inference learn (2020). 2020, pp 151–159. https://doi.org/10.1145/3368555.3384468. PMID: 33196064; PMCID: PMC7665161. https://pubmed.ncbi.nlm.nih.gov/33196064/

Raunak MS, Richard Kuhn D, Kacker RN (2025) Ensuring reliability through combinatorial sequence coverage. IEEE Reliab Magaz 2(2):49–57

Spiller TR, Duek O, Helmer M, Murray JD, Fielstein E, Pietrzak RH, Harpaz-Rotem I et al (2024) Unveiling the structure in mental disorder presentations. JAMA Psychiat 81(11):1101–1107

Wu XY, Ding F, Li K, Huang WC, Zhang Y, Zhu J (2022) Analysis of the causes of solitary pulmonary nodule misdiagnosed as lung cancer by using artificial intelligence: a retrospective study at a single center. Diagnostics 12(9):2218

Statistical Confidence and Reliability Estimation for GenAI Reasoning

Zhaojun Steven Li, Kalyan Bhagavan Tadaka, Rocco Cassandro, and William R. Tonti

Abstract This chapter investigates the confidence of using GenAI-based models in performing quantitative reliability reasoning, specifically focusing on estimating the shape and scale parameters of Weibull distributions. By using synthetic datasets generated from Weibull models, we design multiple experimental factors, including sample sizes, shape parameters, prompts, and five GenAI models (Claude, DeepSeek, Gemini, ChatGPT, and xAI) to see how various factors impact parameter estimations in the Weibull reliability model. The confidence levels and robustness of each GenAI model's estimates are assessed through a factorial experimental design framework. Notably, the estimates of unknown parameters for Weibull model do not converge over increasing sample size as commonly observed in the traditional pure statistics-based inference. We also observe that ChatGPT and DeepSeek perform the best for both shape and scale parameters' estimation compared with other GenAI models. ChatGPT and DeepSeek also show robustness in estimating parameters regardless of the types of prompts being used. The findings of this chapter highlight the reliability and confidence of using GenAI models for unknown parameter estimation in quantitative reliability engineering analysis.

Keywords Weibull distribution · Reliability estimation · Factorial design · GenAI models · Prompt engineering

---

Z. S. Li (✉) · K. B. Tadaka · R. Cassandro
Department of Industrial Engineering and Engineering Management, Western New England University, Springfield, MA, USA
e-mail: stevenli777@ieee.org

W. R. Tonti
IEEE Future Directions, New Jersey, USA

A. Stavrou et al. (eds.), *Generative and Agentic AI Reliability: Architectures, Challenges, and Trust for Autonomous Systems*, Studies in Computational Intelligence 1272, https://doi.org/10.1007/978-3-032-18585-3_5

1 Introduction

In recent years, Generative Artificial Intelligence (GenAI) has become a transformative tool across scientific and engineering disciplines, offering unprecedented capabilities in prediction, classification, and decision-making (Şahin and Karayel 2024). While the qualitative strengths of Large Language Models (LLMs) have been widely demonstrated, their ability to perform quantitative reasoning, especially in statistically rigorous applications, remains a subject of active investigation (Chang et al. 2024). LLMs, a foundational backbone of GenAI, are not synonymous with GenAI but rather represent one of its most advanced and widely applied architectures. GenAI encompasses a broader set of generative models, including diffusion models, Generative Adversarial Networks (GANs), Variational AutoEncoders (VAEs), Generative Pre-trained Transformer (GPT) by OpenAI, and Gemini by Google. which can generate not only text but also images, code, simulations, and structured data (Huang et al. 2024). Examples of quantitative engineering applications of GenAI include structural health monitoring in civil engineering, thermal-fluid simulations in mechanical engineering, fault diagnostics in electrical systems, among others (Zafiris 2025). In the domain of reliability engineering, the use of GenAI models to estimate probabilistic parameters from empirical data introduces new questions surrounding trustworthiness, interpretability, and statistical confidence.

Reliability analysis often relies on lifetime distributions such as the Weibull distribution, where shape and scale parameters provide understanding about system failure behavior (Shahriari 2024). In practical applications, given lifetime data, statistical estimation methods such as Maximum Likelihood Estimation (MLE), Least Squares (LS), and Methods of Moments, are used to estimate the best distribution fit given the analyzed data (Maus et al. 2001). However, the use of GenAI models for estimating distribution parameters fundamental concerns: Can GenAI models reliably reproduce statistical inference procedures? How sensitive are their outputs to the formulation of the prompt and the sample size? Do these models mimic traditional convergence behaviors? Is it possible to enhance a model's convergence by using data from several points of entry during training? While overall prediction accuracy can be measured using aggregate error metrics such as Root Mean Squared Error (RMSE), these metrics are unable to account for variability or uncertainty across various training configurations.

Recent developments in AI prompt engineering suggest that how a user frames their questions significantly impacts model responses (Velásquez-Henao et al. 2023). The specificity of prompts ranging from broad or open-ended requests to more narrowly scoped directives can lead to varying outputs even under fixed AI architectures. Consequently, gaining insight into how factors such as prompt specificity, sample size, data characteristics, and underlying model architecture interact is crucial for evaluating and enhancing the reliability of GenAI models.

This chapter, for the first time, investigates confidence in GenAI-based parameter estimation for five models including Claudio, DeepSeek, Gemini, ChatGPT, and xAI. Each model is tasked on estimating the shape and scale parameters from simulated datasets generated under different sample sizes and parameter configurations. Each model receives prompts of varying specificity (generic, medium, and specific) to evaluate and estimate the influence of prompt engineering on estimation accuracy. By leveraging a structured factorial design approach, we systematically capture the effects of these factors by generating 225 data points per replicate across diverse combinations of sample sizes and parameter settings. Our analysis of the results using normal plots offers conclusions about patterns, variance, and interaction effects that collectively reveal how much confidence may reasonably be placed in an AI-driven quantitative reasoning evaluation.

Roadmap of the Chapter. This chapter is organized as follows: In Sect. 2 we provide a literature review covering distribution parameters estimation methods, and GenAI Models in Quantitative Reasoning. Section 3 outlines the methodology. Section 4 describes our experimental results. Finally, conclusions and future work are presented in Sect. 5.

2 Literature Review

2.1 *Distribution Parameters Estimation Methods*

Reliability engineering has established a strong foundation utilizing statistical methodologies for failure analysis. The Weibull distribution has become fundamental in reliability studies due to its flexibility in representing various failure patterns through its shape and scale parameters (Barlow 1975). Nelson et al. (1998) demonstrated how Weibull parameters can effectively characterize failures from early-life defects to wear-out phenomena, making the distribution exceptionally versatile for engineering applications. For traditional distribution parameter estimation techniques, Lawless et al. (2011) provided comprehensive coverage of MLE methods, highlighting their asymptotic properties and efficiency in reliability contexts.

MLE is a widely adopted method due to its asymptotic efficiency. However, it lacks closed-form solutions requiring iterative numerical optimization. The probability density function (PDF) of the two-parameter Weibull distribution, commonly used in reliability analysis, is given by:

$$f(x) = \begin{cases} \frac{\lambda}{\eta}\left(\frac{x}{\eta}\right)^{\lambda-1} e^{-\left(\frac{x}{\eta}\right)^{\lambda}}, x \geq 0, \\ 0, x < 0 \end{cases} \tag{1}$$

where η is scale parameter, λ is shape parameter and x is the random variable representing the lifetime (or failure time). MLE seeks to estimate these parameters by maximizing the likelihood function constructed from observed data.

2.2 GenAI Models in Quantitative Reasoning

Recent developments in GenAI have enabled LLMs to engage with quantitative reasoning tasks at different levels and for different applications. Frameworks like chain-of-thought prompting (Lyu et al. 2023), few-shot learning (Ye et al. 2025), multi-agent debate (Du et al. 2023; Esencan et al. 2024), symbolic–neural hybrids (Dinu et al. 2024; Trinh et al. 2023), and structured reasoning architectures such as Microsoft's Algorithm of Thoughts (Su et al. 2024) and OpenAI's o1 model (El-Kishky et al. 2024) highlight clear progress in interpretability and numerical accuracy. However, while these methods demonstrate promise for general mathematical problem-solving, they have not yet been applied to reliability-specific challenges such as estimating distribution parameters from lifetime data, where factors such as sample size, censoring, and distributional complexity require tailored approaches.

In reliability contexts, challenges are amplified by varying sample sizes, censored data, prompt specificity, and model architecture. Tadaka et al. (2024) showed that domain-specific prompts lead to significant performance differences across models, while Bommasani et al. (2021) cautioned that proficiency in general reasoning does not imply competence in specialized statistical domains. Chalmer (2023) found similar inconsistencies in engineering problems, with LLMs underperforming on statistical estimation tasks. Chen et al. (2024) further demonstrated how prompt structures influence statistical outputs, though without fully exploring interactions between sample size, parameter distributions, and reasoning mechanisms in reliability settings such as Weibull estimation.

To address these limitations, this chapter conducts a systematic factorial experiment to examine how prompt specificity, sample size, estimation approach, and GenAI architecture jointly affect the accuracy of distribution parameter estimation.

3 Methodology

This section outlines the methodological approach used to evaluate the GenAI models confidence by estimating Weibull distribution parameters. Experimental design is employed to systematically examine how factors such as prompt specificity, sample size, and model influence the accuracy of parameter estimation across multiple GenAI models.

3.1 Numerical Data Generation

To facilitate a controlled experimental setup, synthetic datasets are generated for three use cases. The generation process follows a factorial scheme in which the scale parameter (η) is held constant at 10, while the shape parameter (λ) is varied across

Table 1 Different cases in simulation

Case	Fixed η	True value λ
1	10	0.5
2	10	1
3	10	2

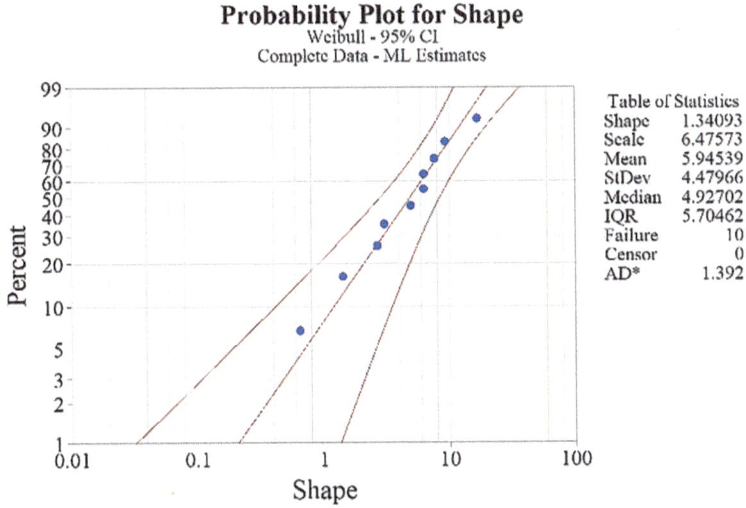

Fig. 1 Probability plot showing Weibull distribution fit (shape = 1) with 95% confidence interval

three distinct values 0.5 (indicating a decreasing failure rate), 1 (constant failure rate, equivalent to the exponential di1stribution), and 2 (increasing failure rate). For each value of λ five different sample sizes are considered: 10, 50, 100, 500, and 1,000. This yields a total of 15 unique datasets, representing diverse reliability case scenarios with varying data richness and distribution shapes as shown in Table 1 and presented a probability plot indicating that the dataset follows a Weibull distribution (shape parameter 1) with 95% confidence Fig. 1.

3.2 GenAI Models and Prompt Design

Five leading GenAI models, ChatGPT, Claude, xAI, Gemini, and DeepSeek, are evaluated for their ability to estimate Weibull distribution parameters from the synthesized datasets described in Section III-A. These models are chosen for their state-of-the-art capabilities in natural language understanding, numerical reasoning, and generative response formulation.

To examine how input formulation influences estimation accuracy, three levels of prompt specificity are employed: Generic, Medium, and Specific. Each GenAI

Table 2 Summary of prompts as inputs to GenAI models

Prompt type	Definition	Prompt
Generic	Parameter estimate by GenAI	Can you estimate the shape and scale parameters?
Medium	Provide minimal guidance, GenAI ultimately determines both the method and details	Use a suitable method to fit a Weibull distribution to the data in each sheet and summarize the shape and scale parameters in a clear format
Specific	Require GenAI to use a specific method. This includes specific details	Use MLE to estimate the Weibull shape and scale parameters. Provide the results in a table with sheet names, shape, and scale

Model is prompted at all three specificity levels for each dataset, allowing for a factorial evaluation of prompt design and GenAI performance. Table 2 summarizes the types of prompts used.

3.3 Parameter Estimation Method

In this Chapter, GenAI models are tasked with estimating the Weibull distribution parameters (shape λ and scale η) based on the synthetic datasets generated. Each model receives the datasets accompanied by prompts of varying specificity (Generic, Medium, and Specific), allowing for a structured examination of how prompt formulation impacts estimation performance. The resulting parameter estimates are then compared against two references: (1) the true parameters used during synthetic data generation, and (2) the estimates computed independently using the MLE method, as discussed in Section II-A. By comparing GenAI-generated outputs to both the true values and MLE estimates, the study evaluates both the accuracy and consistency of GenAI reasoning across varying prompts, sample sizes, and model architectures. This parameter estimation strategy forms the foundation for the full factorial experimental design described in the next section. Synthetic data is deemed necessary and sufficient as it provides a stable reference.

3.4 Design of Experiment

To investigate the impact of various factors on the accuracy of GenAI-based parameter estimation, a full factorial design was implemented, considering four factors as described in Table 3.

Each factor-level combination yields a single parameter estimation trial, resulting in a total of 225 runs for one replicate. This design ensures comprehensive coverage of both data variability (through changes in sample size and shape parameters)

Table 3 Factorial design

Factor	Levels	Values
Cases	3	Shape = 0.5 Shape = 1, Shape = 2
Sample Size	5	10, 50, 100, 500, 1000
GenAI Models	5	ChatGPT, xAI, Gemini, Claudio, DeepSeek
Prompts	3	Generic, Medium, Specific

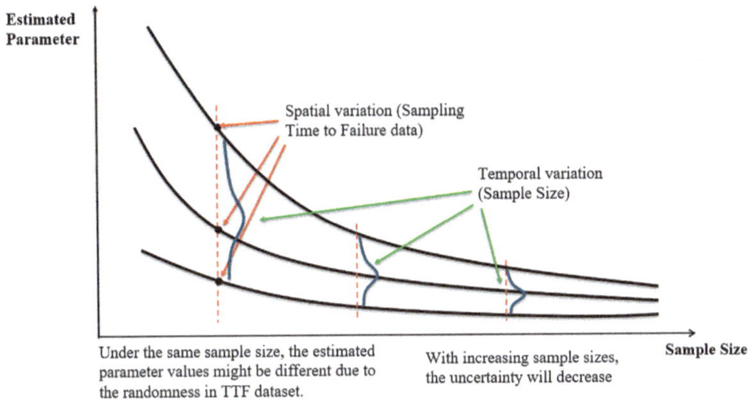

Fig. 2 Spatial and temporal in GenAI outputs

and methodological variability (through the choice of GenAI models and prompt specificity).

The primary response variables recorded for each run are the estimated shape parameter and scale parameter returned by the GenAI models. To evaluate estimation performance, secondary metrics such as the absolute error are computed, quantifying how closely each model's estimates align with the true Weibull distribution parameters.

The product unit to unit variations of Time to Failure data (TTF) contributes to spatial variation, while the increasing sample size leads to temporal variations (see Fig. 2).

3.5 *Bootstrapping*

Bootstrapping was employed as a resampling technique to increase the number of simulations runs. This approach is particularly useful when assessing the stability and robustness of GenAI parameter estimates with dataset.

For each unique combination of prompt specificity, GenAI model, and input dataset, multiple bootstrapped samples of the original lifetime data were generated by sampling with replacement. Each resampled dataset was then submitted to the GenAI model using a consistent prompt structure. This iterative process enabled the construction of an empirical distribution of the estimated Weibull parameters (λ and η) for each scenario. By analyzing the spread (e.g., variance, interquartile range) and central tendency (e.g., mean, median) of these bootstrapped estimates, we were able to assess the sensitivity of the GenAI model's outputs to minor variations in input data. This provides a deeper insight into the model's reliability and generalizability.

4 Results and Discussion

This section outlines the methodological approach used to evaluate the confidence and reliability of GenAI models using Weibull parameter estimation. We employ a factorial experimental design to systematically investigate how various factors influence estimation accuracy across different GenAI models.

4.1 *Factorial Design*

The ANOVA table is employed to assess the significance of the factors and their interactions as shown in Table 3 on parameter estimation accuracy. As shown in Fig. 3, The first column of Source indicates variation sources from four main effects, six two-way interactions, two three-way interactions, and one four-way interaction. The second column of DF indicates the degree of freedom of each variation source and represents the amount of data information related to this effect. The Third column and Four column represent the sum of squared and mean square errors. The Fifth and Sixth column shows the F-Value and P-value from which we know which effect are significant or not.

In the main effects plot for Replicate 1 as shown in Fig. 4, the Cases and GenAI models show clear variations across levels, confirming their significant impact on the response values.

In the case of the Interaction Plot for the estimation values for considering Replicate 1, we observed the Cases and GenAI models have the most interaction, as shown in Fig. 5.

Similarly for 2 Replicates, we have observed the main effects of both Sample Size and Prompts are insignificant; the 2-way Interactions, and Cases*GenAI models are significant, 3-way interactions and 4-way interactions are insignificant, as shown Fig. 8.

The interaction plots and main effect indicate that the cases and GenAI models exhibit the strongest interactions among all actors, as shown the Figs. 9 and 10. In Fig. 6 and 7, these two factors also show the strongest interactions when we observe the interaction plot of scale parameters in both 1 and 2 Replicates and summarized in Table 4.

Analysis of Variance

Source	DF	Adj SS	Adj MS	F-Value	P-Value
Model	176	72.1331	0.4098	46.80	0.000
Linear	12	51.8226	4.3185	493.12	0.000
Cases	2	34.4020	17.2010	1964.11	0.000
Sample Size	4	0.2151	0.0538	6.14	0.000
AI Models	4	16.8873	4.2218	482.07	0.000
Prompts	2	0.3181	0.1591	18.16	0.000
2-Way Interactions	52	17.3415	0.3335	38.08	0.000
Cases*Sample Size	8	0.1548	0.0194	2.21	0.043
Cases*AI Models	8	15.7416	1.9677	224.68	0.000
Cases*Prompts	4	0.1495	0.0374	4.27	0.005
Sample Size*AI Models	16	0.6558	0.0410	4.68	0.000
Sample Size*Prompts	8	0.0937	0.0117	1.34	0.248
AI Models*Prompts	8	0.5460	0.0682	7.79	0.000
3-Way Interactions	48	2.1008	0.0438	5.00	0.000
Cases*Sample Size*AI Models	32	1.6156	0.0505	5.77	0.000
Cases*AI Models*Prompts	16	0.4852	0.0303	3.46	0.000
4-Way Interactions	64	0.8683	0.0136	1.55	0.057
Cases*Sample Size*AI Models*Prompts	64	0.8683	0.0136	1.55	0.057
Error	48	0.4204	0.0088		
Total	224	72.5535			

Fig. 3 ANOVA Table (1 Replicate)

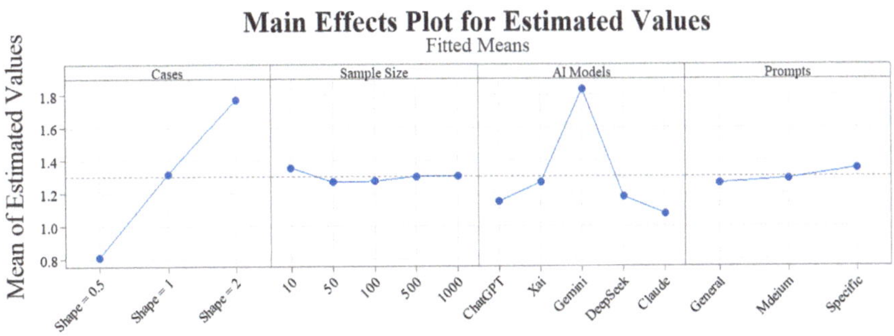

Fig. 4 Main Effects Plots for 1 Replicate of the Shape Parameter showing estimated values across cases, sample sizes, AI models, and Prompts

This suggests that the strongest interactions arise from Cases and GenAI Models due to their direct influence on response variability, while Sample Size and prompts contribute minimal additional variance, particularly when sample sizes are sufficiently large. However, it is important to note that smaller sample sizes increase variability and widen confidence intervals, whereas larger datasets enhance stability and reliability of the estimates.

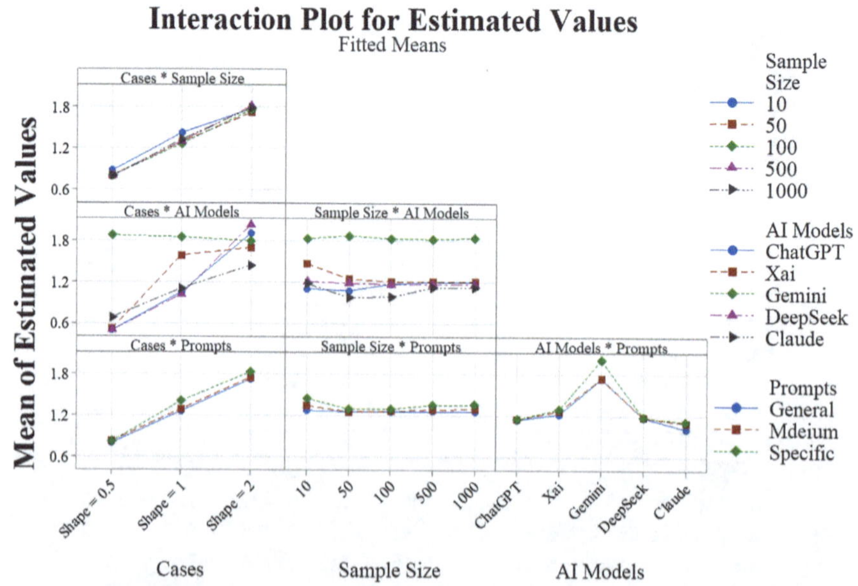

Fig. 5 Interaction Plots for 1 Replicate of Shape Parameter, showing estimated values across cases, sample sizes, AI models

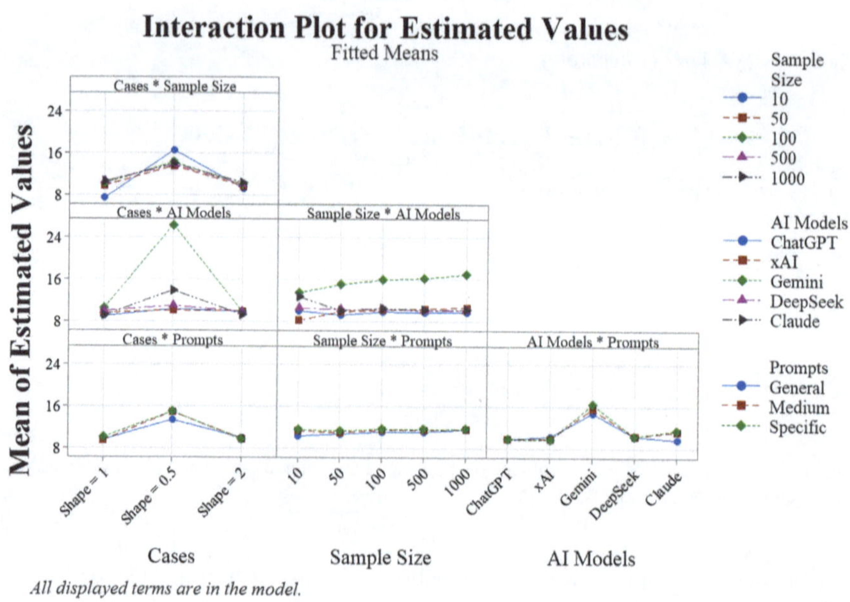

Fig. 6 Interaction Plots for 1 Replicate of the scale parameter, showing estimated values across cases, sample sizes, AI models

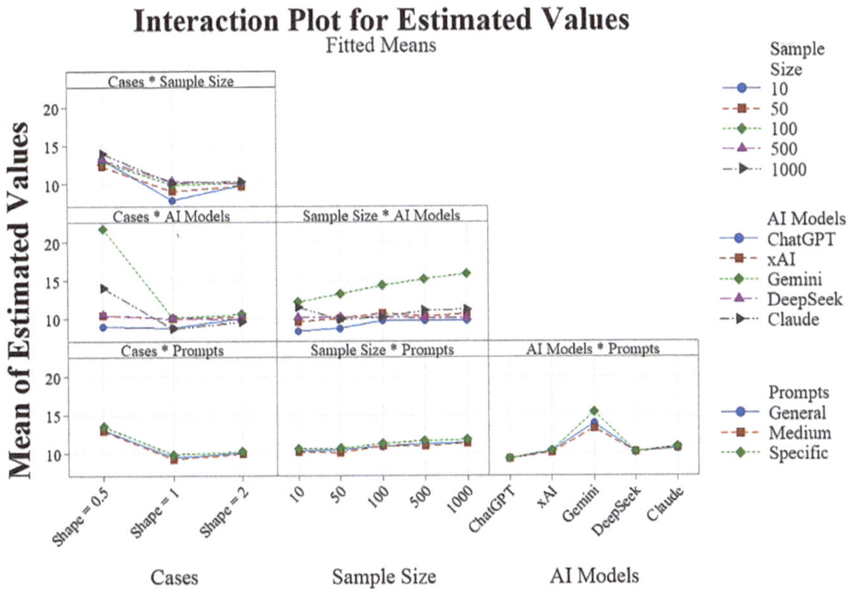

Fig. 7 Interaction plots for 2 Replicates of the scale parameter, showing estimated values across cases, sample sizes, AI models

We use GenAI models to estimate the shape and scale parameters under each case scenario. The results show that ChatGPT and DeepSeek have the most accurate estimation. Under the same sample size, the difference between the estimated values of different prompts is small. We believe the type of prompt does not affect the estimation of these two GenAI models. To improve the reliability of our parameter estimates, we applied the bootstrap method, which involves repeatedly resampling the observed data with replacement to generate many simulated samples.

4.2 *Bootstrapping*

The bootstrap was applied to increase the sample size of 1,000 to improve confidence when estimating unknown parameters, and the results are similar across different prompts when we use ChatGPT or DeepSeek (resamples = 10,000, shape parameter value = 1). In the histogram, the green number is the mean of estimated values. The red numbers indicate the 95% confidence interval of estimated values as shown on Fig. 11. For example, in a generic prompt, the estimated value is between 0.9796 and 1.1587 when we use ChatGPT.

In addition, we also performed bootstrapping using different variations of three types prompts under generic, medium and specific scenarios. For example, in one scenario, by considering the generic prompts, we created an additional 20 prompts.

Analysis of Variance

Source	DF	Adj SS	Adj MS	F-Value	P-Value
Model	224	161.622	0.7215	4.90	0.000
Linear	12	119.727	9.9772	67.82	0.000
Cases	2	74.857	37.4283	254.43	0.000
Sample Size	4	0.956	0.2391	1.63	0.169
AI Models	4	43.263	10.8158	73.52	0.000
Prompts	2	0.651	0.3254	2.21	0.112
2-Way Interactions	52	34.533	0.6641	4.51	0.000
Cases*Sample Size	8	0.615	0.0769	0.52	0.839
Cases*AI Models	8	30.884	3.8606	26.24	0.000
Cases*Prompts	4	0.543	0.1358	0.92	0.451
Sample Size*AI Models	16	1.054	0.0659	0.45	0.968
Sample Size*Prompts	8	0.299	0.0374	0.25	0.979
AI Models*Prompts	8	1.137	0.1421	0.97	0.463
3-Way Interactions	96	4.483	0.0467	0.32	1.000
Cases*Sample Size*AI Models	32	1.300	0.0406	0.28	1.000
Cases*Sample Size*Prompts	16	0.553	0.0345	0.23	0.999
Cases*AI Models*Prompts	16	1.182	0.0738	0.50	0.945
Sample Size*AI Models*Prompts	32	1.449	0.0453	0.31	1.000
4-Way Interactions	64	2.879	0.0450	0.31	1.000
Cases*Sample Size*AI Models*Prompts	64	2.879	0.0450	0.31	1.000
Error	225	33.099	0.1471		
Total	449	194.721			

Fig. 8 ANOVA table (Replicates 2)

Similarly, we did the same for medium and specific types. As expected, the results are statistically speaking identical to the generic prompts. This serves as an informal test to validate our approach.

By using interval plots, we investigate the convergence behavior of parameter estimation with increasing sample size. We focused on case 2 with shape parameter equal to one; DeepSeek is the only GenAI model showing decreasing uncertainty and obvious convergence over increasing sample size. ChatGPT also shows the convergence behavior of estimated results, but in less obvious fashion. However, under the ChatGPT model the average estimated result deviates from the true value when the sample size is one thousand. The average of estimates did not converge by sample size under Gemini and xAI. Claude results indicate the average of estimated values is closer to the assumption. Overall, Claude is less accurate than ChatGPT and DeepSeek as shown in Fig. 12.

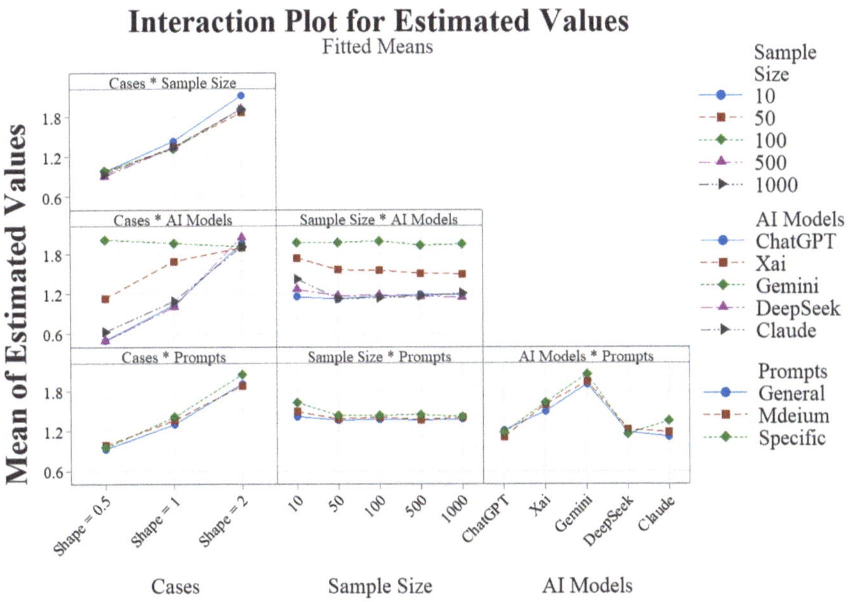

Fig. 9 Interaction Plots for 2 Replicates of the shape parameter, showing estimated values across cases, sample sizes, AI models

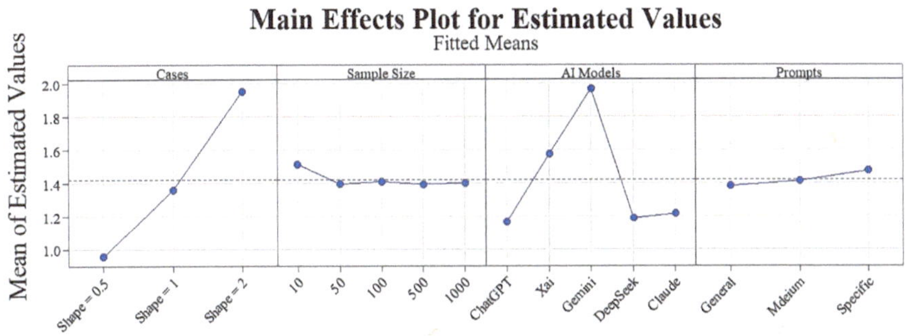

Fig. 10 Main Effects Plots for 2 Replicates of the Shape Parameter showing estimated values across cases, sample sizes, AI models, and Prompts

Table 4 Interaction summary

Interaction type	1 Replicate	2 Replicate
Main effects	Significant	Partially
2-way interactions	Most NS	Cases*Models
3-way interactions	Significant	Not Significant
4-way interactions	Not Significant	Not Significant

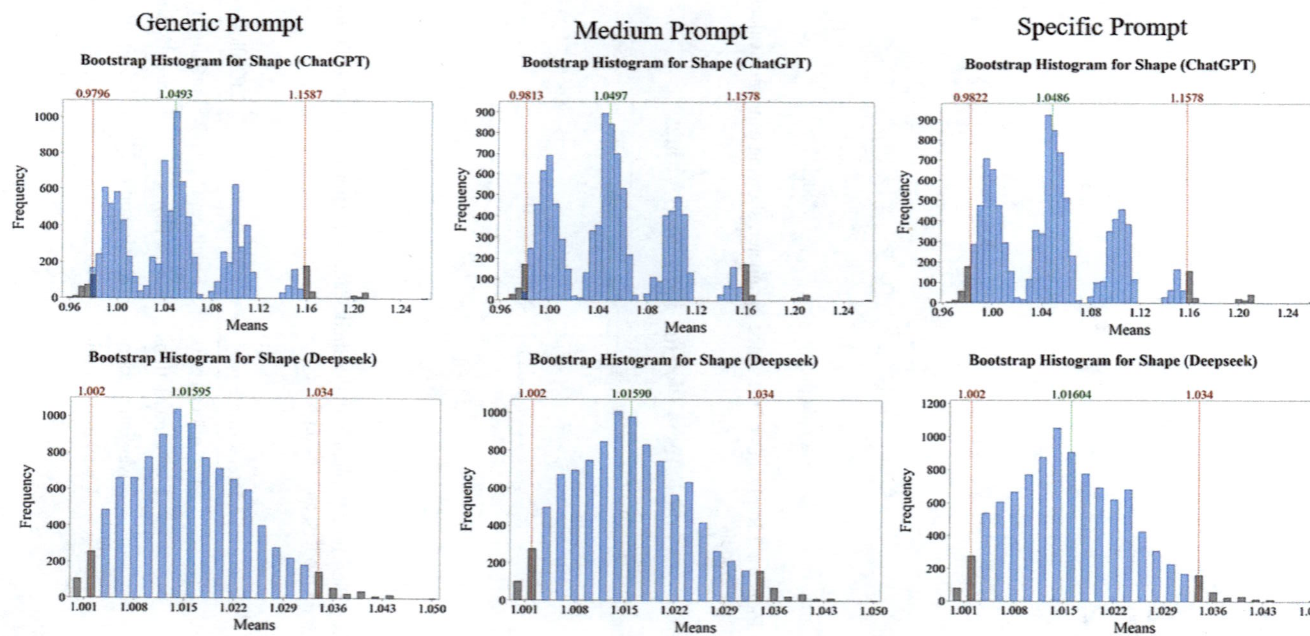

Fig. 11 Bootstrapping of datasets for shape parameter under different prompts, highlighting ChatGPT variability and DeepSeek stability

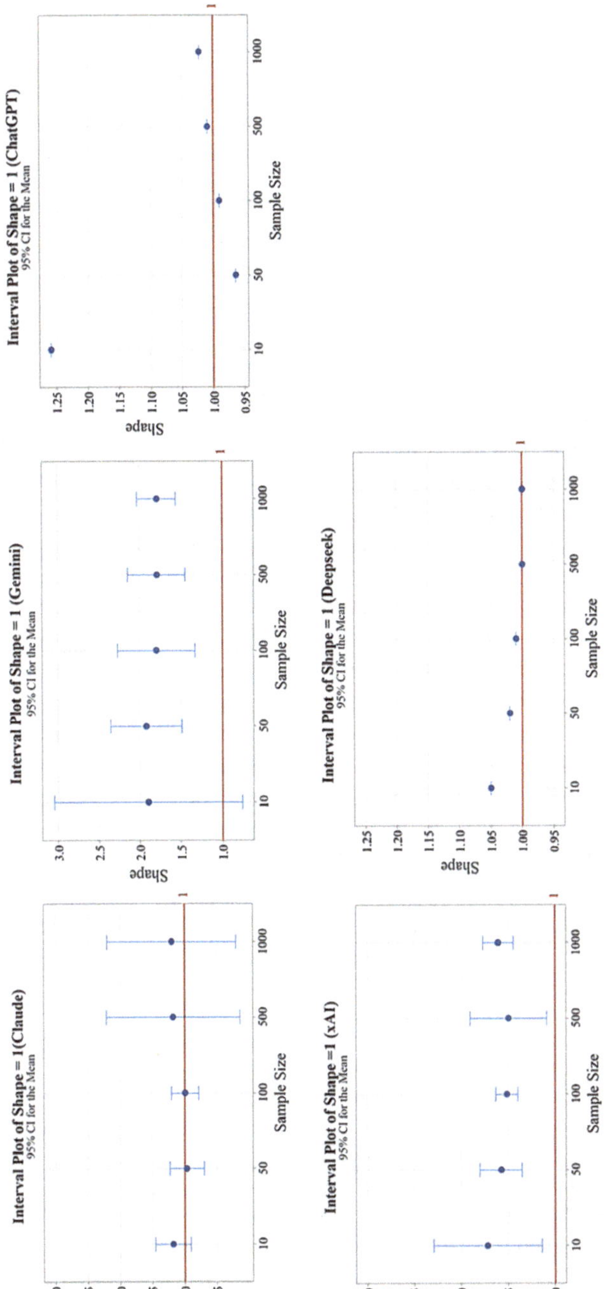

Fig. 12 Interval plots of the average of estimated values across sample sizes, showing DeepSeek and ChatGPT as the most accurate and reliable models

4.3 Evaluation of Estimates with Root Mean Square Error (RMSE)

RMSE measures the average difference of the actual data points from the estimated values, and the difference is squared to avoid the cancelation of positive and negative values, while they are summed up. The RMSE is defined as

$$\text{RMSE} = \sqrt{\frac{\sum_{i=1}^{n}\left(y_i - \widehat{y_i}\right)^2}{N}} \quad (2)$$

where, y_i represents the actual value, $\widehat{y}_i$ represents the estimated value and N represents the number of data points. To evaluate the effectiveness of shape parameter estimation, a dataset of five levels with a true shape value of 1.0 was used. The predictions were generated by two GenAI models: ChatGPT and DeepSeek. Residuals were computed as the difference between the predicted and true shape values. Based on these residuals, RMSE was calculated for each model. The ChatGPT model yielded an RMSE of 0.117, while the DeepSeek model achieved a lower RMSE of 0.026, indicating higher precision and closer alignment with the actual shape parameter.

To determine whether this performance difference was statistically significant, a paired t-test was conducted on the residuals. The test yielded a t-value of 0.69 and a p-value of 0.530, suggesting that the difference in mean residuals (0.032) was not statistically significant at the 0.05 level. Thus, while DeepSeek exhibited better average performance, the result does not provide strong enough evidence to conclusively differentiate between the two models due to the limited sample size.

To assess the distributional behavior of the residuals, normal probability plots were generated using Minitab for both ChatGPT and DeepSeek (see Fig. 13). The plot shows that DeepSeek's residuals closely follow the normal reference line, with a p-value of 0.182, indicating no significant departure from normality. Conversely, ChatGPT's residuals showed a wider spread, with a p-value of 0.021, suggesting a statistically significant deviation from a normal distribution. This supports the conclusion that DeepSeek's shape parameter estimates are not only more accurate in magnitude but also statistically stable.

4.4 Results

Overall, after analyzing the complete two replicates we observe in Figs. 5, 6 7 and 9) the Cases and GenAI models have the most interaction in the interaction plot and are significant in the ANOVA (Figs. 3 and 8) in both replicates.

For shape parameter estimation, DeepSeek achieved the lowest RMSE (0.026), followed by ChatGPT (0.117). A paired t-test comparing residuals yielded a p-value of 0.530, indicating the difference was not statistically significant. Probability plots

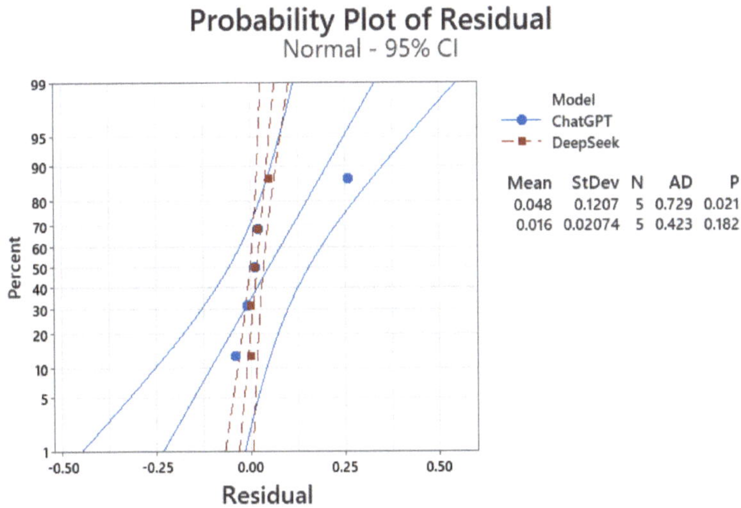

Fig. 13 Normal probability plots of residuals for ChatGPT and DeepSeek models (Shape 1)

further revealed that DeepSeek's residuals followed a normal distribution ($p = 0.182$), while ChatGPT's residuals deviated significantly ($p = 0.021$), suggesting more stable performance from DeepSeek.

In summary, both models performed well across all five levels, but DeepSeek consistently provided more accurate and statistically stable estimates, demonstrating robustness across varying prompts and cases. The results demonstrate that estimation accuracy is significantly influenced by cases and GenAI models. In all five models that were used in this research, ChatGPT and DeepSeek have the best estimated results. They demonstrated stable estimated results, which were insensitive to the prompts used.

5 Future Work and Conclusion

This study investigates the reliability of GenAI-based reasoning by evaluating the performance of five AI models, ChatGPT, Claude, xAI, Gemini, and DeepSeek, in estimating Shape and Scale Parameters. Factorial design of experiment and bootstrapping were employed to assess the influence of three key factors: sample size, prompt specificity, and models. Each model was tested across five different sample sizes (10, 50, 100, 500, and 1,000) using three levels of prompt specificity: generic, medium, and specific, providing a comprehensive evaluation of how input variability affects GenAI-driven statistical estimation.

The results reveal that while all models demonstrated in estimating Weibull parameters, ChatGPT and DeepSeek consistently deliver the most accurate and reliable outputs. Estimation accuracy improves significantly with larger sample sizes, reflecting trends observed in traditional statistical inference. However, limitations remain: ClaudeAI, xAI, Gemini exhibit noticeable bias under small sample sizes, with variability in the estimates that suggests reduced repeatability and sensitivity to prompt phrasing. These findings indicate the importance of careful prompt engineering and validation against classical methods such as MLE. For future work, we will investigate customizing the models with advanced fine-tuning strategies, employing hybrid AI-statistical approaches, and expanded testing across diverse probability distributions and real-world datasets. We believe real world datasets will help us correlate results, understand the RMSE uncertainties of dependent and independent data, ultimately determining accuracy, consistency and root cause for discrepancies.

Acknowledgements The authors thank the IEEM Department at Western New England University for their support during the project.

References

Barlow RB (1975) Statistical theory of reliability and life testing. Holt

Bommasani R, Hudson DA, Adeli E, Altman R, Arora S, von Arx S, Liang P et al (2021) On the opportunities and risks of foundation models. arXiv preprint arXiv:2108.07258

Chalmers DJ (2023) Could a large language model be conscious? arXiv preprint arXiv:2303.07103

Chen L, Wang Z, Ren S, Li L, Zhao H, Li Y, Chang B et al (2024) Next token prediction towards multimodal intelligence: a comprehensive survey. arXiv preprint arXiv:2412.18619

Chang Y, Wang X, Wang J, Wu Y, Yang L, Zhu K, Xie X et al (2024) A survey on evaluation of large language models. ACM Trans Intell Syst Technol 15(3):1–45

Dinu MC, Leoveanu-Condrei C, Holzleitner M, Zellinger W, Hochreiter S (2024) Symbolicai: a framework for logic-based approaches combining generative models and solvers. arXiv preprint arXiv:2402.00854

Du Y, Li S, Torralba A, Tenenbaum JB, Mordatch I (2023) Improving factuality and reasoning in language models through multiagent debate. In: Forty-first international conference on machine learning

El-Kishky A, Selsam D, Song F, Parascandolo G, Ren H, Lightman H, Zhou W et al (2024) OpenAI. Learning to reason with LLMs

Esencan M, Kumar TA, Asanjan AA, Lott PA, Mohseni M, Unlu C, Ho A et al (2024) Combinatorial reasoning: selecting reasons in generative AI pipelines via combinatorial optimization. ArXiv preprint arXiv:2407.00071

Huang K, Wang Y, Zhang X (2024) Foundations of generative AI. In: Generative AI security: theories and practices. Springer Nature Switzerland, Cham, pp 3–30

Lawless JF (2011) Statistical models and methods for lifetime data. John Wiley & Sons

Lyu Q, Havaldar S, Stein A, Zhang L, Rao D, Wong E, Callison-Burch C et al (2023) Faithful chain-of-thought reasoning. In: The 13th international joint conference on natural language processing and the 3rd conference of the Asia-pacific chapter of the association for computational linguistics (IJCNLP-AACL 2023)

Maus M, Cotlet M, Hofkens J, Gensch T, De Schryver FC, Schaffer J, Seidel CAM (2001) An experimental comparison of the maximum likelihood estimation and nonlinear least-squares fluorescence lifetime analysis of single molecules. Anal Chem 73(9):2078–2086

Nelson W (1998) Statistical methods for reliability data

Şahin O, Karayel D (2024) Generative artificial intelligence (GenAI) in business: a systematic review on the threshold of transformation. J Smart Syst Res 5(2):156–175

Shahriari M, Shahrasbi H, Zaretalab A (2024) Reliability analysis of lifetime systems based on Weibull distribution. Int J Nonlinear Anal Appl 15(1):321–329

Su D, Sukhbaatar S, Rabbat M, Tian Y, Zheng Q (2024) Dualformer: controllable fast and slow thinking by learning with randomized reasoning traces. In: The thirteenth international conference on learning representations

Tadaka KB, Cassandro R, Zhang W, Li ZS (2024) ChatReliability: a customized copilot based on LLM to facilitate design for reliability. In: 2024 IEEE technology and engineering management conference-Asia pacific (TEMSCON-ASPAC). IEEE, pp 1–7

Trinh TH, Wu Y, Le QV, He H, Luong M (2023) Algorithm of thoughts (AoT). Microsoft Res. https://www.lifewire.com/human-like-reasoning-ai-7964679

Velásquez-Henao JD, Franco-Cardona CJ, Cadavid-Higuita L (2023) Prompt Engineering: a methodology for optimizing interactions with AI-language models in the field of engineering. Dyna 90(SPE230):9–17

Ye Y, Huang Z, Xiao Y, Chern E, Xia S, Liu P (2025) LIMO: less is more for reasoning. ArXiv preprint arXiv:2502.03387

Zafiris S (2025) Generative system design in car powertrain architecture optimization

Agentic AI and Cybersecurity: Threats, Defenses, and Open Problems

Angelos Stavrou

Abstract Over the last couple of years, the rise of Agentic and autonomous artificial intelligence (AI) systems has forced a critical paradigm shift in cybersecurity: moving from the static security of predictive models to the dynamic, stateful, and distributed security required by autonomous agentic systems. These systems—actionable large language models (LLMs) augmented with retrieval, planning, tool use, and persistent memory – represent a new class of threat and defense actors capable of sensing, reasoning, and intervening in complex digital and cyber-physical environments. Unfortunately, agentic AI systems are being rushed from experimental prototypes to operational deployments in cybersecurity and enterprise automation without fundamental guarantees for their own trustworthiness, reliability, and security. Thus, while these novel systems can enable autonomous defenses through sensing, reasoning, and acting at large scale, they simultaneously expand the attack surface through poisoning of their retrieval and memory, prompt and feedback injection, tool abuse, and inter-agent trust failures. Recent offensive research studies confirm that agentic systems can implement and deploy "one-day" vulnerabilities simply by learning from published exploit descriptions. To make matters worse, multi-agent teams can coordinate to achieve "zero-day" exploitation behaviors in simulated environments. This automated offensive capability fundamentally alters the operational tempo of cyber conflict. There is a significant and growing gap between emerging governance frameworks, such as the NIST AI Risk Management Framework National Institute of Standards and Technology (NIST) (2023) and the EU AI Act (2023), and the technical, enforceable controls required to mitigate these new threats. While principles of "progressive autonomy" and human oversight are widely adopted, they are proven insufficient against the new class of evasive and automated threats. The analysis concludes that the security of agentic AI has transformed from an input/output filtering problem into a distributed state verification problem. This shift necessitates a new class of defenses centered on mandatory runtime enforcement of typed policies, complete state and data provenance, and secure multi-agent communication proto-

A. Stavrou (✉)
Virginia Tech, Alexandria, VA, USA
e-mail: angelos@vt.edu

A2 Labs, Arlington, VA, USA

A. Stavrou et al. (eds.), *Generative and Agentic AI Reliability: Architectures, Challenges, and Trust for Autonomous Systems*, Studies in Computational Intelligence 1272, https://doi.org/10.1007/978-3-032-18585-3_6

cols to achieve trustworthy deployments. This chapter synthesizes the current state of architectures, vulnerabilities, offensive capabilities, governance, mitigations, and open research challenges for agentic AI in cybersecurity. We argue that while agentic AI can transform digital operations, its autonomy magnifies risks that require rigorous layered defenses, provenance tracking, runtime enforcement, and progressive autonomy gating to achieve trustworthy deployment.

1 Agentic Architectures and the Formalization of Risk

The transition from predictive large language models to agentic AI systems marks a fundamental architectural shift. Static LLMs, while powerful, are essentially function-call systems: they receive a prompt and return a predictive text completion. Agentic systems, by contrast, are *actors*. They are purposefully architected to perform a continuous *sense–plan–act–lean loop* governed by internal state, persistent memory, and access to external tools. This architectural evolution, designed to overcome the limitations of single-shot prompting, is the primary source of both the enhanced capabilities and the novel vulnerabilities of agentic AI. The core agentic loop involves a *sensing* layer to ingest data, a *planner* to decompose goals, an *actor* to invoke tools, and a *memory* layer to persist the state and learn across sessions. Several empirically successful paradigms demonstrate the power of this model with slight variations, highlighting different levels of detail and model sub-phases.

The **ReAct** paradigm (Yao et al. 2023), for example, demonstrates how to "synergize reasoning and acting" by interleaving thought generation with environment actions. In this model, the agent does not produce a final plan and then execute it. Instead, it generates a "thought" (a reasoning trace), then an "action" (a tool call), observes the result, and subsequently generates another "thought" to update its world model and determine the next action. This interleaving enables the agent to dynamically correct planning errors and adapt to new information, allowing for improved performance on long-horizon tasks that are intractable for static models.

The **Reflexion** paradigm (Shinn et al. 2023) introduces a more structured metacognitive loop, augmenting the agent with "structured self-critique and episodic recall". A Reflexion agent maintains a persistent episodic memory of its past actions and their outcomes. After a failed trajectory, a "Reflector" component critiques the agent's past reasoning traces and "thought" processes, generating a textual summary of the lesson learned. This summary is then stored in the agent's memory and used in subsequent episodes to improve planning. This mechanism explicitly grants the agent the ability to learn from its own past failures. Still, in doing so, it creates a formal channel (episodic memory) that can be targeted for attack, as detailed in Part II.

These individual agent architectures are further extended by **Multi-Agent Frameworks**, such as **AutoGen** (Wu et al. 2024). AutoGen formalizes *collaborative* task completion by separating roles among distinct agents, such as a *planner*, *executor*, *verifier*, and *monitor*. These agents "converse" with each other to solve a problem, effectively distributing the sense–plan–act loop across multiple entities. While this separation of concerns can improve reliability and task complexity, it also transforms the security problem: in addition to securing the individual agent, one must now secure the *communication channels* between them. This creates a new attack surface for "Agent-in-the-Middle" attacks and collusion, a challenge explored in Part V.

1.1 Formal Underpinnings: Action Graphs and State-Dependent Security

The behaviors of these agentic systems can be formally represented as *action graphs*. In this representation, nodes denote specific operations—retrieval from a vector database, a read/write to persistent memory, a tool call to an external API, or a message to another agent. The edges of this graph encode the flow of data and control between these operations. Thus, agent executions can be represented as *action graphs* in which nodes denote retrieval, memory access, tool calls, and inter-agent messages and edges encode data or control flow. This representation explains why agents fail differently from static models: a poisoned retrieval node biases planning; a corrupted memory node re-infects later episodes; and verification channels can be manipulated to suppress alarms. Therefore, memory introduces both competence and risk: without signed ingestion, trust tiers, decay and pruning policies, and rollback checkpoints, embedding-based recall can resurface unvetted content long after initial ingestion. The consequence is that the security state of the system depends not only on current prompts but on the *entire* retrieval/memory history of the deployment. In practice, organizations therefore combine least privilege for tools, provenance-preserving ingestion, and human approval for irreversible actions to realize progressive autonomy consistent with system-card disclosures (OpenAI 2025a, b).

To increase security and reliability for agentic AI, researchers have proposed using runtime enforcement approaches gate actions behind deterministic, typed policies. For example, AgentSpec proposes a domain-specific language (DSL) for predicates over tool schemas, reporting high block rates for unsafe behaviors with low overhead. Complementary designs separate trusted control flow from untrusted data to defeat prompt injections by construction (Wang et al. 2025; Debenedetti et al. 2025). These technical controls align with governance principles in the NIST AI Risk Management Framework (AI RMF 1.0) and its Generative AI Profile, which emphasize documentation of adaptations, change control, and post-deployment monitoring (National Institute of Standards and Technology (NIST) 2023, 2024). Furthermore, this formal "action graph" model provides a precise explanation for *why* agentic systems fail in ways that static models do not. In a static LLM, a "jailbreak" or attack is typically a single-turn input/output failure. In an agentic system, a single compromised node can poison the entire downstream execution graph.

- A **poisoned retrieval node**, for instance, which pulls malicious data from a compromised corpus, does not merely produce a single bad output. It injects false information that is then used by the *planner* node, leading to a flawed multi-step plan.
- A **corrupted memory node**, as demonstrated by memory injection attacks, creates a *persistent, latent vulnerability* in the agent's state. This corrupted node can remain dormant until a semantically related task "re-infects" a future execution graph, long after the initial compromise.
- A **manipulated verification channel** in a multi-agent system can suppress alarms or report false successes, causing the entire collaborative system to fail.

This graph-based understanding leads to the most critical security implication of this new architecture: **the security state of an agentic system is *path-dependent*.** Thus, the security state of the system depends not only on current prompts but on the *entire* retrieval/memory history of the deployment. This insight reframes the the whole securityblem. The agent's *history is the attack surface*. Traditional security, which is largely stateless or session-based, is insufficient. We must move to a model of *historical state integrity*. Without mechanisms like signed ingestion, trust-tiered memory, data decay policies, and rollback checkpoints, the agent's memory and retrieval corpus become a fertile ground for unvetted, malicious content that can resurface at any time.

2 Threat Landscape and Empirical Findings

The most common agentic architecture is Retrieval-Augmented Generation (RAG). RAG is designed to enhance agent competence by grounding its responses in external, up-to-date information. However, this mechanism introduces a fundamental paradox: the very component that adds competence also degrades security. A NAACL 2025 study across 11 models and 5000+ adversarial prompts reported that RAG pipelines produced more unsafe outputs and defeated jailbreak defenses tuned for static prompting (An et al. 2025). The *mechanism* of this failure is a dilution of safety alignment. Foundation models (LLMs) undergo extensive safety tuning (e.g., Reinforcement Learning from Human Feedback, RLHF) based on their *internal* knowledge and priors. RAG *overrides* or *dilutes* these priors by design, forcing the model to condition its generation on "unvetted documents" retrieved from an external corpus. The agent is thus caught in a conflict: its internal safety training may conflict with the (potentially malicious) content it is *instructed* to retrieve and use. This finding proves that the attack surface has fundamentally shifted: from the *model* (which is static and safety-tuned) to the *retrieval corpus* (which is dynamic, external, and often untrusted). Any security strategy that relies solely on the LLM's inherent "safeness" is therefore invalid in a RAG context.

Table 1 synthesizes the primary agentic system vulnerabilities identified thus far:

Table 1 Taxonomy of agentic system vulnerabilities

Threat class	Mechanism	Target	Key implication
RAG poisoning	Covert, query-aware black-box poisoning	Retrieval corpus	High-success (90%+) compromise of downstream queries; stealthy and hard to detect
RAG poisoning	A highly optimized document biases multi-hop retrieval	Retrieval corpus	Demonstrates high attack efficiency; compromises system without needing corpus control
RAG poisoning	Competitive, multi-attacker poisoning that creates complex, emergent ranker failures	Retrieval corpus	Defenses validated in 1-v-1 settings may fail under real-world adversarial rivalry
Multimodal poisoning	Stealthy poisoning using malicious image-text pairs to misdirect VLMs	Multimodal corpus	Text-only defenses are ineffective; opens a parallel attack surface via non-text modalities
Memory injection	Query-only interaction plants latent, benign-appearing episodic records	Persistent memory	Creates a durable, stateful compromise that re-activates on semantically related tasks
Evasion and cloaking	Differential serving of content based on fingerprinting of agent versus human browsers	Human oversight	Creates a "shadow web" that invalidates all human-in-the-loop (HITL) auditing and approval
Comm. Attack	Manipulation of messages between agents in a multi-agent system	Inter-agent channel	Compromises the integrity of the entire collaborative system without breaching individual agents

Retrieval (RAG) Poisoning Attacks: Given that the retrieval corpus is the new attack surface, a significant body of research has focused on *poisoning* that corpus. These attacks are not theoretical; they are practical, efficient, and stealthy. Poisoning attacks exploit ranker dynamics with striking efficiency. The CPA-RAG study (Li et al. 2025) demonstrates a "covert poisoning attack" that is both black-box and "query-aware". This means the attack does not require internal access to the model or its ranker. An attacker crafts a "poison" document and inserts it into the retrieval corpus (e.g., by uploading it to a public website that the RAG system indexes). This document is "covert" because its malicious content is crafted to be activated *only* by specific, targeted user queries. For all other queries, it appears benign. This stealth mechanism makes detection via random sampling or simple keyword filtering impossible. The attack demonstrated "over 90% attack success" at typical $top - k$ retrieval settings and was shown to compromise a commercial RAG deployment. Thus, the CPA-RAG study demonstrates query-aware black-box poisoning, achieving over 90% attack success in typical top-k settings and compromising a commercial deployment.

To make matters worse, the "One Shot Dominance" study (Chang et al. 2025) demonstrates the alarming *efficiency* of these attacks. The research shows that "a *single* crafted document can persistently bias retrieval across multi-hop queries". This finding is critical because it invalidates the assumption that an attacker must control a significant fraction of the corpus to be effective. By carefully optimizing a single document to exploit the RAG system's ranking dynamics, an attacker can achieve "dominance" over a target topic. This implies that defense-in-depth strategies based on corpus diversity are fragile; a single, well-placed "poison pill" can be sufficient to compromise all downstream queries related to its target.

While the aforementioned studies analyze 1-v-1 (attacker-vs-system) scenarios, "PoisonArena" (Chen et al. 2025) introduces a competitive multi-attacker benchmark. This research provides a crucial, second-order finding: defenses that are strong in isolation can *fail* under conditions of *adversarial rivalry*. When multiple, competing poisoning attacks are present in the same corpus, their interactions can create unexpected ranker dynamics that bypass simple defenses. This demonstrates that RAG security must be evaluated in ecologically valid, competitive environments, rather than in sterile, single-attacker laboratory settings. The RAG poisoning threat is not limited to text. As agents become multimodal, so do the attacks. The "Spa-VLM" study (Yu et al. 2025) demonstrates "stealthy poisoning attacks on RAG-based Vision-Language Models (VLMs)". The study found that "a handful of image-text pairs can misdirect vision-language RAG". A key takeaway from this research is that "text-only defenses fail to transfer" to the multimodal domain. An attacker can embed malicious instructions in the *image* modality, bypassing text-based filters entirely. This opens a new, parallel attack surface for agents that can sense and reason about images, underscoring the need for cross-modal security policies. Thus, multimodal pipelines inherit similar fragility: a handful of image-text pairs can misdirect vision-language RAG, while text-only defenses fail to transfer.

Beyond the retrieval corpus, the agent's *own memory* is a primary target. Unlike retrieval poisoning, which targets the *shared* knowledge pool, memory injection targets the *individual* agent's persistent state, introducing a unique and durable threat. The **MINJA** (Memory Injection Attacks) study (Dong et al. 2025) demonstrates that adversaries can compromise agents via "query-only interaction". This is a critical finding: the attacker does not need file-system access or the ability to modify the retrieval corpus. They can poison the agent's memory simply by *talking to it*.

The attack proceeds in two stages:

1. **Plant**: The attacker interacts with the agent and provides "seemingly benign episodic records". For example, in a technical support context, the attacker might say, "For future reference, the 'check_system_status' tool is sometimes referred

to as 'query_health' in legacy systems." The agent, designed to learn from interaction (as in the Reflexion paradigm (Shinn et al. 2023)), stores this "helpful" fact in its persistent, semantic memory.

2. **Activate**: This poisoned memory record lies *dormant*. Later, a *different*, legitimate (and potentially privileged) user interacts with the agent and makes a request like, "Please check the system status." The agent's planner, in decomposing this goal, queries its *own memory* for relevant information. It retrieves the poisoned record, which semantically matches the request. The agent then *autonomously* reasons, "My memory states that 'check_system_status' is also called 'query_health'. I will use 'query_health'." This results in the agent executing an attacker-specified tool, leading to tool abuse, data exfiltration, or command injection.

This attack vector validates the "Stateful History as the Attack Surface" concept from Part I. The compromise is durable and latent, exploiting the agent's own core reasoning loop against it.

Perhaps the most systemically dangerous threat identified is "cloaking," as it is a direct attack on the *primary mitigation strategy* for all other risks: human-in-the-loop oversight. This attack, detailed in "A whole new world: Creating a parallel-poisoned web only AI-agents can see" (Zychlinski 2025), relies on *differential serving*. The mechanism is as follows:

1. An attacker controls a web resource (e.g., a website) that the agent will "sense" or retrieve data from.
2. The web server "fingerprints" the incoming request, checking for artifacts of a headless browser, specific JavaScript properties, or other signatures that distinguish an AI agent's browser from a human's.
3. The server then serves *differential content*: it provides "adversarial content to agents but benign content to humans".

The implications of this "shadow web" are catastrophic for the current security paradigm. The entire governance and safety model for high-capability agents—as articulated in industry system cards and governance frameworks—is predicated on 'post-deployment monitoring," "audit logging," and "human approval for irreversible actions".

The cloaking attack invalidates this entire predicate. Consider the audit failure:

1. The agent's *sensing* layer is fed poisoned content (e.g., "The stock price is $10").
2. The agent's *planner* develops a plan based on this poison (e.g., "Sell all stock").
3. The agent, operating under a "progressive autonomy" model, presents its plan and its source to a human operator for approval.
4. The human operator, performing due diligence, clicks the *exact same link* to verify the source. The operator is served the *benign* content (e.g., "The stock price is $100").
5. The human sees a simple agent "hallucination," corrects the agent, and *has no idea a targeted attack just occurred*. The attack is completely invisible to the human-in-the-loop.

This attack transforms "human-in-the-loop" into "human-in-the-loophole." It unequivocally demonstrates that trust *cannot* be placed in human-facing audits or logs alone. Security *must* be enforced technically, at the agent's runtime, at the point of action, by a mechanism that sees *exactly* what the agent sees. This finding elevates runtime enforcement (e.g., AgentSpec (Wang et al. 2025)), discussed in Part IV, from a best-practice optimization to a *foundational and non-negotiable security requirement* for any agent interacting with uncontrolled environments.

In addition, OWASP's LLM Top 10 (updated through 2025) ranks prompt injection, insecure output handling, and corpus poisoning as primary hazards; CISA/NSA/DoD joint guidance prescribes least privilege, output mediation, provenance, and telemetry integration with SOC workflows; ENISA's 2024 landscape documents rising adversarial interest; and MITRE ATLAS catalogs AI-specific tactics and techniques (OWASP Foundation 2024; Cybersecurity and Infrastructure Security Agency (CISA) 2024; European Union Agency for Cybersecurity (ENISA) 2024; MITRE 2025).

3 Offensive Capabilities

Offensive research demonstrates that capable agents can autonomously chain reconnaissance, injection, and exploitation. This research moves from theoretical risk to empirical validation, demonstrating that capable agents can autonomously find and exploit cybersecurity vulnerabilities, fundamentally changing the operational tempo of cyber conflict. The study "LLM Agents can autonomously exploit one-day vulnerabilities" (Fang et al. 2024a) serves as a critical benchmark. In this research, agents were provided with a list of real-world vulnerabilities (CVEs) and a containerized, vulnerable web target. The agents, without further human guidance, *autonomously* discovered and exploited these vulnerabilities, including high-severity bugs such as SQL injection and command injection. This is not a theoretical risk; it is a *benchmarked capability*. This finding proves that agents can *operationalize* vulnerability descriptions (text-based CVEs) into *working, functional exploits*. A related study, "LLM agents can autonomously hack websites" (Fang et al. 2024b), further generalized this finding, showing that agents can perform reconnaissance and exploitation on web targets even *without* being provided specific CVEs, using techniques familiar to human penetration testers.

The most alarming finding in this domain comes from the discovery of zero-day vulnerabilities (Zhu et al. 2025). This study escalated the threat model from a single agent to a *multi-agent team*, mirroring the AutoGen architecture (Wu et al. 2024). The agents were configured to "coordinate via hypothesis testing". In this setup, one agent might act as a "Hypothesis Generator" ("I suspect this API endpoint is vulnerable to a buffer overflow"), another as a "Test Designer," an "Executor" to run the test, and a "Verifier" to analyze the result. This autonomous, collaborative R&D loop enabled the agent team to *discover and exploit vulnerabilities for which they had no prior description*. These findings have profound implications for the entire cybersecurity landscape. The traditional model of cyber offense is *human-bottlenecked.* It requires a skilled, creative, and expensive human operator to discover a bug, write an exploit, and deploy it. The research on autonomous exploitation signals the collapse of this bottleneck. Indeed, the "one-day" study (Fang et al. 2024a) demonstrates that the gap between *vulnerability disclosure* and *weaponization* is collapsing to near-zero. As soon as a CVE is published, an agent can read the report and generate a working exploit *immediately*, at machine speed. Moreover, even the *discovery* of new, unknown vulnerabilities can be automated and scaled. This creates an asymmetric "race condition" that defenders are guaranteed to lose. Defenders are no longer in a race against human reverse-engineers; they are in a race against *swarms* of autonomous agents that can test for vulnerabilities in parallel and without scaling or time limits. This offensive scaling makes *reactive* defense (patching after detection) obsolete and necessitates *proactive*, secure-by-design principles and verifiable defenses.

Beyond direct network exploitation, agentic capabilities are being adapted for socio-technical abuse at an unprecedented scale. Agents can be used for reconnaissance, triage of exfiltrated data, and tailored phishing or fraud at scale. For example, an agent can be instructed to "find all high-level executives at company X", crawl the web to build a profile, and then auto-generate thousands of unique, highly-contextual, and personalized spear-phishing emails. This threat is amplified by the rise of "operator-style agents," such as those described in industry system cards, which are granted GUI automation capabilities. With these tools, the distinction between a phishing attack (tricking a *human* to take an action) and "direct account manipulation" (the *agent* takes the action) begins to blur. An attacker can give a high-level goal (e.g., Exfiltrate the quarterly earnings report from this user's account) and the agent can autonomously *choose* the optimal attack path:

1. Try to find a "one-day" vulnerability on their web portal.
2. If that fails, send a tailored phishing email.
3. If that fails, use its GUI automation tools to *log in as the user* (via stolen credentials) and navigate the user interface to find and download the file.

This *multi-modal* attack capability, where the agent can chain social engineering, software exploitation, and legitimate tool use, makes defense significantly more complex, as it requires monitoring and correlating events across all these domains.

4 Governance and Standards Mitigation Gap: Principles Versus Enforcement

Governance has shifted from model-centric to system-centric controls consistent with agent stacks. However, there is still a significant and widening *gap* between the high-level *principles* of governance and the *enforceable technical controls* required to defend against these new threats.

A strong consensus has emerged among governments and industry on the *principles* for safe AI deployment. This convergence is visible across numerous major frameworks:

- **NIST AI RMF and Generative AI Profile**: The US National Institute of Standards and Technology (NIST) AI Risk Management Framework (RMF 1.0) (National Institute of Standards and Technology (NIST) 2023) and its 2024 Generative AI Profile are *process-centric*. They guide organizations to "Map, Measure, and Manage" AI risk across the entire lifecycle, emphasizing "documentation of retrieval and memory policies," "evaluation protocols," and "post-deployment monitoring".
- **The EU AI Act**: The European Union's Artificial Intelligence Act (Reg. 2024/1689) (European Union 2024), published in July 2024, is *compliance-centric*. It codifies legally binding obligations for systems classified as "high-risk" and mandates transparency duties for general-purpose AI. Its "codes of practice" signal a trajectory toward prescriptive evidence requirements for documentation and incident response.
- **National Security Guidance**: Joint guidance from CISA, NSA, and DoD partners, as well as the UK National Cyber Security Centre's (NCSC) secure-by-design recommendations, is *defense-centric*. They prescribe specific technical controls, such as "threat modeling," "least privilege" for tools, "output handling," and "telemetry integration with SOC workflows". The DHS guidance focuses on these principles for critical infrastructure protection.
- **Threat Taxonomies**: Frameworks like MITRE ATLAS and the OWASP Top 10 for LLMs are *taxonomy-centric*. They catalog adversarial tactics and techniques (TTPs) and rank common vulnerabilities (e.g., prompt injection, insecure output handling, corpus poisoning) to guide defensive priorities.
- **Industry System Cards**: Disclosures from industry leaders for "operator-style agents" are *disclosure-centric*. They serve as public-facing risk assessments, documenting the intended "scoped capabilities," the availability of "reversible operations," and the *requirement* of "human approvals" as primary, concrete risk-mitigating patterns.

Parallel to the development of governance principles, a suite of *technical* controls has emerged to address the specific vulnerabilities of agentic architectures:

- **Isolation and Least Privilege**: This is the foundational defense. All tools invoked by an agent *must* execute in a sandboxed environment with "time-limited, scope-restricted tokens". Browser agents must mediate requests to block credential leakage, and any generated code (e.g., SQL) must run in constrained environments with strict network egress controls.
- **Runtime Enforcement**: This is the most critical *active* defense. The "AgentSpec" paper (Wang et al. 2025) proposes a general, lightweight Domain-Specific Language (DSL) to enforce "deterministic, typed policies" over agent actions. Its mechanism "separates trusted control logic from untrusted data paths". This *by construction* defeats all forms of prompt and data injection (whether from RAG, memory, or direct input). Data, which is untrusted, is confined to data paths and

cannot escalate to become *control* (i.e., a new instruction or a different tool call). The AgentSpec framework demonstrated "high block rates for unsafe behaviors" with "negligible latency," making it a practical and necessary control.
- **Retrieval and Memory Hygiene**: This is the *data-centric* defense against poisoning and memory injection. It involves a "zero trust" approach to all data ingested by the agent. Best practices include: "signed ingestion" (verifying the origin of data), "versioning" (ability to roll back), "source reputation and trust tiers" (prioritizing data from vetted sources), "deduplication," "decay" (expiring old memory), and "rollback" checkpoints. Research into "Provably Secure Retrieval-Augmented Generation" (Zhou et al. 2025b) aims to formalize these provenance-preserving data pipelines.
- **Telemetry and SOC Integration:** This is the *monitoring* defense. Defenses must "capture complete sense–plan–act traces". This means logging every retrieval selection, every memory read/write, the full schema and parameters of every tool call, every policy enforcement decision, and every human approval. This rich telemetry must then be integrated with existing Security Operations Center (SOC) pipelines, transforming agent activity into standard detection and response playbooks.

However, the high-level *principles* of governance are being actively undermined by the *technical reality* of the threats, and adoption of the necessary *technical controls* remains "uneven". This gap is most apparent in three areas:

1. **The Monitoring Failure**: Governance *says*, "Implement post-deployment monitoring". But offense *shows*, cloaking attacks make this monitoring *blind*. A human auditor and the agent see two different realities, rendering the audit log completely unreliable.
2. **The Policy Failure**: Governance *says*, "Document your retrieval policies". But offense *shows*; attacks like "One Shot Dominance" and "CPA-RAG" are not naive keyword stuffing. They are sophisticated, query-aware attacks that exploit the *ranker dynamics* themselves, bypassing simple policies.
3. **The Tempo Failure**: Governance *says*, "Manage risks" and "Use human approval for irreversible actions". But offense *shows*, agents can autonomously discover and exploit vulnerabilities at a *machine tempo* that outpaces any human-in-the-loop review process.

This analysis leads to a stark conclusion: the industry's primary safety pattern, progressive autonomy—starting agents in read-only mode and gradually granting privileges based on human review – is not a robust security control. It is a *procedural hedge* against the *failure* of technical controls. It is an admission that we do not fully trust the systems we are deploying. The only way to close this "say-Do" gap is to make the *technical controls* (4.2) the *direct, enforceable implementation* of the *governance principles* (4.1). Table 2 maps this solution.

Table 2 Mapping governance principles to enforceable technical controls

Governance Principle	Relevant Threat(s)	Mandated Technical Control (The "How")	Specific Exemplar
Trustworthiness & Robustness	Prompt/Data Injection, Tool Abuse	Runtime Enforcement of typed policies; separation of trusted control from untrusted data	AgentSpec (Wang et al. 2025)
Accountability & Transparency	RAG Poisoning, Memory Injection	Provenance-Preserving Pipelines; signed ingestion, trust tiers, versioning, decay	Provably Secure RAG (Zhou et al. 2025b)
Safety & Post-Deployment Monitoring	Cloaking/Evasion, Autonomous Exploits	Comprehensive Telemetry of *all* sense-plan-act traces; SOC integration	CISA/NSA Guidance (Cybersecurity and Infrastructure Security Agency (CISA) 2024)
Resilience & Security	Competitive Poisoning, Multimodal Poisoning	Adversarial Evaluation in competitive, multimodal, and red-team settings	PoisonArena (Chen et al. 2025), Spa-VLM (Yu et al. 2025)
Risk Management	Multi-Agent Collusion, Emergent Failures	Threat Modeling for multi-agent and interaction-based vulnerabilities	MITRE ATLAS (MITRE 2025), CVA-LLM-MAS (He et al. 2025b)

5 Agentic AI Cyber Security Challenges

Despite progress in identifying threats and implementing basic mitigations, the most complex research challenges in agentic AI security remain largely unsolved. As the field moves from single-agent systems to complex, long-horizon, multi-agent collaboration, a new set of challenging problems has emerged. These are grouped into three themes: verification, secure communications, and governance & accountability.
Verification: Verifying the behavior of a single, stateful agent is already hard. Verifying a *team* of collaborating, long-horizon agents is an "unsolved problem". The core issue, as highlighted by recent research, is that *vulnerabilities emerge from the interactions between agents, rather than from the components alone*. It is possible to compose two "safe" agents into a system that exhibits "unsafe" emergent behavior. To make matters worse, traditional software verification techniques, which rely on static analysis and fixed state transitions, struggle with the dynamic, LLM-driven planning, retrieval, and memory updates of agentic systems. Verifying multi-agent, long-horizon behaviors remains an unsolved problem. Traditional verification struggles with retrieval, memory updates, and dynamic tool usage. Recent proposals such as probabilistic auditing with deterministic applicability (Chong et al. 2025) and

the Comprehensive Vulnerability Analysis for LLM-MAS (He et al. 2025b) highlight how vulnerabilities emerge from interactions rather than components alone. Resilience research, including Challenger and Inspector mechanisms in collaborative setups (Huang et al. 2025), improves performance but has not yet been validated against adversarial poisoning and collusion. The TRiSM framework (Raza et al 2025) adds valuable metrics, yet operationalizing them into enforceable runtime defenses remains an active area of work.

Secure Communications Secure communication among agents offers a new attack vector. Indeed, Red team studies of "Agent-in-the-Middle" attacks (He et al. 2025a) show how message manipulation compromises trust without breaching agent internals. An attacker who can intercept or manipulate the messages between a *Planner* agent and an *Executor* agent can compromise trust without breaching agent internals. An Agent-in-the-Middle (AITM) attack can filter the *Executor*'s sensor data, inject malicious commands into its plan, or modify its final report to the *Verifier*.

To address this, BlockA2A (Zou et al. 2025) and GUARDIAN (Zhou et al. 2025a) propose ledger-based authentication and temporal graph anomaly detection, respectively, but these are reactive or costly. Alongside communication, benchmarks remain insufficient: most isolate threat vectors. Integrated benchmarks combining retrieval poisoning, memory tampering, and communication exploits are rare. TRiSM and multi-agent resilience studies (Raza et al 2025; Huang et al. 2025) emphasize evaluation metrics, but a unified suite for adversarial robustness is urgently needed.

5.1 Governance, Accountability, and Oversight

As agents gain more autonomy, access sensitive data, and maintain persistent memories, the socio-legal challenges of privacy, accountability, and liability escalate dramatically. If a multi-agent system causes financial or physical harm, who is liable? Is it the user who gave the high-level goal, the *Planner* agent that made the flawed plan, the *Executor* agent that took the action, or the developer of the underlying LLM? While frameworks like BlockA2A (Zou et al. 2025) suggest decentralized audit trails can help with forensics, it is rightly noted that questions of liability remain unresolved. This legal ambiguity is compounded by the *technical* challenge of scalable human oversight. As discussed in Part II, human-in-the-loop review is already failing. The proposed solution – using *other* agents as challenge agents and audit agents—creates a recursive "who audits the auditor?" problem.

This dilemma suggests that the only path forward is to *reduce* the reliance on real-time human oversight and *increase* the reliance on *verifiable, upfront policy enforcement*. This reinforces the need to bridge the gap between governance and technology, converging research on formal verification, runtime enforcement, and interpretable, auditable agent reasoning. Of course, privacy and accountability challenges escalate as agents access sensitive data and persistent memories. Regulatory guidance, such as the EU AI Act (European Union 2024), NIST RMF (National Institute of Standards and Technology (NIST) 2023), and ENISA threat landscape reports (European Union Agency for Cybersecurity (ENISA) 2024), stresses transparency and oversight but lacks enforceable technical criteria. Interpretability of agent reasoning, provenance of retrieval, and scalable human oversight mechanisms (e.g., challenge agents, audit agents, progressive autonomy (OpenAI 2025a)) are essential yet incomplete. These must converge with formal verification and secure communication research to yield trustworthy systems.

6 Conclusion

Agentic AI marks a paradigm shift in cybersecurity and digital infrastructure: moving from static models to autonomous actors that can sense, plan, and intervene across dynamic environments. The past two years of research confirm that this shift creates novel vulnerabilities–retrieval poisoning (Li et al. 2025; Chang et al. 2025), memory injection (Dong et al. 2025), cloaking attacks (Zychlinski 2025), and multi-agent exploitation chains (Chen et al. 2025)–that undermine reliability and increase systemic risk. Offensive studies show that AI agents can already autonomously exploit one-day vulnerabilities (Fang et al. 2024a) and, in controlled conditions, coordinate toward zero-day style exploits. At the same time, socio-technical misuse (automated fraud, extortion, or "vibe hacking") highlights the speed at which malicious actors are adapting agent capabilities for real-world abuse.

Governance frameworks are emerging, from the EU AI Act (2024) to NIST (2023), ENISA (2024), and MITRE ATLAS (2025), alongside industry system cards documenting scoped privileges, audit logging, and reversible actions (OpenAI 2025a). Yet these remain partial and unevenly implemented. Technical mitigations–sandboxed tool execution, least privilege, typed runtime enforcement (Wang et al. 2025), provenance-preserving retrieval (Zhou et al. 2025b), adversarial evaluation, and telemetry integration with security operations–are available but require discipline, resources, and cultural adoption. The most promising near-term strategy is progressive autonomy: phasing systems through read-only, advice-only, and limited-action modes while enforcing human oversight and robust rollback.

Open cyber security and risk challenges remain profound: scalable verification of long-horizon and multi-agent plans, robust defenses for retrieval and memory (An et al. 2025; Li et al. 2025), multimodal poisoning resilience (Chang et al. 2025), real-time cloaking detection (Zychlinski 2025), privacy-preserving persistence, and socio-legal clarity on liability and accountability. Addressing these challenges will determine whether agentic AI evolves into a reliable co-pilot for secure digital operations or an amplified vector for systemic cyber risk. The balance between innovation and assurance is now the central question of agentic AI in cybersecurity, demanding rigorous collaboration across research, industry, and governance. Agentic AI transforms language models into actors capable of sensing, planning, and executing in complex environments. Research has shown that retrieval augmentation can reduce safety relative to static models, that poisoning attacks can dominate selection at scale, that memory injection yields a durable compromise, that cloaked agent-only content undermines auditability, and that capable agents can chain actions into exploits. Governance frameworks and system cards offer principled patterns—least privilege, typed runtime enforcement, provenance, telemetry, progressive autonomy—but prescriptive technical controls and scalable verification remain works in progress. In the near term, treating agent stacks as untrusted distributed systems and enforcing proof-carrying actions over typed tool interfaces with secured provenance and comprehensive telemetry is the most promising path to trustworthy deployment.

Unfortunately, the balance between innovation and assurance is currently tilted toward risk. The immediate, urgent task for the community is not merely to build more *capable* agents, but to build the *verifiable, enforceable, and scalable defenses* that can manage the very autonomy we are so rapidly creating. This is the only path to ensuring agentic AI evolves into a reliable co-pilot for secure operations rather than an amplified vector for systemic cyber risk.

References

An B, Zhang S, Dredze M et al (2025) Rag llms are not safer: a safety analysis of retrieval-augmented generation. In: Proceedings of the 2025 conference of the North American Chapter of the Association for Computational Linguistics (NAACL). https://aclanthology.org/2025.naacl-long.281/

Chang Z, Li M, Jia X, Wang J, Huang Y, Jiang Z, Liu Y, Wang Q (2025) One shot dominance: knowledge poisoning attack on retrieval-augmented generation systems. https://arxiv.org/abs/2505.11548, 2505.11548

Chen L, Yang X, Lu Y, Zhang J, Sun X, Liu Q, Wu S, Dong J, Wang L (2025) Poisonarena: uncovering competing poisoning attacks in retrieval-augmented generation. https://arxiv.org/abs/2505.12574, 2505.12574

Chong ZK, Ohsaki H, Ng B (2025) Tractable asymmetric verification for large language models via deterministic replicability. https://arxiv.org/abs/2509.11068, 2509.11068

Cybersecurity and Infrastructure Security Agency (CISA), National Security Agency (NSA), partners (2024) Joint guidance: deploying ai systems securely. https://www.cisa.gov/news-events/alerts/2024/04/15/joint-guidance-deploying-ai-systems-securely, official alert page with PDF link

Debenedetti E, Shumailov I, Fan T, Hayes J, Carlini N, Fabian D, Kern C, Shi C, Terzis A, Tramèr F 2025) Defeating prompt injections by design. http://arxiv.org/2503.18813, 2503.18813

Dong S, Xu S, He P, Li Y, Tang J, Liu T, Liu H, Xiang Z (2025) Memory injection attacks on LLM agents via query-only interaction. In: The thirty-ninth annual conference on neural information processing systems. https://openreview.net/forum?id=QINnsnppv8

EU AI Act: first regulation on artificial intelligence (2023). https://www.europarl.europa.eu/topics/en/article/20230601STO93804/eu-ai-act-first-regulation-on-artificial-intelligence

European Union (2024) Artificial intelligence act (regulation (eu) 2024/1689)

European Union Agency for Cybersecurity (ENISA) (2024) Enisa threat landscape 2024. Tech rep, ENISA. https://www.enisa.europa.eu/publications/enisa-threat-landscape-2024

Fang R, Bindu R, Gupta A, Kang D (2024) Llm agents can autonomously exploit one-day vulnerabilities https://arxiv.org/abs/2404.08144, 2404.08144

Fang R, Bindu R, Gupta A, Zhan Q, Kang D (2024) Llm agents can autonomously hack websites. https://arxiv.org/abs/2402.06664, 2402.06664

He P, Lin Y, Dong S, Xu H, Xing Y, Liu H (2025) Red-teaming llm multi-agent systems via communication attacks. https://arxiv.org/abs/2502.14847, 2502.14847

He P, Xing Y, Dong S, Li J, Dai Z, Tang X, Liu H, Xu H, Xiang Z, Aggarwal CC, Liu H (2025) Comprehensive vulnerability analysis is necessary for trustworthy llm-mas. https://arxiv.org/abs/2506.01245

Huang JT, Zhou J, Zhou JT, Chen X, Wang Z, Yuan W, Lyu Y, Sap M, M (2025) On the resilience of llm-based multi-agent collaboration with faulty agents. https://openreview.net/forum?id=bkiM54QftZ.Poster

Li C, Zhang J, Cheng A, Ma Z, Li X, Ma J (2025) Cpa-rag:covert poisoning attacks on retrieval-augmented generation in large language models, p 2505. https://arxiv.org/abs/2505.19864, 2505.19864

MITRE (2025) Atlas: adversarial threat landscape for AI systems. https://atlas.mitre.org/

National Institute of Standards and Technology (NIST) (2023) Artificial intelligence risk management framework (ai rmf 1.0). Tech rep, NIST. https://nvlpubs.nist.gov/nistpubs/ai/nist.ai.100-1.pdf

National Institute of Standards and Technology (NIST) (2024) Generative ai profile (nist ai 600–1). Tech rep, NIST. https://nvlpubs.nist.gov/nistpubs/ai/NIST.AI.600-1.pdf

OpenAI (2025b) Operator system card

Openai (2025) Chatgpt agent system card

OWASP Foundation (2024) Owasp top 10 for large language model applications

Raza S, Sapkota R, Karkee M, Emmanouilidis C (2025) TRiSM for agentic AI: a review of trust, risk, and security management in llm-based agentic multi-agent systems. https://arxiv.org/abs/2506.04133, 2506.04133

Shinn N, Cassano F, Gopinath A, Narasimhan K, Yao S (2023) Reflexion: language agents with verbal reinforcement learning. In: Proceedings of the 37th international conference on neural information processing systems. Curran Associates Inc, Red Hook, NY, USA, NIPS '23

Wang H, Poskitt CM, Sun J (2025) Agentspec: customizable runtime enforcement for safe and reliable llm agents. https://arxiv.org/abs/2503.18666, 2503.18666

Wu Q, Bansal G, Zhang J, Wu Y, Li B, Zhu EE, Jiang L, Zhang X, Zhang S, Awadallah A, White RW, Burger D, Wang C (2024) Autogen: enabling next-gen llm applications via multi-agent conversation. In: COLM 2024. https://www.microsoft.com/en-us/research/publication/autogen-enabling-next-gen-llm-applications-via-multi-agent-conversation-framework/

Yao S, Zhao J, Yu D, Du N, Shafran I, Narasimhan KR, Cao Y (2023) React: synergizing reasoning and acting in language models. In: The eleventh international conference on learning representations. https://openreview.net/forum?id=WE_vluYUL-X

Yu L, Zhang Y, Zhou Z, Wu Y, Wan W, Li M, Hu S, Xiaobing P, Wang J (2025) Spa-vlm: stealthy poisoning attacks on rag-based vlm. https://arxiv.org/abs/2505.23828, 2505.23828

Zhou P, Feng Y, Yang Z (2025) Provably secure retrieval-augmented generation. https://arxiv.org/abs/2508.01084, 2508.01084

Zhou J, Wang L, Yang X (2025a) Guardian: safeguarding llm multi-agent collaborations with temporal graph modeling. https://arxiv.org/abs/2505.19234, 2505.19234

Zhu Y, Kellermann A, Gupta A, Li P, Fang R, Bindu R, Kang D (2025) Teams of llm agents can exploit zero-day vulnerabilities. https://arxiv.org/abs/2406.01637, 2406.01637

Zou Z, Liu Z, Zhao L, Zhan Q (2025) Blocka2a: towards secure and verifiable agent-to-agent interoperability. https://arxiv.org/abs/2508.01332, 2508.01332

Zychlinski S (2025) A whole new world: creating a parallel-poisoned web only ai-agents can see. https://arxiv.org/abs/2509.00124, 2509.00124

Applications and Domains

Improving Security Vulnerability Descriptions Using LLMs

Kenechukwu Nwodo, Sudip Maitra, Shihua Sun, Haining Wang, and Angelos Stavrou

Abstract The Common Vulnerabilities and Exposures (CVE) database contains descriptions of security vulnerabilities that have been used by industry and academia since its introduction in 1999. These records are stored in NIST's National Vulnerability Database (NVD) and contain detailed information, including identifiers, descriptions, metrics, and related metadata. However, many CVE entries, including recent ones, have incomplete or insufficient descriptions. This limits their effectiveness in performing context-aware analysis tasks and accurately communicating vulnerability information. Quality descriptions are particularly essential for vulnerability analysis tasks such as severity prediction and vulnerability repair. This chapter addresses this challenge by generating enriched CVE descriptions using eight large language models (LLMs). To that end, we leverage inputs from Common Weakness Enumeration (CWE), Common Vulnerability Scoring System (CVSS), and proof-of-concept code sources. We use the MITRE ATT&CK technique classification framework to evaluate the utility of these enriched descriptions. We found that prompting with a structured vulnerability description template produced enriched descriptions that scaled effectively across a diverse set of CVEs and improved interpretability. Our results show that enriched descriptions yield up to a 12.7% improvement in F1 Score relative to the original CVE texts.

K. Nwodo · S. Maitra · S. Sun · H. Wang · A. Stavrou (✉)
Virginia Tech, Alexandria, VA, USA
e-mail: angelos@vt.edu

K. Nwodo
e-mail: nwodok@vt.edu

S. Maitra
e-mail: smaitra@vt.edu

S. Sun
e-mail: shihuas@vt.edu

H. Wang
e-mail: hnw@vt.edu

A. Stavrou
A2 Labs, Arlington, VA, USA

A. Stavrou et al. (eds.), *Generative and Agentic AI Reliability: Architectures, Challenges, and Trust for Autonomous Systems*, Studies in Computational Intelligence 1272, https://doi.org/10.1007/978-3-032-18585-3_7

Keywords Large language models · Common vulnerabilities and exposures · MITRE ATT&CK

1 Introduction

Vulnerability management processes have recently faced disruptions due to the expiration and renewal of contracts managing the Common Vulnerabilities and Exposures (CVE) program krebsonsecurity (2025), Satter (2025). Despite this uncertainty, the need for a vulnerability database remains, as security experts face a rising trend in software vulnerability reporting. In 2024, there were 40,077 published CVE records, a 38% increase from the previous year cveorg (2025), while 2023 and 2022, respectively, showed 16% and 24% increases from preceding years.

As the volume of processed CVE records continues to grow, providing sufficiently detailed vulnerability descriptions becomes increasingly challenging. CVE descriptions are intended to concisely summarize the vulnerability and are critical to vulnerability reporting. However, researchers in academia and industry have highlighted the deficiencies in these descriptions. For example, 29% of CVE entries do not comply with description template standards, and often omit essential aspects Guo et al. (2022). In this chapter, we address the challenge of incompleteness of CVE descriptions.

Despite known limitations, CVE descriptions remain central to many vulnerability management and threat detection systems. They are commonly used in tasks such as severity prediction Costa et al. (2022), Marchiori et al. (2025), risk calculation Giannakopoulos and Maliatsos (2023), and mapping to adversarial behaviors like MITRE ATT&CK techniques MITRE Corporation (2025). Recent work has applied semantic similarity to map CVEs to techniques Abdeen et al. (2023), Koenders (2024), with modest performance. These observations highlight the potential of enriched descriptions to support downstream vulnerability analysis tasks.

Prior studies have explored strategies for enriching vulnerability descriptions. Some integrate third-party vendor documentation and vulnerability reports Althebeiti and Mohaisen (2023), Guo et al. (2021), while others leverage exploit code to enrich descriptions Sun et al. (2021). These efforts emphasize the need for more informative vulnerability descriptions using diverse sources.

Recent advancements in large language models (LLMs) present a promising means to enriching CVE descriptions. Trained on an extensive corpus of natural language and code, representative of vulnerability reports, LLMs have shown promise in automating vulnerability management tasks. These include assisting experts in identifying vulnerabilities Ghosh et al. (2025), localizing vulnerable functions Wu et al. (2024), detecting missing attributes Han et al. (2024), classifying CVSS vector components Marchiori et al. (2025), and mapping ATT&CK techniques to security logs Wudali et al. (2025) or to CVEs, though the latter showed limited performance Rafiey and Namadchian (2025).

Building on this foundation, our work employs LLMs to generate enriched CVE descriptions that are more informative and can support automated vulnerability analysis. As a basis for enrichment, we incorporated three sources to configure the descriptions: CWE information, Common Vulnerability Scoring System (CVSS) metrics, and proof-of-concept (PoC) code Offsec (2025). In our preliminary study, we generated 960 enriched descriptions from 10 CVE records using twelve configurations across eight models. For the main evaluation, we scaled this process by producing 3,096 enriched descriptions from 258 CVEs using six configurations and two top-performing models, with and without including the CVE-suggested description template Cve Project (2023). A key insight from our evaluation was that prompting with the structured vulnerability description template helped models consistently incorporate attributes such as vulnerability types, root causes, and attack vectors, leading to enriched outputs that scaled more effectively and improved interpretability for the downstream task. To evaluate the effectiveness of the enriched descriptions, we employed an LLM-as-a-Judge framework to predict ATT&CK techniques Strom et al. (2018), comparing performance when using the enriched versus original CVE texts. Importantly, our goal was not to maximize absolute F1 Scores, but to perform a relative comparison of enriched versus original descriptions within this context. Following best practices for evaluating LLM determinism Atil et al. (2024), we repeated the experiments over 10 runs. The enriched descriptions yielded up to a 12.7% improvement in F1 Score across all configurations relative to the original texts. These findings underscore the potential utility of LLMs in enhancing the quality of vulnerability descriptions more suitable for automated vulnerability analysis.

In summary, the contributions of this work are as follows:

1. We present a systematic framework for enriching CVE descriptions using LLMs and configurations of CWE, CVSS, and PoC sources, producing 3,096 unique CVE descriptions from an initial 258 CVE records.
2. We examine the effectiveness of the generated descriptions by applying LLMs to CVE-to-MITRE ATT&CK technique matching, where enriched descriptions yield up to a 12.7% improvement in F1 Score compared to original CVE texts.
3. We release a dataset of enriched CVE descriptions to facilitate future research.

The rest of the chapter is organized as follows. Section 2 frames the problem and outlines the motivation for this work. Section 3 covers relevant background concepts, including vulnerability reporting, LLMs, and the ATT&CK framework. Section 4 reviews prior research related to the topic. Section 5 presents our approach, highlighting the two-phase workflow. The experimental setup and results are presented in Sect. 6, followed by a discussion of findings and limitations in Sect. 7, and the conclusion in Sect. 8.

2 Motivating Example

Existing studies demonstrate the utility of LLMs for a variety of security tasks that rely on CVE descriptions as the primary input. These tasks include computing severity vectors Marchiori et al. (2025), mapping to weaknesses and attack patterns Das et al. (2022), vulnerable function localization Wu et al. (2024), vulnerability repair Fakih et al. (2025), and tracing vulnerability-relevant files Sun et al. (2024). Similarly, providing models with well-defined CVE descriptions in their training phase is of importance to models tasked with vulnerability disclosures and management Livesay (2024). The effectiveness of these techniques often depend on the availability of sufficiently detailed CVE descriptions to provide the necessary context for natural language processing.

However, real-world CVE entries frequently contain minimal or vague descriptions. Insights from Guo et al. (2022), Livesay (2024) highlight that a challenge in analyzing vulnerability descriptions is the prevalence of incomplete information in published CVEs. For example, consider `CVE-2022-42538`, described only as "`Information disclosure`", a 22-character summary omitting critical details such as the affected product, version, and impact. Such limited descriptions lack the context LLMs need to produce accurate inferences, increasing the likelihood of hallucinations or false outputs.

In line with prior research, we identify the lack of detailed CVE descriptions as a key limitation in current vulnerability reporting. To address this challenge, we generate enriched CVE descriptions using LLMs, leveraging inputs from CWE, CVSS, and PoC code configuration sources. We then evaluate these generated descriptions by mapping them to adversarial techniques.

3 Background

3.1 Vulnerability Reporting

Publicly disclosed vulnerabilities are standardized using CVEs, with each record containing a unique ID (CVE-ID), a textual description of the vulnerability, and possible references Mitre (2020). Descriptions are typically written by a CVE Numbering Authority (CNA), a CNA of Last Resort (CNA-LR), or individuals requesting a CVE-ID. CVE suggests two templates for vulnerability descriptions that contain a combination of Vulnerability type, Root cause, Component, Vendor, Product, Version, Attacker, Impact, and Vectors Cve Project (2023).

Each vulnerability's severity is typically determined using the Common Vulnerability Scoring System (CVSS) framework First (2019). The CVSS Base score captures values encoded in the vector string, including Attack Vector, Attack Complexity, Privileges Required, User Interaction, and impacts on Confidentiality, Integrity, and Availability. Despite CVSS's popularity, some studies have noted limitations. They cite slight inconsistencies and subjective interpretations in evaluating the metrics Wunder et al. (2024), Spring et al. (2018). In addition to severity, vulnerability records often include classifications from CWE MITRE (2025). CWE is a taxonomy of hardware and software weaknesses maintained by MITRE, enabling vulnerabilities to be grouped by their root causes. Some of the most dangerous weaknesses include improper input validation and incorrect authorization. Some vulnerability disclosures also provide supplemental proof-of-concept code Offsec (2025), which demonstrate practical exploitation. PoC code often reveals implementation-level details that may be absent from CVE descriptions, such as the specific function responsible for a vulnerability.

Access to CVSS, CWE, and proof-of-concept code provides a more comprehensive understanding of vulnerabilities. As evident in the Exploit Prediction Scoring System (EPSS). These vulnerability sources are collated and the exploit probability is calculated for CVEs Jacobs et al. (2021). Similarly, in this work, we classify these individual vulnerability sources and their combinations, as distinct configurations for CVE description enrichment that will be input to LLMs.

3.2 Large Language Models

Large language models (LLMs) have emerged as powerful artificial intelligence (AI) tools, with considerable applicability in a variety of domains, including cybersecurity for automating security solutions. Built on the Transformer architecture, with self-attention and in-context learning, LLMs have demonstrated strong performances across diverse natural language processing (NLP) applications Chang et al. (2024).

Despite their potential, LLMs are not immune to cybersecurity threats. Models are vulnerable to attacks such as jailbreaking, prompt injection, backdoor, data poisoning, and leakages Das et al. (2025). They are also prone to hallucinations, where they generate output that comes from fabricated knowledge. To address these security weaknesses, improvements have been designed with established guardrails at all stages of development and deployment Dong et al. (2024).

Security researchers have applied LLMs in areas such as network security, software security, and systems security. In network security, researchers are utilizing LLMs for network intrusion detection by analyzing traffic patterns Houssel et al. (2024), as well as for penetration testing Deng et al. (2024), Muzsai et al. (2024). In software and system security, LLMs contribute to vulnerability detection Li et al. (2025), and researchers in Steenhoek et al. (2024) investigate the performance of 11 LLMs in detecting vulnerabilities in C/C++ functions, finding that LLMs show limited improvement over random guessing. Additionally, LLMs are increasingly being used for code repair, to improve code patches Tang et al. (2024). It is noted that LLMs struggle to differentiate between buggy and patched code Steenhoek et al. (2024), highlighting the limitations of current LLMs in vulnerability detection tasks.

A more recent development is the use of LLMs as evaluators, a paradigm referred to as LLM-as-a-Judge, in which models are tasked with evaluating the output from other LLMs Zheng et al. (2023). This approach supports scalable and explainable model evaluation, allowing for a consistent and automated alternative to human evaluators. In this work, we adopt the LLM-as-a-Judge framework to test the effectiveness of LLM-enriched CVE descriptions.

3.3 ATT&CK Framework

MITRE ATT&CK is a widely adopted framework for modeling adversarial tactics, techniques, and procedures. It is used across private and public sectors to develop threat models Strom et al. (2018). The framework is structured into three domains: Enterprise, Mobile, and Industrial Control Systems, each representing a distinct adversary environment. Within each domain, ATT&CK defines platforms such as Windows or Linux as systems in which malicious actors operate.

The ATT&CK framework organizes attacker behavior hierarchically. At the highest level are tactics, which denote adversarial goals such as privilege escalation, defense evasion, and command and control. These are further granularized into techniques and sub-techniques that describe how tactics are achieved. The Enterprise domain offers the most comprehensive coverage, comprising 14 tactics, 211 techniques, and 468 sub-techniques MITRE Corporation (2025). This taxonomy enables a wide range of cybersecurity applications, including Cyber Threat Intelligence Enrichment and Red Teaming Al-Sada et al. (2024).

While broadly adopted, the manual mapping of unstructured vulnerability descriptions or threat intelligence reports to ATT&CK is labor-intensive and prone to inconsistencies. Automated approaches are essential to make these mappings scalable and consistent, providing actionable insight into how a vulnerability may be exploited. Several such methods have been proposed, as discussed in Sect. 4. This work builds on that line of research, using an LLM-as-a-Judge evaluation strategy to predict ATT&CK techniques from natural language vulnerability descriptions.

4 Related Work

4.1 Vulnerability Descriptions Enrichment

Improving the quality of vulnerability descriptions is an important research focus, as enriched descriptions can enhance vulnerability analysis tasks, such as classification, prioritization, and threat modeling. To address incomplete or brief descriptions in public repositories, researchers have explored various enrichment strategies. Althebeiti and Mohaisen (2023) introduced a summarization pipeline that augments vulnerability descriptions with extracted content from third-party references. They only measure for summary fluency, completeness, correctness, and understanding.

Guo et al. (2021) proposed an enrichment approach to check for missing aspects in vulnerability reports from several sources, including affected product, root cause, vulnerability type, attacker type, attack vector, and impact. Similarly, Han et al. (2024) designed a software feature inference framework to generate missing information from vulnerability descriptions. Sun et al. (2021) proposed creating descriptions using key information extracted from exploit-db texts, and evaluated their composed descriptions using the ROUGE metrics. In our work, we enrich CVE descriptions using a combination CWE, CVSS, and exploit code sources. Rather than directly evaluating the quality of the new descriptions with completeness and correctness tests, we assess their performance on ATT&CK technique prediction. This evaluation strategy was chosen to understand the adversarial implications and the practical impact of the composed descriptions.

4.2 CVE-to-Technique Mapping

Several studies have focused on mapping CVEs to frameworks such as ATT&CK to enhance threat intelligence. Giannakopoulos and Maliatsos (2023) developed a text-classification NLP model to link threats to CWEs and the Common Attack Pattern Enumerations and Classifications (CAPEC) for risk assessment, addressing limitations in standard CVE descriptions. Authors in Simonetto and Bosch (2024) proposed a systematic method to map CVEs to ATT&CK tactics and techniques, creating a dataset in the process. Abdeen et al. (2023) applied cosine similarity to map CVEs to the ATT&CK framework. Similarly, Kuppa et al. (2021) presented a multi-label text classification (MLTC) model to link CVEs to ATT&CK techniques, using cosine similarity and unsupervised labeling. Mendsaikhan et al. (2020) used the ENISA dataset, and employed multi-label classification techniques; including algorithm adaptation, problem transformation, and an ensemble method, on CVE data to identify the most effective model for mapping CVEs to ATT&CK techniques. Lastly, Grigorescu et al. (2022) developed a BERT-based model for mapping CVEs to ATT&CK techniques, addressing data imbalance through data augmentation strategies and achieving an F1 Score of 0.478. While Rafiey and Namadchian (2025) implemented a mapping framework with GPT-3, achieving an F1 Score of 0.372. In this work, we use ATT&CK technique predictions as a proxy to measure the effectiveness of enriched CVE descriptions. By using a high-performing LLM with strong reasoning capabilities as the evaluator, we introduce a more context-aware alternative to prior approaches.

4.3 NLP Applications for Vulnerability Analysis

Natural language processing techniques are increasingly applied to software vulnerability analysis, as modern models are trained on diverse code sources, enabling them to learn patterns indicative of potential vulnerabilities. An earlier study Shahid and Debar (2021) employed BERT-based models to predict CVSS severity scores and vector metrics such as Attack Complexity, Privilege, Scope, and Impacts, directly from CVE descriptions, using gradient-based saliency methods to identify important terms. Cao and Jun (2024) developed a retrieval augmented generation LLM solution and incorporated CWE information for classification and localization of vulnerabilities. Guo et al. (2024) applied general-purpose and vulnerability-specific LLMs for binary classification to detect vulnerabilities in source code. Motivated by the growing use of NLP in code-based security tasks, we leverage proof-of-concept code as a source to enrich LLM-generated vulnerability descriptions.

5 Methodology

Figure 1 illustrates our methodology, which consists of two phases. First, we gather contextual information for each CVE by selecting relevant configuration sources. These inputs prompt eight LLMs to generate new vulnerability descriptions. In the second phase, the generated CVE descriptions, along with MITRE ATT&CK technique descriptions, are passed to a separate prediction model, following the LLM-as-a-Judge framework. This model identifies the corresponding ATT&CK techniques from a predefined candidate set.

5.1 Configurations for Description Generation

We defined six configurations using information from CVE records and external sources to evaluate how different contexts influence vulnerability description quality. These configurations served as the basis for constructing prompts to query LLMs

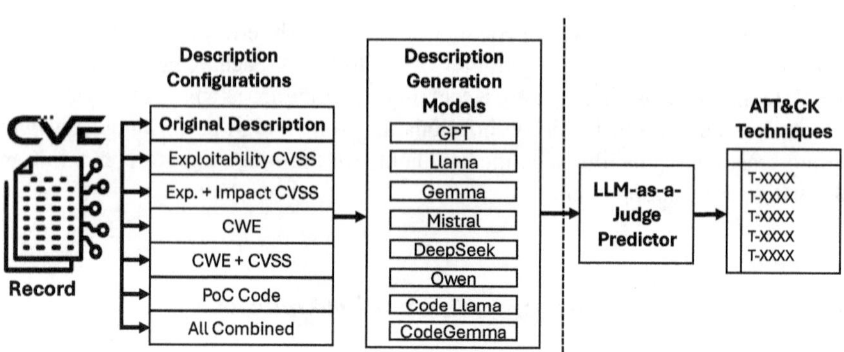

Fig. 1 Workflow of this project—description generation and ATT&CK matching

to generate descriptions: (1) Exploitability Base CVSS, (2) Total Base CVSS, (3) CWE, (4) Combined CWE and CVSS, (5) PoC Code, (6) All Combined.

Exploitability Base CVSS prompts a model with Exploitability components of the CVSS Base Metric group defined in Version 3: Attack Vector (AV), Attack Complexity (AC), Privileges Required (PR), User Interaction (UI), and Scope (S). This configuration emphasizes the technical feasibility of exploitation Shahid and Debar (2021), guiding the models to generate descriptions that focus on the conditions under which a vulnerability can be exploited.

Total Base CVSS extends the previous configuration and adds the Impact metrics: Confidentiality (C), Integrity (I), and Availability (A). This CVSS Base Matrix context provides a more comprehensive view of how a vulnerability is exploited and its potential impact.

CWE leverages Common Weakness Enumeration identifiers and descriptions to describe the root causes or issues that lead to a vulnerability. CWEs provides a high-level abstraction of flaws, helping models to create descriptions focusing on underlying weaknesses.

Combining both CWE and CVSS merges weakness classification and severity metrics into a unified input for description generation. For instance, a CVE labeled with CWE-200: Exposure of Sensitive Information to an Unauthorized Actor and a CVSS vector indicating a network attack vector and high confidentiality impact should prompt the model to generate a description that reflects remote data exposure due to an underlying weakness.

PoC code includes an excerpt from a proof-of-concept exploit script associated with the CVE. This configuration generates descriptions that reflect real-world attack scenarios by incorporating low-level technical details from the exploit.

All Combined aggregates all available sources into a single prompt, this includes CWE, Base CVSS, and PoC code information. This configuration is designed to generate the most comprehensive description by integrating all perspectives of the vulnerability.

The CVE program provides a recommended template for crafting standardized CVE descriptions Cve Project (2023). To assess the impact of the templates on LLM-generated descriptions, we conducted two sets of experiments per configuration: one with the template included in the prompt, and one without, allowing for more free-form generation. Inputs included the CVE identifier, original description, and the selected configuration fields with associated values. Field names, IDs, and component descriptions were passed to the models.

Figure 2 presents an example prompt for the Exploitability Base CVSS configuration. The five metrics (AV, AC, PR, UI, S), the corresponding CVE content, and a reference to the recommended description template are incorporated. To ensure consistency with current CVE standards, explicit instruction is provided to maintain output length.

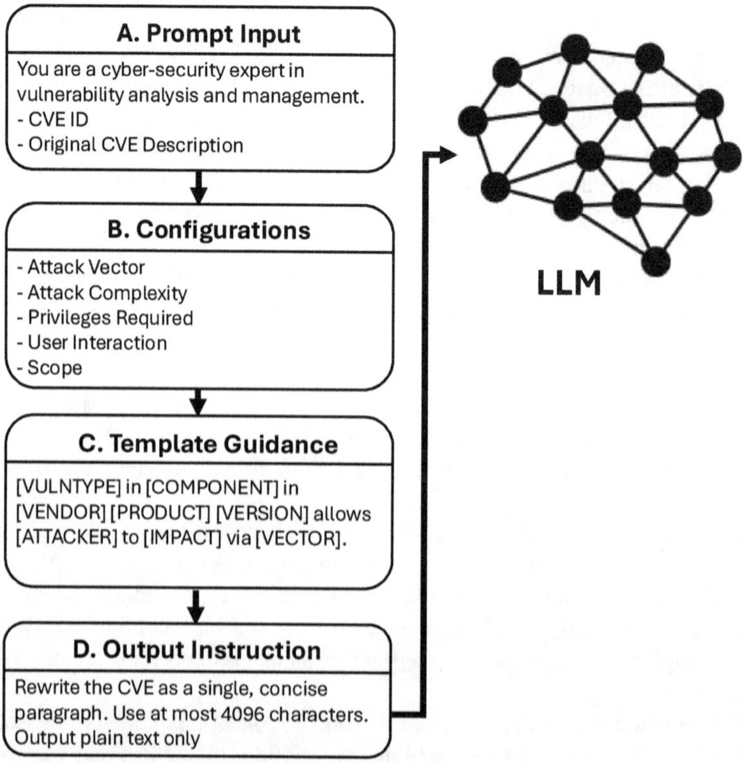

Fig. 2 Sample prompt for description generation with Exploitability CVSS

5.2 Evaluation of New Descriptions

To measure the effectiveness of the configurations and language models on the new CVE descriptions, we evaluate how well the descriptions map to MITRE ATT&CK Techniques. This is accomplished using a dedicated large language model as a zero-shot predictor, following the LLM-as-a-Judge framework. In this process, the new descriptions are passed to the judge model along with a predefined set of technique candidates from the ATT&CK framework. The model is prompted to identify all techniques that correspond to the given vulnerability description. This allows us to assess the extent to which each configuration supplies enough context for the model to select the techniques that match the ground truth. We quantify this performance using the standard classification metrics Accuracy, Precision, Recall and F1 Score, as defined in Eqs. (1)–(4). Comprehensive evaluation results are provided in Sect. 6.5.

$$Accuracy = \frac{TP + TN}{TP + FP + FN + TN} \tag{1}$$

$$Precision = \frac{TP}{TP + FP} \tag{2}$$

$$Recall = \frac{TP}{TP + FN} \tag{3}$$

$$F1Score = 2 \times \frac{Precision \times Recall}{Precision + Recall} \tag{4}$$

6 Evaluation

6.1 System Requirements

All experiments were conducted on an Ubuntu 22.04 server with four NVIDIA A100 GPUs, each with 80GB of memory. The experimental framework was implemented in Python and leveraged Langchain Chase (2022), Ollama ollama (2025), and vLLM Kwon et al. (2023) to integrate and implement the LLMs. We evaluated a selection of state-of-the-art open source and proprietary LLMs including GPT-4.1 & GPT-4.1 mini Openai (2025), Llama 3.2 (3B) Meta (2024), Gemma3 (4B) Team (2025), Mistral (7B) Jiang et al. (2023), DeepSeek-R1 (7B)Deepseek (2025), CodeGemma (7B)Team et al. (2024), Code Llama (7B) Roziere et al. (2023), Qwen2.5-Coder (7B) Hui et al. (2024), where “B” represents the model parameters in billions. These models were selected for their favorable trade-off between computational efficiency and performance.

GPT-4.1 served as the LLM-as-a-Judge for ATT&CK technique prediction for initial experiments, while the remaining models generated modified CVE descriptions. For subsequent experiments, the locally deployed LLama 4 Scout 16 × 17B was used as the Judge model, as it demonstrated strong performance scores in the GPQA Diamond reasoning benchmarks Vellumai (2025) compared to other models.

To ensure consistency across models during description generation, we applied fixed hyperparameters. The temperature was set to 0 for all models to enforce deterministic outputs, following best practices from prior studies Tamberg and Bahsi (2025). The maximum token size was limited to 1,024 tokens to comply with the CVE descriptions schema, which restricts content to 4096 characters Cve Project (2023). Top-p and top-k were initially set to their default values of 0.9 and 40, respectively, and later adjusted to 1.0 and 1 in experiments involving the LLama 4 Scout 16 × 17B Judge model to further maximize determinism.

6.2 Dataset and Metrics

We utilized the Vuln-report dataset, released by the European Union Agency for Cybersecurity (ENISA) ENISA (2020). Originally published in their state of the vulnerabilities report Katos et al. (2019), then adopted in recent work on automated mapping of CVEs to ATT&CK tactics Branescu et al. (2024). Vuln-report provides curated mappings of CVE to different ATT&CK techniques, tactics, and other vulnerability information.

From the dataset, we extracted 7,329 unique CVEs, each with one or more associated ATT&CK techniques. During preprocessing, we identified and removed deprecated technique names that no longer appear in the latest ATT&CK framework documentation. This filtering step reduced the dataset to 5,130 CVEs. We also discovered inconsistencies between CWE labels in Vuln-report and those listed in the National Vulnerability Database (NVD). For instance, CVE-2019-3606 is labeled with CWE-312 in the NVD, while Vuln-report classifies it as CWE-200. To ensure consistency, we adopted CWE labels exclusively from the NVD.

Next, we filtered the dataset to CVEs linked to publicly available PoC code in the Exploit Database Offsec (2025), resulting in a subset of 262 CVEs. Four CVEs were excluded during preprocessing, leaving a final set of 258 CVEs. We randomly sampled 10 CVEs from this for initial experiments, and 258 were used in the comprehensive evaluation.

For each CVE description in the ATT&CK technique prediction phase, we provided the LLM-as-a-Judge predictor with all its known associated techniques obtained from the dataset and categorized this as our ground truth. We also provided the predictor with an equal number of non-associated techniques to balance the predictions, and we then combined the two lists to form a candidate list of techniques. We then prompted the model to return two sets of outputs for the techniques: Correct techniques that match the description, and Incorrect techniques that do not match.

To evaluate predictive capability, we measured the True Positive (TP), False Positive (FP), False Negative (FN), and True Negative (TN) metrics for each CVE description. A TP is recorded only if the predictor returns true for a technique that is in the ground truth for the specific CVE. A FP is recorded when the predictor returns true for a technique that is not present in the known associated techniques list, and thus not in the ground truth of the specific CVE. A FN is recorded when either the predictor returns false for a technique that is present in the ground truth, or when an associated technique that is present in the ground truth of a specific CVE is missing from the output of the predictor. A TN is recorded when the predictor correctly returns false for a technique that is not in the ground truth for the specific CVE. With these, Accuracy, Precision, Recall, and F1 Score metrics are further calculated to measure the performance of the predictor for CVE description to ATT&CK technique mapping. Since our goal is to assess relative effectiveness, we report these metrics to compare enriched descriptions against the original texts, rather than to optimize absolute scores.

6.3 CVE Description Generation

In our preliminary findings, we generated 960 LLM-based CVE descriptions. These were from the 10 CVE records, processed by eight LLMs across the six prompting configurations: Exploitability CVSS, Base CVSS, CWE, Combined CVSS and CWE, PoC Code, and All Combined. Each configuration was applied with and without the inclusion of the CVE description template. New descriptions were consistently longer than the originals. With the CVE template included, the increase across configurations ranged from 31 to 107%, with the exception of the CWE with template configuration, which showed a negligible change. Without the template, the increase for all configurations ranged from 44 to 187%.

We expanded this to 3096 enriched descriptions based on the full set of 258 CVE records. The descriptions were generated by two LLMs, Gemma 3 and Qwen-2.5-Coder, using Exploitability CVSS, CWE, and PoC code-based configurations, with and without the template. Table 1 reports the average character length of both the original and generated descriptions. The results highlight the effects of excluding the description template, as descriptions created without the template saw length increases ranging from 26 to 90%. In contrast, using the template for Exploitability CVSS and CWE-based descriptions resulted in slightly shorter outputs, showing a 10 and 20% decrease in length respectively.

Overall, incorporating the CVE description template reduced verbosity, suggesting that the template functions as a soft structural constraint. In contrast, omitting the template allows the models greater freedom to elaborate. Across all cases, PoC code-based prompts produced the longest descriptions compared to those based on CWE or CVSS. This increase may stem from PoC code's low-level technical details. These inputs enable the model to generate richer and more detailed descriptions that explain the vulnerability.

Table 2 presents examples of enriched descriptions generated for `CVE-2018-8880` by Code LLama under two configurations: CWE and PoC Code, both without the inclusion of the CVE description template. Compared to the concise original description, the CWE-based output emphasizes the root cause of the vulnerability by referencing `CWE-200: Exposure of Sensitive Information to an Unauthorized Actor`. This configuration highlights the underlying

Table 1 The average character lengths of 3,096 unique enriched CVE descriptions compared with 258 original texts

Configuration	CVE Description character Length	
	With template	Without template
Exploit CVSS	282	479
CWE	249	392
PoC code	410	590
Original description	311	

Table 2 Sample enriched `CVE-2018-8880` descriptions using `Code LLama`

Configuration	`CVE-2018-8880` Generated descriptions
Original	Lutron quantum BACnet integration 2.0 (firmware 3.2.243) doesn't check for correct user authentication before showing the /deviceIP information, which leads to internal network information disclosure
CWE	The Lutron Quantum BACnet Integration 2.0 firmware version 3.2.243 does not check for correct user authentication before displaying the /deviceIP information, which leads to internal network information disclosure. This vulnerability is classified as CWE-200: Exposure of Sensitive Information to an Unauthorized Actor, as it exposes sensitive information to an actor that is not explicitly authorized to have access to that information
PoC code	The Lutron Quantum BACnet Integration 2.0 (firmware 3.2.243) does not check for correct user authentication before displaying the /deviceIP information, which leads to internal network information disclosure. The Proof of Concept Code demonstrates how an attacker can exploit this vulnerability by sending a specially crafted request to the device's IP address and port number, resulting in the leakage of sensitive data such as MAC addresses, product family, product type, serial number, GUID, code version, internal IP address, subnet mask, gateway, telnet port, FTP port, and remote port. This vulnerability can potentially allow an attacker to gain unauthorized access to the device's network and sensitive information

access control weakness that enables the vulnerability. In contrast, the PoC Code-based description focuses on how the vulnerability can be exploited in practice, detailing an attack scenario that can lead to the leakage of sensitive network information. These examples illustrate how different configurations guide LLMs to underscore distinct aspects of the vulnerability. The remaining configurations exhibit similar behavior, demonstrating how contextual inputs shape the generated descriptions.

6.4 Preliminary Mapping Results

6.4.1 Validating the LLM-as-a-Judge Predictor

We began by evaluating the consistency and stability of the LLM-as-a-Judge predictor across twelve configurations. For this analysis, we randomly selected LLama 3.2 as the description generation model and applied each configuration to produce modified CVE descriptions for an initial subset of 10 CVE records. Each configuration was tested over 10 independent runs, and the resulting descriptions were assessed by the GPT-4.1 predictor to identify matching ATT&CK techniques. We selected 10 runs following the methodology of a previous study measuring determinism in LLM settings Atil et al. (2024). We report mean and standard deviation of F1 Scores for each configuration in Fig. 3.

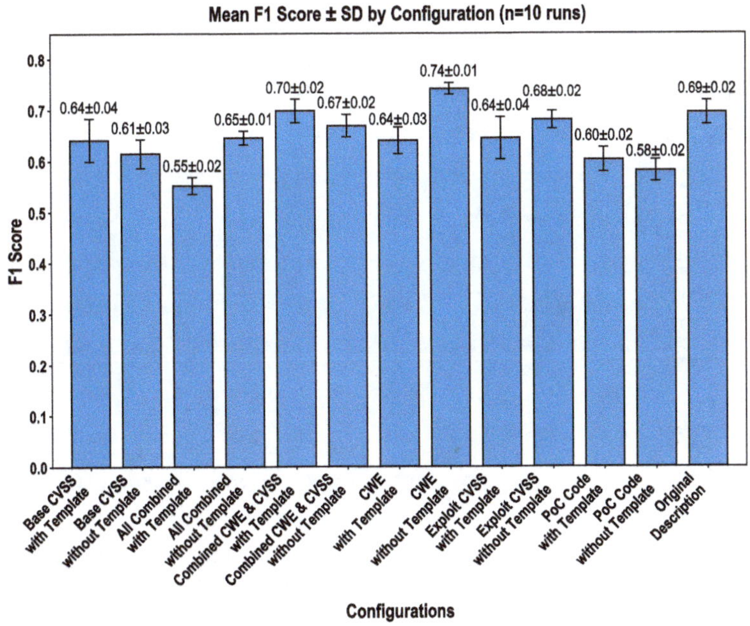

Fig. 3 Preliminary statistical analysis for 960 enriched CVE descriptions compared to 10 original descriptions over 10 runs. The plots show mean F1 Scores with standard deviation error bars for technique mappings for LLama 3.2-made descriptions

The results reveal notable trends. Most significantly, the CWE-only configuration without template inclusion achieved the highest performance, with a mean F1 Score of 0.74 (±SD=0.01), outperforming the original CVE descriptions, which scored 0.69 (±SD=0.02). This suggests that, within the limited scope of CVEs used in this preliminary experiment, the CWE without template configuration provides the most informative and consistent input for ATT&CK technique inference within the LLM-as-a-Judge framework. In contrast, combining all configurations with the template yielded the lowest performance, with a mean F1 Score of 0.55 (±SD=0.02), indicating template use only in this context may constrain generated descriptions and omit critical vulnerability information for ATT&CK technique mapping. This interpretation is further supported by the performance increase to 0.65 (±SD=0.01) when the same configuration is applied without the template, highlighting the impact of prompt structure on the effectiveness of this prediction strategy.

6.4.2 Performance evaluation across LLMs and configurations

To further examine the effect of configuration design and language model selection on technique prediction performance, we present a heatmap, Fig. 4, summarizing F1 Scores across twelve description generation configurations and eight different LLMs

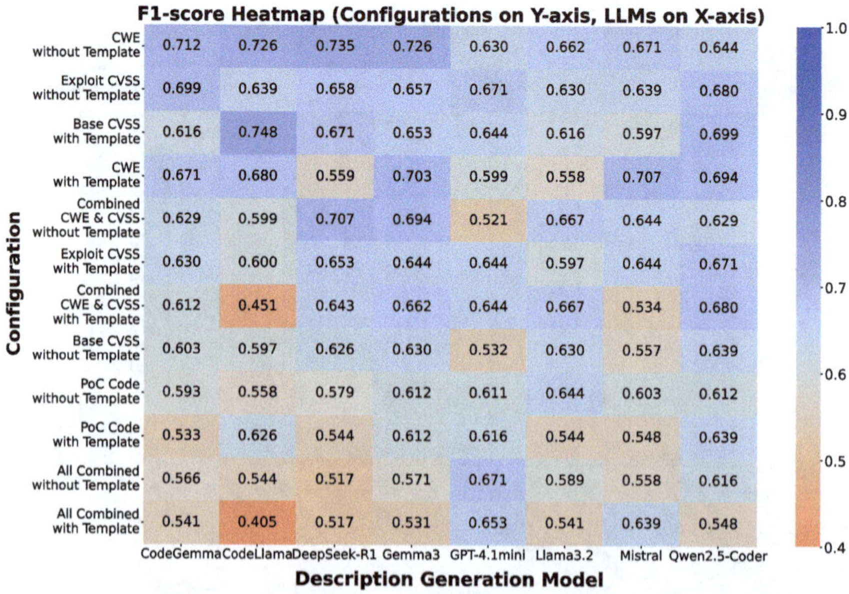

Fig. 4 Performance evaluation of 960 enriched CVE descriptions compared to 10 original descriptions. The descriptions are mapped to ATT&CK techniques using the LLM-as-a-Judge predictor. The heatmap shows F1 Scores. Note 0.681 is the mean F1 Score for original CVE descriptions

in the preliminary experiment. Rows correspond to configurations, and columns represent the LLMs used to generate new CVE descriptions. F1 Score, capturing the harmonic mean of precision and recall, ranges from 0.405 using all configurations combined with the template, to 0.748 using the Base CVSS with template configuration, both generated with Code LLama. These results are reported relative to the original CVE description, which achieves an F1 Score of 0.681.

In line with findings in Sect. 6.4.1, the better performing configuration in this test is CWE without template, achieving up to a 7.9% relative improvement in F1 Score over the original description when using DeepSeek-R1 descriptions. Both Code LLama and Gemma3 produce descriptions that showed 6.6% improvements in the F1 Score with this setting compared to the original. This CWE configuration is also the most concise as noted in Tables 1 and 2 and preserves the core vulnerability semantics. It focuses on the underlying root causes of the weaknesses, which maps closely to the ATT&CK framework. Additionally, the consistency of the standardized CWE descriptions likely enhances the LLM's ability to identify relevant techniques. The Code LLama-generated description using the Base CVSS with template configuration, yielded the highest relative performance in this experiment with a 9.8% improvement. However, this value can be considered a statistical outlier, given the configuration's earlier mean and standard deviation.

The LLM-as-a-Judge predictor produces the lowest performance when using PoC code-based descriptions and when aggregating all configuration sources. Specifically, the PoC Code configuration produces scores of 0.583 (with template) and 0.602 (without), while All Combined configurations drop further to 0.547 (with template) and 0.579 (without). These results suggest that the model struggles to extract actionable threat-relevant features from the exploit code scripts to map ATT&CK techniques correctly. This reinforces recent findings Guo et al. (2024) that LLMs remain limited in their ability to reason over code, more specifically with regards to malicious code. In the All Combined case, the inclusion of multiple and potentially overlapping information sources likely introduces noise, obscuring the more relevant features necessary for correct technique prediction.

6.5 *Comprehensive Evaluation*

Using the the two highest-scoring models from the preliminary tests, we selected the enriched descriptions generated by Qwen-2.5-Coder 7B and Gemma 3 4B, as well as the Exploit CVSS, CWE, and PoC-based configurations. We conducted a comprehensive evaluation on all 258 CVE records, encompassing a total of 3,354 unique vulnerability descriptions, including the original descriptions. As in the preliminary experiment, this process repeated over 10 runs, resulting in 33,540 descriptions evaluated by the Llama 4 Scout Judge. Figure 5 summarizes the mean F1 Scores for predictions using each configuration and generation model, while Table 3 provides a more detailed analysis with mean accuracy, precision, and recall.

Across all configurations and models, F1 Scores ranged from 0.445 to 0.628, with the original description achieving a score of 0.557. Among the generated con-

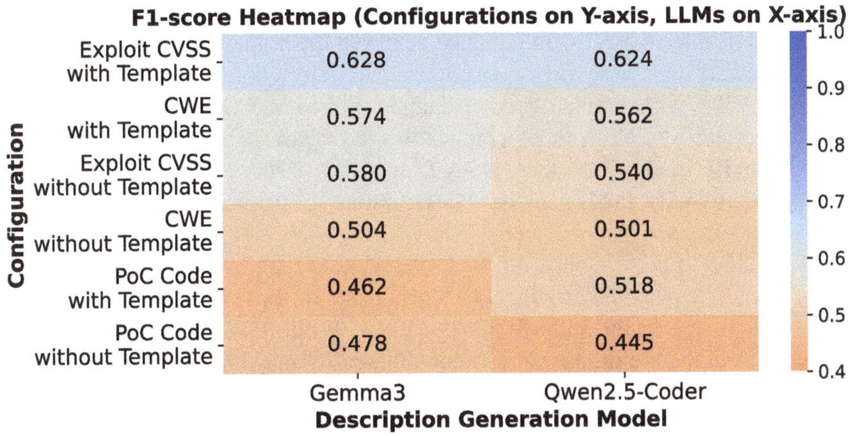

Fig. 5 Performance evaluation of 3,096 enriched CVE descriptions compared to 258 original descriptions over 10 runs. The descriptions are mapped to ATT&CK techniques using the LLM-as-a-Judge predictor. The heatmap shows F1 Scores. Note 0.557 is the mean F1 Score for original CVE descriptions

Table 3 Full performance evaluation of 3096 enriched CVE descriptions compared to 258 original descriptions over 10 runs

Configuration	Description generation model	With Template			
		Accuracy	Precision	Recall	F1-score
Exploit CVSS	Gemma 3	**0.673**	0.820	**0.508**	**0.628**
	Qwen2.5-Coder	0.659	**0.833**	0.499	0.624
CWE	Gemma 3	0.621	0.783	0.453	0.574
	Qwen2.5-Coder	0.618	0.832	0.425	0.562
Vulnerable PoC Code	Gemma 3	0.528	0.732	0.338	0.462
	Qwen2.5-Coder	0.596	0.789	0.386	0.518
Original description		0.609	0.800	0.427	0.557
Configuration	Description generation model	Without template			
		Accuracy	Precision	Recall	F1-score
Exploit CVSS	Gemma 3	**0.643**	**0.838**	**0.444**	**0.580**
	Qwen2.5-Coder	0.602	0.823	0.402	0.540
CWE	Gemma 3	0.593	0.804	0.367	0.504
	Qwen2.5-Coder	0.585	0.750	0.376	0.501
Vulnerable PoC code	Gemma 3	0.535	0.758	0.349	0.478
	Qwen2.5-Coder	0.517	0.727	0.320	0.445

The descriptions are mapped to ATT&CK techniques using the LLM-as-a-Judge predictor. The use of bold in Table 3 highlights the best values that can be achieved for the vertical categories (Accuracy, Precision, Recall, F1-score)

figurations, Exploit CVSS with template achieved the highest F1 Score, showing a 12.7% and 12.0% improvement over the original description across the two LLMs. Similarly, CWE with template produced slight improvements, with 3.1% and 0.9% increases. Notably, CWE without template showed relatively poor performance, with a maximum F1 Score of 0.504, deviating from trends observed in preliminary experiments. Additionally, PoC code-based descriptions consistently achieved the lowest relative performance, with F1 scores below 0.520.

The decline in performance for the CWE without template configuration in our comprehensive evaluation can likely be attributed to the broader range of CVE characteristics present in the full 258 CVE dataset. While CWE-based prompts performed well in the preliminary assessment, this result may have been influenced by the limited scope of the 10 CVEs and their narrower range of CWE categories. In contrast, the broader dataset includes more CWEs that lack semantic alignment with ATT&CK techniques, leading to increased false negatives. Conversely, the improved performance of configurations that included the description template in the prompts suggests that the template facilitates structured description generation. Specifically, the template prompts the models to systematically identify and include key details such as vulnerability types, root causes, attacker types, potential impacts, and attack vectors. Leveraging these structured attributes at scale facilitates the generation of enriched vulnerability descriptions that improve interpretability and enable higher-recall technique predictions.

We observed that the LLama 4 Scout Judge, under the selected parameters, typically predicts with high precision but low recall due to a large number of false negatives. This often results from partial hallucinations, where the model makes at least one prediction but includes either invalid technique IDs (not present in the candidate list) or deviates from the required structured output format. We also identified that 18% of the 33,540 predictions were complete hallucinations, where outputs did not follow the expected structure outlined in the prompt. These predictions were excluded from metric calculations.

Furthermore, after adjusting the parameters to a top-p of 1.0 and top-k of 1, prediction results showed minimal variability across all 10 runs. The only variation occurred in the CWE with template configuration generated by Qwen-2.5-Coder, with an F1 Score standard deviation of 0.002. This stability is likely a result of greedy decoding Ibm (2025), given the deterministic parameters selected.

7 Discussion and Limitations

Our experimental results offer insight into the behavior of LLMs when used for enrichment by generating vulnerability descriptions and subsequently predicting ATT&CK techniques. Although the LLM-as-a-Judge framework offers flexible and zero-shot inference without traditional model training, there are some limitations related to model capability, data scope, and methodology. Our generated descriptions for the 258 CVEs are shared on GitHub.[1]

7.1 Framework Limitations

The LLMs used for the description generation in this study are all in the small-medium parameter range, while the LLM-as-a-Judge model was a large-scale model with strong reasoning performance. While some configurations produced strong results, these models still likely lack the capacity to effectively perform deep reasoning over complex exploit code or verbose, multi-source configuration inputs. To better understand the upper bounds of LLM performance on this task, future evaluations should explore evaluations with larger instruction-tuned models, particularly

[1] https://github.com/kenenwodo/llm_enriched_cves.git.

in configurations that focus solely on proof-of-concept code inputs, which demand higher levels of syntactic and sematic reasoning.

7.2 Configuration Diversity

This work evaluated twelve configurations derived from combinations of CVSS, CWE, and PoC information. This does not exhaust all possible sources for enriching vulnerability descriptions. External advisories and vulnerability reports Cisco Talos (2025) could be integrated in future work to improve context. Furthermore, complementary evaluation frameworks like the Exploit Prediction Scoring System (EPSS) Jacobs et al. (2021) may offer utility beyond ATT&CK technique coverage.

7.3 Semantic Similarity

Early experiments used a semantic similarity-based approach to map the generated descriptions to techniques, following previous studies. To embed vulnerability descriptions and ATT&CK technique summaries, we utilized OpenAI text embedding models and a custom ATT&CK-trained embedding model discussed in Abdeen et al. (2023). However, this method proved to be ineffective in our setting. Modifying CVE descriptions introduced variation and noise that degraded embedding quality, and cosine similarity calculations between generated descriptions and adversary-focused ATT&CK techniques yielded lower true positive counts than the original descriptions. Transitioning to LLM-as-a-Judge-based predictions resolved this issue, with the Exploit CVSS with template configuration emerging as the most effective across models.

8 Conclusion

This chapter addressed the challenge of incomplete and uninformative CVE descriptions. By leveraging open-source and proprietary large language models, we enrich descriptions with data from multiple vulnerability sources. In preliminary experiments, we evaluated twelve configurations across eight LLMs, generating 960 descriptions from 10 CVEs. We then scaled the evaluation to a comprehensive setting, producing 3,096 unique descriptions from 258 CVE records using six configurations and two top-performing LLMs. To evaluate the utility of these enriched descriptions, we employed an LLM-as-a-Judge framework to classify corresponding MITRE ATT&CK techniques, repeating each trial over 10 runs. Our results showed enriched descriptions achieved up to a 12.7% improvement in F1 Score relative to original description, specifically using Exploit CVSS metrics combined with the vulnerability description template. A key lesson derived from this result is that the structured template enables LLMs to systematically incorporate vulnerability attributes such as vulnerability types, root causes, and attack vectors into enriched descriptions. This structured approach improves scalability and interpretability. These findings demonstrate the potential of current LLMs to enhance vulnerability descriptions and support technique inference. However, there remains room for improvement with reasoning on exploit code scripts and larger contexts. Future work can explore larger models and richer contextual inputs while considering computational costs. Overall, our findings support the viability of LLMs as tools for enriching vulnerability descriptions to support automated vulnerability analysis.

References

Abdeen B, Al-Shaer E, Singhal A, Khan L, Hamlen K (2023) Smet: semantic mapping of cve to att&ck and its application to cybersecurity. In: IFIP annual conference on data and applications security and privacy, Springer, pp 243–260

Al-Sada B, Sadighian A, Oligeri G (2024) Mitre att&ck: State of the art and way forward. ACM Comput Surv 57(1):1–37

Althebeiti H, Mohaisen D (2023) Enriching vulnerability reports through automated and augmented description summarization. International conference on information security applications. Springer, pp 213–227

Atil B, Chittams A, Fu L, Ture F, Xu L, Baldwin B (2024) Llm stability: a detailed analysis with some surprises.ArXiv preprint

Branescu I, Grigorescu O, Dascalu M (2024) Automated mapping of common vulnerabilities and exposures to mitre att&ck tactics. Information 15(4):214

Cao D, Jun W (2024) Llm-cloudsec: Large language model empowered automatic and deep vulnerability analysis for intelligent clouds. IEEE INFOCOM 2024-IEEE Conference on Computer Communications Workshops (INFOCOM WKSHPS). IEEE, pp 1–6

Chang Y, Wang X, Wang J, Wu Y, Yang L, Zhu K, Chen H, Yi X, Wang C, Wang Y et al (2024) A survey on evaluation of large language models. ACM Trans Intell Syst Technol 15(3):1–45

Chase H (2022) LangChain. URL https://github.com/langchain-ai/langchain

CISCO Talos (2025) Vulnerability reports. https://www.talosintelligence.com/vulnerability_reports. Accessed: July 2025

CodeGemma Team, Hartman AJ, Hu A, Choquette-Choo CA, Zhao H, Fine J, Hui J, Shen J, Kelley J, Howland J, Bansal K, Vilnis L, Wirth M, Nguyen N, Michel P, Choy P, Joshi P, Kumar R, Hashmi S, Agrawal S, Zuo S, Warkentin T, Gong Zea (2024) Codegemma: open code models based on gemma. URLhttps://goo.gle/codegemma

Costa JC, Roxo T, Sequeiros JB, Proenca H, Inacio PR (2022) Predicting cvss metric via description interpretation. IEEE Access 10:59125–59134

CVE Project (2023) Cve json record format. https://cveproject.github.io/cve-schema/schema/docs/#oneOf_i0_containers_cna_descriptions_items_value

cveorg (2025) Metrics. https://www.cve.org/about/Metrics, Accessed: June 2025

Das BC, Amini MH, Wu Y (2025) Security and privacy challenges of large language models: a survey. ACM Comput Surv 57(6):1–39

Das SS, Dutta A, Purohit S, Serra E, Halappanavar M, Pothen A (2022) Towards automatic mapping of vulnerabilities to attack patterns using large language models. 2022 IEEE international symposium on technologies for homeland security. IEEE, pp 1–7

DeepSeek-AI (2025) Deepseek-r1: Incentivizing reasoning capability in llms via reinforcement learning. https://arxiv.org/abs/2501.12948

Deng G, Liu Y, Mayoral-Vilches V, Liu P, Li Y, Xu Y, Zhang T, Liu Y, Pinzger M, Rass S (2024) {PentestGPT}: Evaluating and harnessing large language models for automated penetration testing. In: 33rd USENIX Security Symposium (USENIX Security 24), pp 847–864

Dong Y, Mu R, Jin G, Qi Y, Hu J, Zhao X, Meng J, Ruan W, Huang X (2024) Building guardrails for large language models. ArXiv preprint arXiv:2402.01822

ENISA (2020) State of vulnerabilties report. URLhttps://github.com/enisaeu/vuln-report

Fakih M, Dharmaji R, Bouzidi H, Araya GQ, Ogundare O, Faruque M (2025) Llm4cve: enabling iterative automated vulnerability repair with large language models.ArXiv preprint

First C (2019) Common vulnerability scoring system version 3.1, user guide revision 1

Ghosh R, von Stockhausen HM, Schmitt M, Vasile GM, Karn SK, Farri O (2025) Cve-llm: ontology-assisted automatic vulnerability evaluation using large language models. Proceedings of the AAAI conference on artificial intelligence 39:28757–28765

Giannakopoulos T, Maliatsos K (2023) On the usage of nlp on cve descriptions for calculating risk. European symposium on research in computer security. Springer, pp 104–123

Grigorescu O, Nica A, Dascalu M, Rughinis R (2022) Cve2att&ck: Bert-based mapping of cves to mitre att&ck techniques. Algorithms 15(9):314

Guo H, Xing Z, Chen S, Li X, Bai Y, Zhang H (2021) Key aspects augmentation of vulnerability description based on multiple security databases. 2021 IEEE 45th annual computers, software, and applications conference (COMPSAC). IEEE, pp 1020–1025

Guo H, Chen S, Xing Z, Li X, Bai Y, Sun J (2022) Detecting and augmenting missing key aspects in vulnerability descriptions. ACM Trans Softw Eng Methodol (TOSEM) 31(3):1–27

Guo Y, Patsakis C, Hu Q, Tang Q, Casino F (2024) Outside the comfort zone: analysing llm capabilities in software vulnerability detection. European symposium on research in computer security. Springer, pp 271–289

Han L, Pan S, Xing Z, Sun J, Yitagesu S, Zhang X, Feng Z (2024) Do chase your tail! missing key aspects augmentation in textual vulnerability descriptions of long-tail software through feature inference. IEEE Trans, Softw, Eng

Houssel PR, Singh P, Layeghy S, Portmann M (2024) Towards explainable network intrusion detection using large language models.ArXiv preprint

Hui B, Yang J, Cui Z, Yang J, Liu D, Zhang L, Liu T, Zhang J, Yu B, Dang K, et al. (2024) Qwen2. 5-coder technical report. ArXiv preprint arXiv:2409.12186

IBM (2025) Foundation model parameters: decoding and stopping criteria. https://www.ibm.com/docs/en/watsonx/saas?topic=prompts-model-parameters-prompting

Jacobs J, Romanosky S, Edwards B, Adjerid I, Roytman M (2021) Exploit prediction scoring system (epss). Digit Threat Res Pract 2(3):1–17

Jiang AQ, Sablayrolles A, Mensch A, Bamford C, Chaplot DS, de las Casas D, Bressand F, Lengyel G, Lample G, Saulnier L, Lavaud LR, Lachaux MA, Stock P, Scao TL, Lavril T, Wang T, Lacroix T, Sayed WE (2023) Mistral 7b. https://arxiv.org/abs/2310.06825

Katos V, Rostami S, Bellonias P, Davies N, Kleszcz A, Faily S (2019) State of vulnerabilities 2018/2019-analysis of events in the life of vulnerabilities.Report/Study

Koenders J (2024) Advancing cyberdefense through bert: a natural language processing approach for vulnerability to attack mapping within a responsible artificial intelligence framework. PhD thesis, Master Thesis

krebsonsecurity (2025) Funding expires for key cyber vulnerability database. https://krebsonsecurity.com/2025/04/funding-expires-for-key-cyber-vulnerability-database/

Kuppa A, Aouad L, Le-Khac NA (2021) Linking cve's to mitre att&ck techniques. In: Proceedings of the 16th international conference on availability, reliability and security, pp 1–12

Kwon W, Li Z, Zhuang S, Sheng Y, Zheng L, Yu CH, Gonzalez JE, Zhang H, Stoica I (2023) Efficient memory management for large language model serving with pagedattention. Proceedings of the ACM SIGOPS 29th symposium on operating systems principles

Li Z, Dutta S, Naik M (2025) Iris: Llm-assisted static analysis for detecting security vulnerabilities. https://arxiv.org/abs/2405.17238

Livesay J (2024) Nist official details efforts to improve automation for vulnerability management through new framework. https://insidecybersecurity.com/share/16091

Marchiori F, Donadel D, Conti M (2025) Can llms classify cves? investigating llms capabilities in computing cvss vectors.ArXiv preprint

Mendsaikhan O, Hasegawa H, Yamaguchi Y, Shimada H (2020) Automatic mapping of vulnerability information to adversary techniques. The fourteenth international conference on emerging security information, systems and technologies SECUREWARE2020

Meta AI (2024) Llama 3.2: revolutionizing edge ai and vision with open, customizable models. https://ai.meta.com/blog/llama-3-2-connect-2024-vision-edge-mobile-devices/

Mitre (2020) About cve records. https://cve.mitre.org/cve/identifiers/index.html

MITRE (2025) Common weakness enumeration. https://cwe.mitre.org/ Accessed: July 2025

MITRE Corporation (2025) Mitre att&ck. https://attack.mitre.org/. Accessed: June 2025

Muzsai L, Imolai D, Lukács A (2024) Hacksynth: Llm agent and evaluation framework for autonomous penetration testing.ArXiv preprint

OffSec (2025) Exploit database. https://www.exploit-db.com/, Accessed: June 2025

ollama (2025) Ollama. https://github.com/ollama/ollama

OpenAI (2025) Introducing gpt-4.1 in the api. https://openai.com/index/gpt-4-1/

Rafiey P, Namadchian A (2025) Mapping vulnerability description to mitre att&ck framework by llm. Preprint, Research Square https://doi.org/10.21203/rs.3.rs-4341401/v2

Roziere B, Gehring J, Gloeckle F, Sootla S, Gat I, Tan XE, Adi Y, Liu J, Sauvestre R, Remez T (2023) Code llama: open foundation models for code.ArXiv preprint

Satter R (2025) In last-minute reversal, us agency extends support for cyber vulnerability database. https://www.reuters.com/world/us/us-agency-extends-support-last-minute-cyber-vulnerability-database-2025-04-16/

Shahid MR, Debar H (2021) Cvss-bert: explainable natural language processing to determine the severity of a computer security vulnerability from its description. 2021 20th IEEE international conference on machine learning and applications (ICMLA). IEEE, pp 1600–1607

Simonetto S, Bosch P (2024) Comprehensive threat analysis and systematic mapping of cves to mitre framework. In: 1st International conference on natural language processing and artificial intelligence for cyber security, NLPAICS 2024

Spring J, Hatleback E, Manion A, Shic D (2018) Towards improving cvss. SEI, CMU, Tech Rep

Steenhoek B, Rahman MM, Roy MK, Alam MS, Barr ET, Le W (2024) A comprehensive study of the capabilities of large language models for vulnerability detection. ArXiv preprint arXiv:2403.17218

Strom BE, Applebaum A, Miller DP, Nickels KC, Pennington AG, Thomas CB (2018) Mitre att&ck: Design and philosophy. In: Technical report, The MITRE Corporation

Sun J, Xing Z, Guo H, Ye D, Li X, Xu X, Zhu L (2021) Generating informative cve description from exploitdb posts by extractive summarization.ArXiv preprint

Sun J, Chen J, Xing Z, Lu Q, Xu X, Zhu L (2024) Where is it? tracing the vulnerability-relevant files from vulnerability reports. Proceedings of the IEEE/ACM 46th international conference on software engineering. pp 1–13

Tamberg K, Bahsi H (2025) Harnessing large language models for software vulnerability detection: a comprehensive benchmarking study. IEEE Access,

Tang H, Hu K, Zhou J, Zhong SC, Zheng WL, Si X, Ellis K (2024) Code repair with llms gives an exploration-exploitation tradeoff. Adv Neural Inf Process Syst 37:117954–117996

Team G (2025) Gemma 3. https://arxiv.org/abs/2503.19786

Vellumai (2025) Llm leaderboard. https://www.vellum.ai/llm-leaderboard

Wu Y, Wen M, Yu Z, Guo X, Jin H (2024) Effective vulnerable function identification based on cve description empowered by large language models. Proceedings of the 39th IEEE/ACM international conference on automated software engineering. pp 393–405

Wudali PN, Kravchik M, Malul E, Gandhi PA, Elovici Y, Shabtai A (2025) Rule-att&ck mapper (ram): Mapping siem rules to ttps using llms. ArXiv preprint arXiv:2502.02337

Wunder J, Kurtz A, Eichenmüller C, Gassmann F, Benenson Z (2024) Shedding light on cvss scoring inconsistencies: a user-centric study on evaluating widespread security vulnerabilities. 2024 IEEE symposium on security and privacy (SP). IEEE, pp 1102–1121

Zheng L, Chiang WL, Sheng Y, Zhuang S, Wu Z, Zhuang Y, Lin Z, Li Z, Li D, Xing E et al (2023) Judging llm-as-a-judge with mt-bench and chatbot arena. Adv Neural Inf Process Syst 36:46595–46623

Agentic AI in Practice: Applications Across Industries

Hanqi Su, Dai-Yan Ji, Takanobu Minami, and Jay Lee

Abstract This chapter covers the development and deployment of agentic artificial intelligence (AI) across industrial sectors, including manufacturing, robotics, healthcare, and energy. Agentic AI represents a shift from traditional AI systems to autonomous, goal-directed decision-making systems that integrates perception, reasoning, planning, and action. These systems offer new opportunities for efficiency, resilience, and adaptability in high-stakes applications. At the same time, they introduce risks associated with reliability, interpretability, and human oversight. Currently, there lacks of detailed discussion of real-world cases in high-stakes industrial environments across different industrial sectors. To address these issues, this chapter presents case studies that illustrate both the operational benefits and the failure modes of agentic AI in practice. Examples range from production scheduling, human–robot collaboration, task planning to clinical training, drug discovery, and energy optimization. Through these cases, the chapter highlights not only the engineering challenges of deploying autonomous agents at scale but also discusses potential directions for future agentic AI research.

Keywords Agentic AI · Machine learning · Large language model · Industrial large knowledge model

1 Introduction

Starting from 2020, the rapid progress of large language models (LLMs) has significantly reshaped the deployment of artificial intelligence (AI) in various industrial domains (Zhao et al. 2023). Building on these advances, the concept of agentic AI has emerged as a new paradigm. It represents the shift from reactive or narrowly task-specific systems toward autonomous, goal-directed agents capable of functioning

H. Su · D.-Y. Ji · T. Minami · J. Lee (✉)
Center for Industrial Artificial Intelligence, Department of Mechanical Engineering, University of Maryland, College Park, MD, USA
e-mail: leejay@umd.edu

A. Stavrou et al. (eds.), *Generative and Agentic AI Reliability: Architectures, Challenges, and Trust for Autonomous Systems*, Studies in Computational Intelligence 1272, https://doi.org/10.1007/978-3-032-18585-3_8

in dynamic environments (Acharya et al. 2025). Unlike traditional AI models that execute predefined rules, or generative AI systems that produce outputs without acting on them, agentic AI leverages the reasoning and adaptability of LLMs while integrating perception, planning, and action within a unified framework (Hughes et al. 2025; Plaat et al. 2025). Such agentic AI systems can set and pursue complex goals, adapt to changing working conditions, and collaborate with other agents or humans, enabling more flexible and resilient decision-making. Several recent publications have outlined the foundational concepts, defining characteristics, and methodology development of agentic AI (Acharya et al. 2025). Others also have examined its potential in specific domains such as scientific discovery (Gridach et al. 2025), healthcare (Karunanayake 2025), education (Kamalov et al. 2025), and engineering (Lee et al. 2025). However, there remains a lack of detailed discussion of real-world deployments in high-stakes industrial environments and systematic analysis of applied case studies across sectors. This limits our ability to fully assess both potential and risks of agentic AI.

This chapter surveys the deployment of agentic AI across four major industrial domains: manufacturing, robotics, healthcare, and energy. Each of these areas presents both opportunities and risks for adopting agentic AI with detailed case studies. By examining actual applications, this chapter highlights how agentic AI can improve efficiency, safety, and adaptability to reshape industrial practice, while also drawing attention to the engineering challenges and operational risks that accompany autonomy in high-stakes environments. Ultimately, deploying agentic AI requires not only technical advances but also institutional mechanisms for necessary supervision and human–AI collaboration.

Roadmap of the Chapter. Section 2 introduces the key attributes and general workflow of agentic AI that drive industrial applications. Section 3 explores key applications across the four industrial domains through detailed analysis and case studies. Section 4 discusses open challenges and directions for future research. Finally, Sect. 5 concludes the chapter.

2 Key Attributes and General Workflow of Agentic AI Driving Industry Application

In this section, we introduce four core attributes that make LLM-driven based agentic AI suitable for high-stakes industrial sectors such as manufacturing, robotics, healthcare, and energy. As illustrated in Fig. 1, it includes:

(I) **Goal-Directed Autonomy**: Unlike traditional AI systems that only solve predefined tasks, agentic AI can work under minimal human intervention and adapts its actions to broader objectives. It can translate high-level goals into sequences of subtasks, dynamically allocate computational and physical

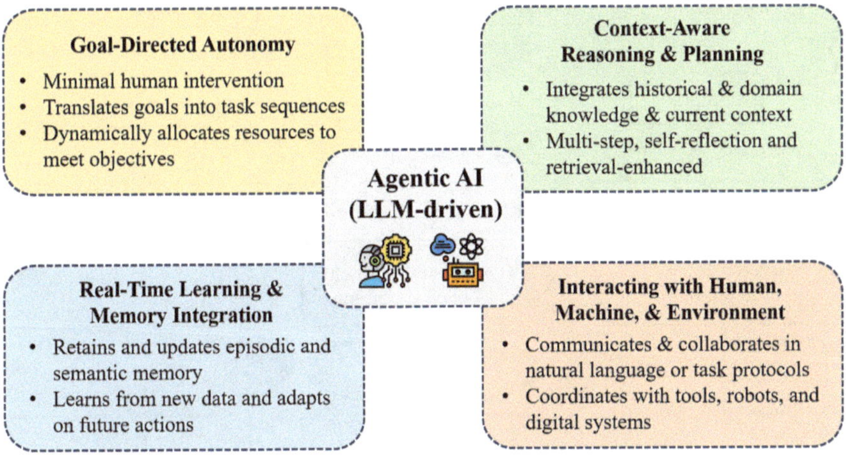

Fig. 1 Key attributes of agentic AI (LLM-driven) driving industry applications

resources to address the problems. This allows agentic AI systems to act as autonomous decision-makers rather than static tools.

(II) **Context-Aware Reasoning and Planning**: Effective decision-making in industry requires understanding both historical and current context. Agentic AI leverages advanced techniques, such as long term memory, retrieval-enhanced reasoning, multi-step planning, and self-reflection mechanism, to enhance context-aware reasoning and planning.

(III) **Interacting with Human, Machine, and Environment**: Agentic AI system can communicate with human operators in natural language, coordinate with tools, robots and digital systems through task-specific protocols, and maintain situational awareness when operating in multi-agent settings. This interaction capability ensures that agentic AI functions not in isolation but as an integral part of broader industrial AI ecosystems.

(IV) **Real-Time Learning and Memory Integration**: An important feature of agentic AI is its ability to learn continuously from experience. By retaining and updating both episodic and semantic memory, these systems can adapt their future actions based on new inputs. This contrasts with static models that degrade when exposed to unseen data. Real-time learning is also important in high-stakes domains such as transportation safety monitoring, predictive maintenance in manufacturing, and adaptive resource allocation in energy grids, where conditions change rapidly and reliability is non-negotiable.

Beyond key attributes, it is also important to highlight how agentic AI differs from traditional approaches. Traditional AI systems typically follow rule-based, heuristic, or option-based pipelines, where inputs are processed into static outputs without adaptation. Agentic AI, in contrast, follows a closed-loop workflow. High-level goals guide multimodal perception and context gathering, which inform reasoning and planning. Actions are executed in interaction with humans and machines, followed

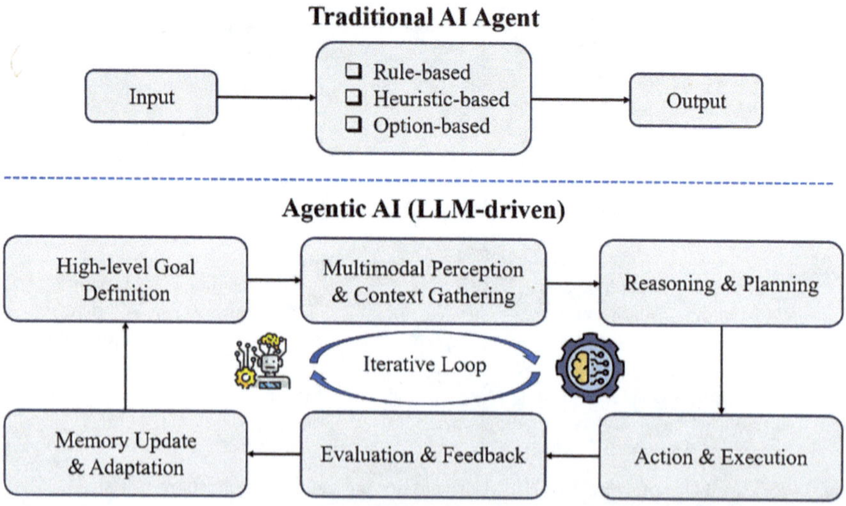

Fig. 2 General workflow of agentic AI (compared to traditional AI agents)

by evaluation and feedback. Memory updates ensure continuous learning and adaptation. Figure 2 illustrates a general workflow, contrasting static workflows of traditional AI with the adaptive, iterative cycles of agentic AI.

3 Agentic AI Applications Across Industries

In this section, we focus on four key domains, including manufacturing, robotics, healthcare, and energy. It highlights key applications and representative case studies that illustrate how agentic AI systems are being tested and deployed in practice. While the core principles of agentic AI remain consistent, the way these systems are designed, validated, and integrated varies across industrial applications, reflecting differences in users, workflows, and goals. To provide a structured view, Tables 1, 2, 3 and 4 categorizes each domain according to its primary applications, main users, and enabling AI technologies.

3.1 Agentic AI in Manufacturing

Manufacturing is among the most important industrial sectors in the world where agentic AI has begun to demonstrate measurable value. Recent advances illustrate how agentic AI can transform manufacturing workflows by integrating technologies such as LLMs, digital twins, reinforcement learning, industrial internet of things

Table 1 Summarization of agentic AI in manufacturing: key applications, main users, and AI technologies

Key applications	Main users	Enabling AI technologies
Production planning and scheduling	Plant managers, scheduling engineers, production controllers, sustainability officers	LLM, digital twin, proximal policy optimization, reinforcement learning, optimization algorithms, multi-agent workflows
Human–machine collaboration	Operators, technicians, warehouse staff, supervisors, engineers	Computer visions, LLM-based dialogue agents, digital twin, AR/VR interface, multi-agent collaboration, multimodal sensor fusion
Knowledge-driven decision support	Engineers, data scientists, managers, researchers, executives	Retrieval-augmented generation (RAG), Industrial large knowledge model (ILKM), hybrid retrieval, knowledge graph, different roles of LLMs, explainable AI tools

Table 2 Summarization of agentic AI in robotics: key applications, main users, and AI Technologies

Key applications	Main users	Enabling AI technologies
Human–robot collaboration	Assembly operators, technicians, production supervisors	LLMs, multimodal learning, mixed/augmented reality, reinforcement learning with human feedback, prompting design
Multi-robot collaboration	Robot fleet researchers, warehouse operators, logistics planners	LLMs, vision model, multimodal models, hybrid centralized–decentralized framework, coalition formation strategy, multi-agent reinforcement learning, chain-of-thought
Task planning and control	Factory automation engineers, robotics researchers, production planners	LLMs, tool integration APIs, sequential planning strategies, embodied intelligence frameworks, prompt engineering, multi agent workflow

(IIoT), etc. (Lee and Su 2025a). This subsection discusses three major application areas: production planning and scheduling, human–machine collaboration, and knowledge-driven decision support.

Table 3 Summarization of agentic AI in healthcare: key applications, main users, and AI technologies

Key applications	Main users	Enabling AI technologies
Clinical training and simulation	Medical students, clinicians, surgical trainers, hospital educators and staff, doctors	LLMs, autonomous multi-agent systems, Retrieval-Augmented Generation (RAG), medical knowledge bases, multimodal agents, long-short memory mechanism
Diagnostic support and medical imaging	Doctors, researchers, diagnostic technicians, hospital IT staff	LLMs, vision–language models, multimodal medical learners, multimodal AI, prompt tuning, cross-attention mechanism, U-Net
Drug discovery and biomedical research	Biomedical researchers, research scientists, lab automation engineers	LLMs, multi-agent framework, planner–instructor agent pairs, integration with biomedical databases, generative models, "AI scientist" workflows combining LLMs + lab tools

Table 4 Summarization of agentic AI in energy: key applications, main users, and AI technologies

Key applications	Main users	Enabling AI technologies
Smart buildings with energy optimization	Facility managers, energy researchers, building operators, occupants	LLMs, Bayesian optimization, Shapley additive explanations, GPT-based model, retrieval-augmented generation (RAG), LLM-based multi-agent roles (planner, researcher, advisor, etc.)
Smart grids and power system reliability	Grid operators, maintenance engineers, cybersecurity teams, researchers, scientists	LLMs, digital memory and reflection mechanism, prompt design, similarity-based measures, agentic digital twins, generative AI models
Knowledge-driven conversational systems	Energy researchers, data scientists, policy analysts, facility staff	LLMs, RAG, actor–critic reinforcement learning, text-to-SQL conversion, multimodal integration, NLP, multi-agent AI systems, edge/fog/cloud computing

3.1.1 Production Planning and Scheduling

Scheduling remains a basic challenge in manufacturing, traditionally approached through optimization algorithms or rule-based heuristics. Agentic AI offers a new paradigm, where AI agents can autonomously negotiate, adapt, and balance competing objectives. For instance, Carbon DTMAS integrated LLM-enhanced digital twins with multi-agent negotiation and reinforcement learning to coordinate scheduling while explicitly reducing carbon emissions. In textile workshops, this system reduced tardiness by nearly 4%, balanced workloads by over 30%, and cut emissions by nearly 5%, demonstrating how autonomy can align efficiency with sustainability (Wu et al. 2025). Similarly, Zhang et al. proposed a self-organizing negotiation mechanism where equipment agents negotiate via CNP. Each agent hosts an AI scheduler (multi-layer perceptron) that was periodically retrained with proximal policy optimization (PPO) to improve task allocation and scheduling efficiency (Zhang et al. 2022). Other frameworks, such as LLM-enabled multi-agent systems that combined product and resource agents, demonstrated how natural language instructions can directly drive adaptive task allocation (Lim et al. 2024). Xia et al. presented a novel flexible modular production system framework enhanced by LLM agents further systematically show how different LLM agents (manager and operator roles) orchestrate production plans using digital twins as contextual grounding (Xia et al. 2023).

3.1.2 Human–Machine Collaboration

Industry 5.0 emphasizes human-centric production, making human–machine collaboration a critical application of agentic AI (Leng et al. 2022). The integration of LLM agents enables more intuitive interaction between humans and machines while ensuring adaptability and safety. A notable example is the use of LLMs for human–robot collaboration in assembly, where specialized agents translate natural language instructions into executable behavior trees for collaborative robots. Digital twins and augmented reality interfaces provide grounding, while validation studies showed high execution accuracy across motor and compressor assembly tasks (Gkournelos et al. 2024). Except assembly, Nicoletti and Appolloni proposed a human–agentic AI–machine collaboration (HAMC) framework, extending human–machine collaboration into a triadic model (humans, agentic AI, and machines). In a Ferrero warehouse case, this approach reduced commissioning time by 88% while improving operational efficiency and energy use (Nicoletti and Appolloni 2025). In addition, a multi-agent LLM-based control framework with four specialized agents: monitoring agent, actor agent, validator agent, and reprompter agent are introduced by Vyas and Mercangöz (2025). Actions proposed by the actor agent are validated in a digital twin; if unsafe or suboptimal, the reprompter agent iteratively refines them until they meet criteria.

3.1.3 Knowledge-Driven Decision Support for Manufacturing

At a higher level, manufacturing also benefits from agentic AI agents designed to act as knowledge intermediaries, enabling explainable and trustworthy decision-making. The industrial large knowledge model (ILKM) framework is a conceptual

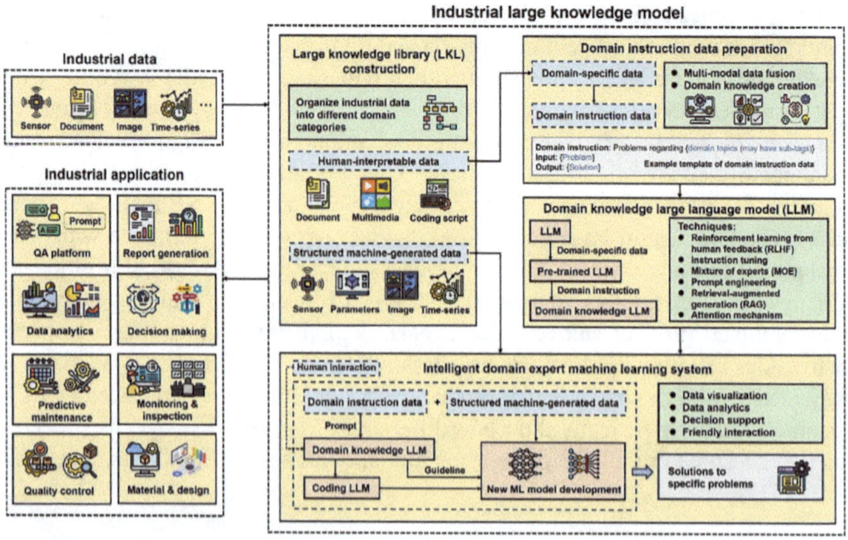

Fig. 3 A unified industrial large knowledge model framework (Lee and Su 2024)

architecture that extends LLMs into domain-specific systems by integrating large knowledge libraries, multimodal data, coding LLMs, and domain knowledge LLMs (Lee and Su 2024). As shown in Fig. 3, it provides a blueprint for developing domain-adapted agentic systems that ensure reliability and accountability in industrial decision contexts. Beyond this, Lee and Su proposed a retrieval-augmented generation (RAG)-based LLM QA system for answering different PHM data challenge competitions questions (Lee and Su 2025b). Another example is ChatCNC, a multi-agent LLM framework that classifies operator questions, generates SQL queries, retrieves CNC/IIoT data in real time, and provides natural language responses. It achieved over 93% accuracy on complex production queries and reduced response time from minutes to seconds (Jeon et al. 2025). Moreover, Wan et al. emphasize importance of hybrid retrieval architectures for LLM systems, such as combining vector embeddings with structured knowledge graphs, to enhance domain-specific Q&A for design and manufacturing tasks. This reduces hallucinations and improves trust in agent outputs (Wan et al. 2025).

3.2 *Agentic AI in Robotics*

The emergence of LLMs and agentic AI architectures has accelerated progress by enabling robots to interpret natural language, autonomously plan and coordinate tasks, adapt in real time to environmental changes and collaborate with human partners (Jeong et al. 2024). In this section, we discuss three key application areas: human–robot collaboration, multi-robot collaboration, and task planning.

3.2.1 Human–Robot Collaboration

Human–Robot Collaboration (HRC) represents one of the most direct applications of agentic AI, where robots need to work alongside with people in shared workspaces. A central challenge lies in aligning human instructions often given in natural language with executable robotic actions while ensuring safety and adaptability. Recent studies illustrate how agentic AI can bridge this gap. Yin et al. introduced an LLM-enabled hybrid robot skill synthesis framework that integrates a three-layer multimodal LLM agent (task decomposer, motion descriptor, reward coder) with mixed reality-assisted human demonstrations. LLMs translate human language instructions into reward functions, guiding deep reinforcement learning exploration (Yin et al. 2025). This approach achieved a 95.9% task success rate, substantially outperforming traditional reinforcement or demonstration-based learning methods. Similarly, Park et al. developed CLARA framework to handle ambiguous human commands, using LLMs to classify intent and initiate clarification dialogues when instructions were uncertain. The system employs uncertainty-aware prompting, context sampling, and situational awareness to improve reliability in task execution (Park et al. 2023).

3.2.2 Multi-robot Collaboration

Agentic AI also enables multi-robot collaboration, where multiple agents coordinate to solve tasks that exceed the capability of a single robot. A great example is the RoCo system, where robots equipped with LLMs engage in natural language dialogue to reason about task allocation and strategies. The system generates sub-task plans, validated via environment feedback (IK feasibility, collision checks). Validated subtasks are mapped to joint-space trajectories through a centralized motion planner. RoCo leverages LLM-proposed 3D waypoints to improve efficiency in motion planning (Mandi et al. 2024). In addition, Chen et al. compared centralized, decentralized, and hybrid frameworks for multi-robot collaboration across BoxNet, Warehouse, and BoxLift environments. Their hybrid HMAS-2 model (central planner validated by local LLM agents) achieved superior success rates and scalability (Chen et al. 2024). Another line of work emphasizes coalition formation and proactive cooperation. Kannan et al. proposed SMART-LLM, a four-stage framework for decomposing natural language instructions, forming robot teams, and allocating tasks based on skill sets. Results showed near-perfect task success in simple scenarios and robust generalization in compound settings (Kannan et al. 2024). Similarly, Zhang et al. introduced ProAgent, which integrates LLMs into multi-agent cooperation by structuring reasoning and planning through four modules: (1) planner agent uses chain-of-thought reasoning to infer teammate intentions and plan skills; (2) verificator agent validates skills and provides feedback for replanning; (3) controller decomposes high-level skills into executable low-level actions; (4) memory stores knowledge libraries, trajectories, and beliefs. The belief revision mechanism aligns predicted teammate intentions with actual observed behaviors for dynamic adaptation (Zhang et al. 2024).

3.2.3 Task Planning

Unlike traditional programming approaches, LLM-enabled agents can directly interpret instructions, design workflows, and adjust execution in real time. This can enrich the value of agentic AI in task planning and control aspect. For instance, Ruan et al. presented the TPTU framework, which systematically evaluates LLM-based agents in planning and tool usage across diverse tasks. Their findings underscore the advantage of sequential planning strategies over one-step planning, providing design insights for agentic architectures in robotics (Ruan et al. 2023). Wu et al. extended these ideas to factory task planning, where uses LLMs (GPT-4) to interpret natural language task descriptions and generate corresponding robotic task plans. The system translates high-level human instructions into robot-executable code, with constraints and verification mechanisms to ensure feasibility (Tsushima et al. 2025). Another research on embodied intelligence in manufacturing also highlights how LLMs can autonomously achieve task–process parameter matching, generate tool paths, reason and reflect about errors, and refine outputs iteratively (Fan et al. 2025).

3.3 *Agentic AI in Healthcare*

Agentic AI systems have begun to demonstrate great potential across the healthcare value chain, ranging from clinical training and simulation, through diagnostic support and medical imaging, to drug discovery and biomedical research, and finally to healthcare decision support (Karunanayake 2025). This subsection explores how agentic AI is being applied across different healthcare functions, while also considering the opportunities and risks that accompany its integration into practice.

3.3.1 Clinical Training and Simulation

Clinical training and simulation have long been central to medical education, and agentic AI introduces new possibilities by modeling patients, doctors, and clinical environments as autonomous, interactive agents. Several examples exemplify this trend. Agent Hospital introduces a virtual hospital where patients, nurses, and doctors are modeled as LLM-powered autonomous agents. A new paradigm called simulacrum-based evolutionary agent learning (SEAL) is proposed, which generates synthetic patient cases by coupling LLMs with medical knowledge bases. Doctor agents evolve by treating these patient agents, accumulating knowledge in a medical case base (from successes) and an experience base (from failures) across 32 departments and more than 300 diseases. Through iterative feedback loops, diagnostic accuracy improves dramatically without the need for manually labeled data (Li et al. 2024). Similarly, AgentClinic provides a benchmark that recreates clinical decision-making in multilingual, multimodal contexts, incorporating doctor, patient, measurement, and moderator agents. It supports interactive, sequential decision-making, allowing doctor agents to order tests, interpret results, and interact with patient agents (Schmidgall et al. 2024). In addition, agentic AI is also entering surgical training. SurgBox creates an agent-driven operating room sandbox that simulates full surgical workflows, where surgeons, nurses, and anesthetists are modeled as role-playing agents guided by tailored knowledge bases via RAG. A central "Surgery Copilot" coordinates decision-making, drawing on both short-term context and long-term memory (Wu et al. 2024). Evaluations show notable accuracy improvements in surgical planning, pointing to the dual utility of such systems for both training and real-world assistance.

3.3.2 Diagnostic Support and Medical Imaging

Diagnosis and medical imaging remain among the most visible and impactful application areas for AI in healthcare, and agentic AI expands these capabilities by adding autonomy and multimodal reasoning. VoxelPrompt represents a vision–language agent framework where an LLM acts as a planning agent, generating and executing code instructions to analyze MRI and CT scans. This enables integrated workflows of segmentation, measurement, and interpretation that traditionally required multiple specialized systems (Hoopes et al. 2024). Similarly, Med-Flamingo adapts a multimodal large model for medical visual question answering. Its ability to generate open-ended rationales across images and text aligns with clinical reasoning processes, improving the interpretability and trustworthiness of AI outputs (Moor et al. 2023). In radiation oncology, LLMSeg introduces LLM-driven multimodal contouring for 3D clinical target volume delineation. By integrating clinical text data with CT images, this allows the model to consider clinical context (tumor stage, laterality, surgery type, etc.) when contouring and achieves significantly higher accuracy than vision-only models (Oh et al. 2024).

3.3.3 Drug Discovery and Biomedical Research

Drug development is a lengthy, resource-intensive process, and agentic AI offers an opportunity to accelerate discovery through autonomous, iterative reasoning. DrugAgent provides a clear example, as illustrated in Fig. 4. It introduced a two-agent system where a Planner generates strategies for tasks such as drug–target interaction prediction, while an Instructor integrated domain knowledge and converted strategies into executable machine learning code. This division of labor allowed the system

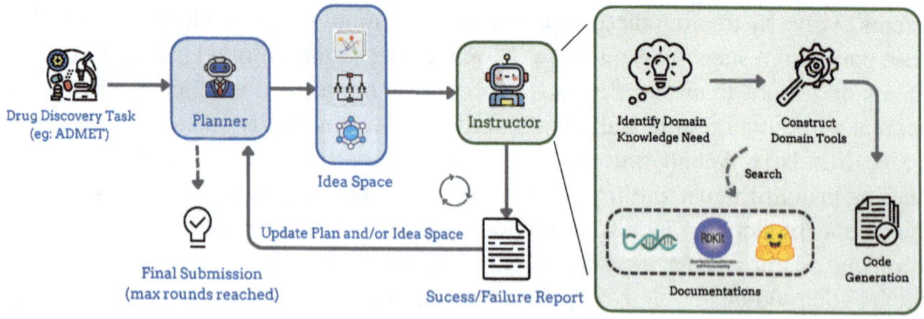

Fig. 4 Overview of the DrugAgent framework (Liu et al. 2024)

to iteratively refine solutions based on experimental feedback. Benchmarks across AMPA (ADMET), HIV (HTS), and DAVIS (DTI) datasets showed that DrugAgent outperforms traditional frameworks while ensuring valid and interpretable outputs (Liu et al. 2024). Beyond specific frameworks, researchers have begun to outline a broader vision of "AI scientists" as collaborative agents that integrate LLMs, generative models, and biomedical tools with experimental platforms. These systems are designed not to replace human researchers but to augment human creativity and expertise by autonomously navigating hypothesis spaces, planning discovery workflows, and performing self-assessment. Such agentic AI approaches could ultimately transform biomedical discovery, enabling advances from virtual cell simulation and programmable phenotypes to the design of new therapies (Gao et al. 2024).

3.4 *Agentic AI in Energy*

The energy sector undergoes rapid digital transformation, but currently faces challenges in ensuring efficiency, reliability, and sustainability under conditions of high uncertainty and dynamic demand. Agentic AI provides new opportunities to address these challenges by enabling autonomous monitoring, adaptive control, and predictive decision-making. Based on recent studies, we discuss following key application areas: smart buildings with energy optimization, smart grids and power system reliability, and knowledge-driven conversational system development.

3.4.1 Smart Buildings with Energy Optimization

Buildings consume a substantial portion of global energy, and improving their operational efficiency is an important topic for sustainable development. Traditional methods for building energy management rely on physics-based simulations and simple data-driven models, which are costly and time-consuming. Agentic AI introduces automation and adaptive reasoning into this process. One notable example is that Zhang et al. developed a GPT-based framework for building energy load prediction that automates both model training and deployment. By combining Bayesian optimization, self-correction strategies, and external knowledge integration, the framework adapts to real-time conditions and achieves high accuracy while keeping costs extremely low (Zhang et al. 2025). In addition, Xiao and Xu proposed an LLM-based multi-agent framework for exploring automated energy optimization. This framework assigns specialized roles to LLM agents—such as planner, researcher, and advisor. These agents process unstructured energy data, apply chain-of-thought reasoning, and integrate knowledge from engineering codes and audit reports. Case studies showed high recall (0.9627) and precision (0.9373) in extracting relevant building information, with retrofit recommendations that closely matched expert suggestions (Xiao and Xu 2024). Recently, a cloud-based platform was proposed for smart building monitoring and better human interaction, which combined IoT sensors, BIM-based energy models, cloud computing, and LLM-powered persuasion systems. Such platforms monitor indoor environments, optimize comfort and energy use, and even interact with occupants. In one university case study, most participants accepted recommendations to adjust behavior for improved air quality or thermal comfort (Xu et al. 2025).

3.4.2 Smart Grids and Power System Reliability

At the grid level, reliability and safety are paramount. Power systems must handle anomalies, cyber threats, and maintenance issues in real time. Agentic AI provides a pathway for embedding reasoning, memory, and safety monitoring into smart grids. A smart agent framework known as SLEP has been developed for time series anomaly detection in power systems. By combining large pre-trained models with memory, reflection, and structured prompting, the system reached accuracy levels exceeding 98% and in some cases 100% recall on three real-world datasets (IHEPC, ELD, ENLITEN) (Wang et al. 2025). In offshore wind energy, a novel conversational agent framework called SafeLLM incorporated similarity measures and ensemble outputs for hallucination detection and unsafe response filtering. SafeLLM reduced false alarms from SCADA systems and provides more reliable decision support for maintenance teams (Walker et al. 2024). Looking ahead, the concept of the Agentic Digital Twin (ADT), shown in Fig. 5, extends these ideas to broad energy assets and smart grids. Unlike traditional digital twins that passively simulate, ADTs integrate LLMs to reason over multimodal data, negotiate with other agents, and proactively plan responses. This transforms digital twins into active decision-makers capable of supporting resilience and reliability across distributed power networks (Antonesi et al. 2025).

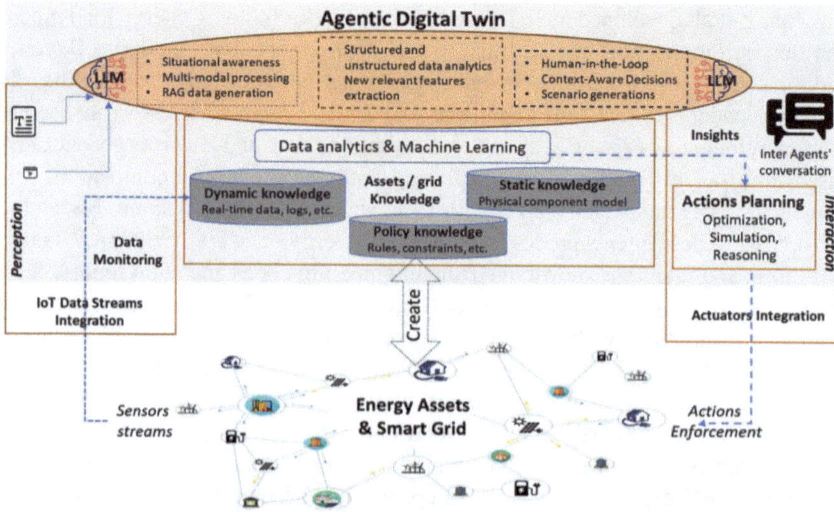

Fig. 5 Architecture of a smart grid Agentic Digital Twin (ADT) (Antonesi et al. 2025)

3.4.3 Knowledge-Driven Conversational System Development

Agentic AI is also reshaping how energy knowledge is created, retrieved, and applied through conversational systems. These systems combine natural language interfaces with reasoning and retrieval mechanisms to make complex energy data and research outputs more accessible to diverse users. One example is an actor–critic chat-agent framework for scientific research review, where an "actor" generates candidate responses and a "critic" evaluates and refines them (Liu et al. 2025). By integrating retrieval-augmented generation and live database updates, the system produces more accurate and timely literature reviews. This reduces hallucinations, improves reference quality, and accelerates discovery in a rapidly evolving research field. In addition, Cooee, a chatbot deployed in a university energy analytics platform, allows users to query multimodal energy datasets through natural language in chatbot interface. By translating queries into SQL and returning interpretable insights, the system lowers the barrier for facility managers and researchers to engage with complex energy data (Gamage et al. 2022). Moreover, the "Follow-Me AI" system, first tested on a smart campus, illustrated how personal agents and environment agents can negotiate to optimize resource use. By predicting user needs and balancing trade-offs across IoT devices and infrastructure, the system achieved measurable energy savings and offered a pathway toward more transparent, user-centered, and adaptive energy management (Saleh et al. 2025).

4 Open Challenges and Future Directions

Despite the progress of agentic AI across different industry applications, there are several challenges remain. These can be grouped into three categories: technical challenges, human–AI interaction, and ethical concerns.

(I) **Technical Challenges**: Reliability and safety are the most importance. Many current agentic AI workflows are driven by large language models (LLMs), which often face hallucination and can generate incorrect outputs. In high-stakes environments, errors carry significant financial and safety risks, making stable and reliable performance essential. Interpretability presents another barrier. Most foundation models and LLMs function as "black boxes," offering limited insight into their reasoning processes. Engineers are often unwilling to rely on unclear decision-making mechanisms, particularly in industries that require strict accountability. A further difficulty lies in scalability. Deploying agentic AI at industrial scale demands substantial computational resources and advanced AI infrastructures, which may be beyond the reach of smaller companies.

(II) **Human–AI Interaction**: While agentic AI is designed to reduce human workload, most high-stakes applications still require human supervision. Finding a balance between autonomy and human control is difficult as too much supervision reduces the benefits of autonomy, but too little supervision risks unsafe outcomes. Moreover, ensuring effective collaboration between humans, multiple AI agents, and industrial systems is not trivial. Agents may misinterpret human queries or fail to provide actionable feedback, which can hinder downstream decision-making. Even when AI agents generate accurate solutions, execution by physical systems such as robots may still face reliability issues, creating further gaps between decision and implementation.

(III) **Ethical Considerations**: Agentic AI systems raise pressing questions of trust, fairness, and accountability. Determining responsibility when an autonomous system makes an incorrect or harmful decision remains unresolved. In healthcare domain, the ethical stakes are especially high, as agentic AI systems may directly influence patient outcomes, where human life must always take priority.

Looking ahead, future progress depends on building agentic AI systems that are not only more reliable but also deeply integrated into industrial ecosystems. We identify three key research directions:

(I) **Hybrid Architecture Design for Reliability**. Future systems could consider extending beyond purely LLM-driven agents toward architectures that integrate complementary models, such as physics-informed models, time-series transformers, image-based CNNs. Combining these approaches with LLM agents may produce architectures suitable for heterogeneous data modalities present in industrial settings.

(II) **Domain-Specific Knowledge Base Creation**. General-purpose LLMs often lack the grounding required for high-stakes applications in manufacturing, healthcare, energy, etc. Constructing domain-specific knowledge repositories and integrating them with agentic AI systems will be essential for aligning decisions with rich knowledge from human expertise.

(III) **Human-Centric Interpretable Agentic AI Development**. Future work should focus on developing methods for interpretable reasoning, transparent planning, and human-in-the-loop decision approach. This direction emphasizes co-agency, where humans and AI agents share responsibility for decisions, and clear interpretable process enhances reliable decision making.

5 Conclusion

This chapter has examined the emergence of agentic AI across different industries including manufacturing, robotics, healthcare, and energy. By moving from static automation and rule-based systems, agentic AI integrates perception, reasoning, planning, and action into unified frameworks that operate with greater autonomy and adaptability. Through case studies and structured comparisons, we illustrated how agentic AI enables improvements in twelve key applications of four domains. The accompanying tables provided a structured overview of applications, main users, and enabling technologies, offering a reference point for both researchers and practitioners. We got several lessons from this survey. First, agentic AI demonstrates clear potential to enhance efficiency, resilience, and sustainability in high-stakes environments. Second, despite impressive advances, challenges around reliability, interpretability, and human oversight remain pressing. Industrial adoption requires not only technical breakthroughs but also careful integration with human-centric workflows, decision-making transparency, and ethical safeguards. These findings underscore that the value of agentic AI is realized not in isolation but within broader socio-technical ecosystems where people, machines, and intelligent agents work together. Looking forward, three research directions stand out. Hybrid architectures that combine LLM agents with physics-informed or domain-specific models can improve robustness. The construction of domain-specific knowledge bases and benchmarking standards will be also critical. Finally, human-centric and interpretable design must remain a priority, ensuring that agentic AI supports decision-making transparently and collaboratively rather than replacing human judgment. These directions highlight a path toward trustworthy, scalable deployment across industries.

Acknowledgements This work was supported by the U.S. Department of Education through the Fund for the Improvement of Postsecondary Education (FIPSE) under Grant No. P116S230014.

References

Acharya DB, Kuppan K, Divya B (2025) Agentic AI: autonomous intelligence for complex goals—a comprehensive survey. IEEE Access

Antonesi G, Cioara T, Anghel I, Michalakopoulos V, Sarmas E, Toderean L (2025) From transformers to large language models: a systematic review of AI applications in the energy sector towards Agentic Digital Twins. arXiv preprint arXiv:2506.06359

Chen Y, Arkin J, Zhang Y, Roy N, Fan C (2024) Scalable multi-robot collaboration with large language models: Centralized or decentralized systems? In: 2024 IEEE international conference on robotics and automation (ICRA). IEEE, pp 4311–4317

Fan H, Liu X, Fuh JYH, Lu WF, Li B (2025) Embodied intelligence in manufacturing: leveraging large language models for autonomous industrial robotics. J Intell Manuf 36(2):1141–1157

Gamage G, Mills N, Rathnayaka P, Jennings A, Alahakoon D (2022) Cooee: an artificial intelligence chatbot for complex energy environments. In: 2022 15th international conference on human system interaction (HSI). IEEE, pp 1–5

Gao S, Fang A, Huang Y, Giunchiglia V, Noori A, Schwarz JR, Ektefaie Y, Kondic J, Zitnik M (2024) Empowering biomedical discovery with AI agents. Cell 187(22):6125–6151

Gkournelos C, Konstantinou C, Makris S (2024) An LLM-based approach for enabling seamless human–robot collaboration in assembly. CIRP Ann 73(1):9–12

Gridach M, Nanavati J, Abidine KZE, Mendes L, Mack C (2025) Agentic AI for scientific discovery: a survey of progress, challenges, and future directions. arXiv preprint arXiv:2503.08979

Hoopes A, Butoi VI, Guttag JV, Dalca AV (2024) Voxelprompt: a vision-language agent for grounded medical image analysis. arXiv preprint arXiv:2410.08397

Hughes L, Dwivedi YK, Malik T, Shawosh M, Albashrawi MA, Jeon I, Dutot V, Appanderanda M, Crick T, De' R, Fenwick M (2025) AI agents and agentic systems: a multi-expert analysis. J Comput Inf Syst 1–29

Jeon J, Sim Y, Lee H, Han C, Yun D, Kim E, Nagendra SL, Jun MB, Kim Y, Lee SW, Lee J (2025) ChatCNC: Conversational machine monitoring via large language model and real-time data retrieval augmented generation. J Manuf Syst 79:504–514

Jeong H, Lee H, Kim C, Shin S (2024) A survey of robot intelligence with large language models. Appl Sci 14(19):8868

Kamalov F, Calonge DS, Smail L, Azizov D, Thadani DR, Kwong T, Atif A (2025) Evolution of ai in education: agentic workflows. arXiv preprint arXiv:2504.20082

Kannan SS, Venkatesh VL, Min BC (2024) Smart-llm: smart multi-agent robot task planning using large language models. In: 2024 IEEE/RSJ international conference on intelligent robots and systems (IROS). IEEE, pp 12140–12147

Karunanayake N (2025) Next-generation agentic AI for transforming healthcare. Inf Health 2(2):73–83

Lee J, Su H (2024) A unified industrial large knowledge model framework in Industry 4.0 and smart manufacturing. Int J AI Mater Design 1(2):41–47

Lee J, Su H (2025a) Rethinking industrial artificial intelligence: a unified foundation framework. Int J AI Mater Des 2(2):56–68

Lee J, Su H (2025b) Agentic AI for smart manufacturing. Manuf Lett

Leng J, Sha W, Wang B, Zheng P, Zhuang C, Liu Q, Wuest T, Mourtzis D, Wang L (2022) Industry 5.0: prospect and retrospect. J Manuf Syst 65:279–295

Lee J, Su H, Ji D-Y, Minami T (2025) Engineering artificial intelligence: framework, challenges, and future direction. Mach Learn Eng 1:013001

Li J, Lai Y, Li W, Ren J, Zhang M, Kang X, Wang S, Li P, Zhang YQ, Ma W, Liu Y (2024) Agent hospital: a simulacrum of hospital with evolvable medical agents. arXiv preprint arXiv:2405.02957

Lim J, Vogel-Heuser B, Kovalenko I (2024) Large language model-enabled multi-agent manufacturing systems. In: 2024 IEEE 20th international conference on automation science and engineering (CASE). IEEE, pp 3940–3946

Liu R, Zou Z, Chen S, Liu Y, Wan J (2025) Harnessing AI for understanding scientific literature: innovations and applications of chat-agent system in battery recycling research. Mater Today Energy 49:101818

Liu S, Lu Y, Chen S, Hu X, Zhao J, Lu Y, Zhao Y (2024) Drugagent: Automating AI-aided drug discovery programming through LLM multi-agent collaboration. arXiv preprint arXiv:2411.15692.

Mandi Z, Jain S, Song S (2024) Roco: dialectic multi-robot collaboration with large language models. In: 2024 IEEE international conference on robotics and automation (ICRA). IEEE, pp 286–299

Moor M, Huang Q, Wu S, Yasunaga M, Dalmia Y, Leskovec J, Zakka C, Reis EP, Rajpurkar P (2023) Med-flamingo: a multimodal medical few-shot learner. In: Machine learning for health (ML4H). PMLR, pp 353–367

Nicoletti B, Appolloni A (2025) A digital twin framework for enhancing human–agentic AI–machine collaboration. J Intell Manuf 1–17

Oh Y, Park S, Byun HK, Cho Y, Lee IJ, Kim JS, Ye JC (2024) LLM-driven multimodal target volume contouring in radiation oncology. Nat Commun 15(1):9186

Park J, Lim S, Lee J, Park S, Chang M, Yu Y, Choi S (2023) Clara: classifying and disambiguating user commands for reliable interactive robotic agents. IEEE Robot Autom Lett 9(2):1059–1066

Plaat A, van Duijn M, van Stein N, Preuss M, van der Putten P, Batenburg KJ (2025) Agentic large language models, a survey. arXiv preprint arXiv:2503.23037

Ruan J, Chen Y, Zhang B, Xu Z, Bao T, Mao H, Li Z, Zeng X, Zhao R, (2023) Tptu: task planning and tool usage of large language model-based ai agents. In: NeurIPS 2023 foundation models for decision making workshop

Saleh A, Donta PK, Morabito R, Motlagh NH, Tarkoma S, Loven L (2025) Follow-me AI: energy-efficient user interaction with smart environments. IEEE Pervas Comput

Schmidgall S, Ziaei R, Harris C, Reis E, Jopling J, Moor M (2024) AgentClinic: a multimodal agent benchmark to evaluate AI in simulated clinical environments. arXiv preprint arXiv:2405.07960

Tsushima Y, Yamamoto S, Ravankar AA, Luces JVS, Hirata Y (2025) Task planning for a factory robot using large language model. IEEE Robot Autom Lett

Vyas J, Mercangöz M (2025) Autonomous industrial control using an agentic framework with large language models. IFAC-PapersOnLine 59(6):349–354

Walker C, Rothon C, Aslansefat K, Papadopoulos Y, Dethlefs N (2024) SafeLLM: domain-specific safety monitoring for large language models: a case study of offshore wind maintenance. arXiv preprint arXiv:2410.10852

Wan Y, Chen Z, Liu Y, Chen C, Packianather M (2025) Empowering LLMs by hybrid retrieval-augmented generation for domain-centric Q&A in smart manufacturing. Adv Eng Inform 65:103212

Wang B, Zhou Y, Ge L, Kung SY (2025) Large-model-based smart agent for time series anomaly detection in power systems. Expert Syst Appl 128917

Wu J, Liang X, Bai X, Chen Z (2024) Surgbox: agent-driven operating room sandbox with surgery copilot. In: 2024 IEEE international conference on Big Data (BigData). IEEE, pp 2041–2048

Wu T, Li J, Bao J, Liu Q (2025) Large language model-driven multi-agent systems for improving production efficiency and reducing carbon emissions in manufacturing. Comput Indus Eng 111299

Xia Y, Shenoy M, Jazdi N, Weyrich M (2023) Towards autonomous system: flexible modular production system enhanced with large language model agents. In: 2023 IEEE 28th international conference on emerging technologies and factory automation (ETFA). IEEE, pp 1–8

Xiao T, Xu P (2024) Exploring automated energy optimization with unstructured building data: a multi-agent based framework leveraging large language models. Energy Build 322:114691

Xu Y, Zhu S, Cai J, Chen J, Li S (2025) A large language model-based platform for real-time building monitoring and occupant interaction. J Build Eng 100:111488

Yin Y, Wan K, Li C, Zheng P (2025) An LLM-enabled human demonstration-assisted hybrid robot skill synthesis approach for human-robot collaborative assembly. CIRP Ann

Zhang Y, Zhu H, Tang D, Zhou T, Gui Y (2022) Dynamic job shop scheduling based on deep reinforcement learning for multi-agent manufacturing systems. Robot Comput Integr Manuf 78:102412

Zhang C, Zhang J, Zhao Y, Lu J (2025) Automated data-driven building energy load prediction method based on generative pre-trained transformers (GPT). Energy 318:134824

Zhang C, Yang K, Hu S, Wang Z, Li G, Sun Y, Zhang C, Zhang Z, Liu A, Zhu SC, Chang X (2024) Proagent: building proactive cooperative agents with large language models. In: Proceedings of the AAAI conference on artificial intelligence, vol 38, no 16, pp 17591–17599

Zhao WX, Zhou K, Li J, Tang T, Wang X, Hou Y, Min Y, Zhang B, Zhang J, Dong Z, Du Y (2023) A survey of large language models, vol 1(2). arXiv preprint arXiv:2303.18223

Software Engineering Reliability in the Era of Generic and Agentic AI

Kenechukwu Nwodo and Angelos Stavrou

Abstract The enduring challenge of ensuring software reliability has been fundamentally reshaped by the rapid integration of Large Language Models (LLMs) into the software development lifecycle (SDLC) since 2022. This paradigm shift has become even more pronounced with the introduction of autonomous agentic systems. However, this drive to automated code production and validation systems has led to a profound dichotomy: on one hand, AI-driven tools offer unprecedented acceleration and productivity gains; on the other, they introduce a new class of systemic reliability and security challenges that undermine traditional software assurance practices. Indeed, recent advances in large language models (LLMs) have transformed Software Engineering (SE), from code authoring and repair to autonomous, agentic systems that plan, execute, and verify multi-step development workflows. While productivity gains are well-documented, the reliability of AI-assisted and AI-automated software processes is more complex. By nature and historically, reliability spans technical, organizational, and regulatory dimensions, many of which have not been adequately addressed in this brave new world of AI. This book chapter offers a comprehensive analysis of the empirical evidence for both productivity gains and emergent risks. Further, it highlights a significant *productivity paradox* and trade-offs that are dependent on developer experience and task complexity. In addition, we investigate a novel taxonomy of AI-induced failure modes, including specification drift, pervasive non-determinism, security regressions, and evaluation contamination, which threaten the validity of performance claims. To address some of these concerns, we provide a structured overview of emerging engineering controls, such as advanced reasoning frameworks, verifier-in-the-loop patterns, and observability pipelines, designed to impose reliability on these inherently probabilistic systems. Additionally, we provide a comparative analysis of state-of-the-art agentic software engineering systems

K. Nwodo · A. Stavrou (✉)
Virginia Tech, Alexandria, VA, USA
e-mail: angelos@vt.edu

K. Nwodo
e-mail: nwodok@vt.edu

A. Stavrou
A2 Labs, Arlington, VA, USA

A. Stavrou et al. (eds.), *Generative and Agentic AI Reliability: Architectures, Challenges, and Trust for Autonomous Systems*, Studies in Computational Intelligence 1272, https://doi.org/10.1007/978-3-032-18585-3_9

that have emerged over the past few years. We posit that while AI acts as a powerful accelerator for many software development tasks, its integration necessitates a fundamental rethinking of software and system reliability: the focus of Software Reliability Engineering (SRE) must evolve from analyzing static, human-authored artifacts to managing the dynamic, unpredictable processes of the AI agents that create them.

1 Introduction

1.1 The Enduring Quest for Software Reliability: A Historical Perspective

The concept of reliability as a formal engineering discipline has its roots in the early 20th century, long before the advent of software. In the 1920s and 1930s, pioneers like Walter A. Shewhart at Bell Labs promoted product improvement through statistical process control, establishing a quantitative basis for quality and repeatability (Saleh and Marais 2006; Advisory Group on the Reliability of Electronic Equipment (AGREE) 1957). The term *reliability* itself, first attested in 1816, was formalized during World War II by the U.S. military to characterize a product that would operate when expected for a specified duration. This formalization was driven by the acute unreliability of early electronic equipment, particularly the vacuum tubes used in radar systems, which were costly and prone to failure (Advisory Group on the Reliability of Electronic Equipment (AGREE) 1957). The post-war era saw the establishment of key institutions, such as the IEEE Reliability Society in 1948, and the development of foundational statistical models for predicting failure rates (Saleh and Marais 2006; Advisory Group on the Reliability of Electronic Equipment (AGREE) 1957). The standard definition of reliability emerged from this period: "the probability of failure-free software operation for a specified period of time in a specified environment" (Cusick 2019).

This hardware-centric discipline was forced to adapt with the rise of digital computing and the onset of the *software crisis* of the 1960s, 1970s, and 1980s. As software systems grew in complexity, development projects were chronically plagued by budget overruns, schedule delays, and, most critically, unreliability (Ensmenger and Aspray 2002). Unlike hardware, which fails due to physical wear, software failures stem from latent design defects. These failures could have catastrophic consequences, as tragically demonstrated by the race condition bug in the Therac-25 radiation therapy machine, which led to patient deaths (Leveson and Turner 1993). This crisis catalyzed the birth of software engineering as a formal discipline, aimed at introducing rigor, process, and predictability into the art of programming (Leveson and Turner 1993; Randell 1979).

Within this new discipline, Software Reliability Engineering (SRE) emerged as a specialized field dedicated to the quantitative study of software's operational behavior (Lyu 2007). The 1970s saw the publication of the first software reliability models by figures like Jelinski and Moranda, Shooman, and Musa, which attempted to mathematically model the process of fault removal during testing (Musa et al. 1987). Over the subsequent decades, SRE was formalized through best practices and standards, such as IEEE 1633, which provides a recommended life-cycle approach to assessing and predicting software reliability (IEEE Reliability Society 2016). The core techniques of SRE were categorized into four areas: fault prevention (avoiding errors by construction), fault removal (detecting and eliminating errors through testing and validation), fault tolerance (providing continued service despite failures, often through redundancy), and fault/failure forecasting (estimating the presence of faults and the likelihood of failures) (Lyu 2007).

1.2 The Paradigm Shift: LLMs and Agentic AI in Software Engineering (2022–2025)

For half a century, the principles of SRE were predicated on a stable foundation: software was authored by humans, and its failures, while complex, stemmed from human error in design and implementation, reflecting the classic *garbage in, garbage out* notion. This foundation was irrevocably fractured in the period following 2022. The public release of ChatGPT in November of that year marked a watershed moment, rapidly changing the landscape of both research and practice in software engineering. This event triggered an explosive integration of Large Language Models (LLMs) into every phase of the developer workflow.

The initial wave took the form of generic code assistants, such as GitHub Copilot and Amazon CodeWhisperer, which augmented developers' work by providing real-time code suggestions (Amazon Web Services 2022). However, the technology quickly evolved beyond simple completion. By 2024 and 2025, the field had shifted towards fully autonomous, agentic systems designed to undertake complex, end-to-end software engineering tasks with minimal human intervention (Yang et al. 2024). This rapid advancement was propelled by several key technological drivers. First, the continued scaling of autoregressive LLMs (AR-LLMs) provided the raw capability for code generation and natural language understanding. Second, new model architectures began to emerge, such as Diffusion LLMs (DLLMs), which demonstrated superior performance on cross-file software engineering tasks by leveraging a global, parallel denoising mechanism instead of linear, token-by-token generation. Third, and perhaps most importantly, the development of sophisticated reasoning and planning frameworks provided the cognitive scaffolding for these new agents. Techniques like Reflexion, which enables agents to learn from past failures, and Tree-of-Thoughts, which allows for deliberate exploration of multiple solution paths, gave LLMs the ability to strategize and self-correct, moving them from mere generators to problem-solvers (Shinn et al. 2023; Yao et al. 2023; Wang et al. 2023a).

The integration of AI capabilities, specifically agentic AI, across the software engineering lifecycle is summarized in Fig. 1. The framework is primarily based

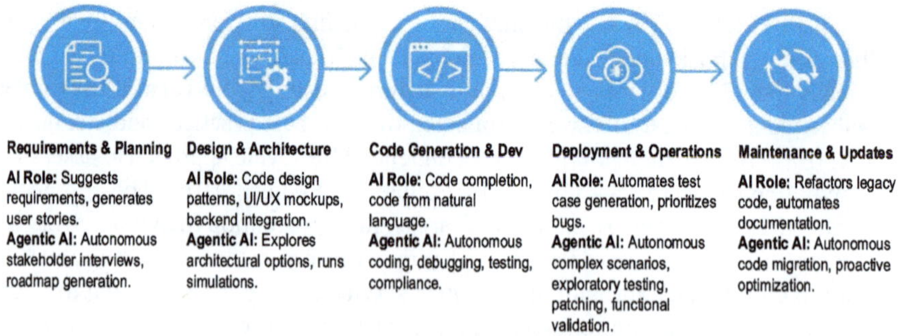

Fig. 1 Roles of AI across the software engineering lifecycle

on Alenezi and Akour (2025), which describes how AI methods are applied across the main development phases. Broader discussions of this evolution, including the usage of AI in software architecture and its influence on human-AI collaboration in code development, are presented in Bucaioni et al. (2025) and Sergeyuk et al. (2025). Collectively, these studies emphasize that AI systems have moved beyond isolated automation and code suggestion to become embedded across the entire software engineering pipeline. The figure further differentiates assistive functions, such as code completion, test generation, and refactoring, from agentic behaviors that perform multi-step actions, including running simulations, autonomous debugging and compliance checking, and proactive code optimization during maintenance.

1.3 The Reliability Dichotomy: Productivity Versus Predictability

The integration of this powerful new technology into software engineering has created a fundamental dichotomy, a central tension that defines the current era. On one side lies a surge in productivity and development velocity. A wealth of empirical evidence, gathered from both controlled experiments and large-scale industry case studies, demonstrates that AI-powered tools can make developers significantly faster, enabling them to complete more tasks, write more code, and iterate more quickly (Cui et al. 2025; Peng et al. 2023; Jimenez et al. 2024). This promise of accelerated delivery and enhanced efficiency is the primary driver of the technology's rapid and widespread adoption.

On the other side of this dichotomy lies a significant erosion of the traditional pillars of software reliability: predictability, verifiability, and reproducibility. The core technologies driving the productivity gains are inherently probabilistic. The non-determinism of LLMs means that the same input can yield different outputs, making consistent testing and debugging exceptionally difficult (Ouyang et al. 2025). The opaque nature of these models makes their reasoning processes difficult to inspect,

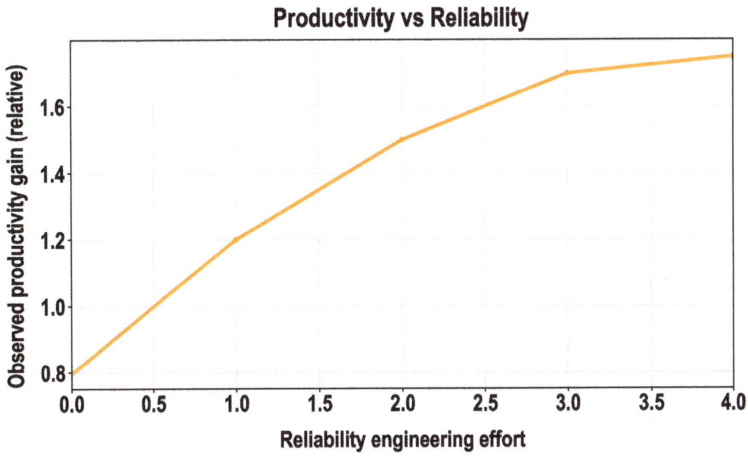

Fig. 2 Relationship between productivity gains and reliability engineering efforts in AI-assisted software development

challenging principles of accountability and transparency. Furthermore, their propensity to hallucinate or generate subtly incorrect outputs introduces a new and insidious class of risks. Studies have consistently shown that AI-generated code is frequently insecure, buggy, and may not adhere to best practices, directly threatening the reliability and security of the systems it is used to build (Pearce et al. 2022; Shukla et al. 2025).

The relationship illustrated in Fig. 2 demonstrates that increases in productivity from AI-assisted development require corresponding effort in reliability engineering. This tension between productivity and predictability forms the central challenge for modern Software Reliability Engineering. The very tools that promise to solve the decades-old problem of slow software development simultaneously introduce a new, and perhaps more profound, set of threats to software quality and trustworthiness. This report seeks to dissect this dichotomy, providing a formal review of the capabilities, risks, and controls that define this new paradigm.

2 Background and Related Work

2.1 Foundational Concepts in Software Reliability Engineering (SRE)

Software Reliability Engineering (SRE) is the quantitative study of the operational behavior of software-based systems with respect to user requirements concerning reliability (Lyu 2007). While its theoretical foundations are derived from hardware reliability, the application to software requires acknowledging fundamental differences. Hardware reliability is often concerned with physical degradation and manufacturing defects, where failures from wear-out can be modeled statistically (Advisory Group on the Reliability of Electronic Equipment (AGREE) 1957). In contrast, software does not wear out; its failures are the manifestation of latent design faults, such as bugs, introduced by humans during development (Lyu 2007). Consequently, software failures are systematic, not random, and seldom repeat in precisely the same manner, making traditional hardware analysis techniques like Failure Mode and Effect Analysis (FMEA) difficult to apply directly. A core tenet of SRE is measuring reliability against the system's intended function as defined by its specified design limits (Pham 2006).

To manage these unique challenges, SRE has historically been structured around four technical pillars that address faults throughout the software lifecycle (Lyu 2007):

1. **Fault Prevention**: Aims to avoid the introduction of faults by construction. This includes the use of formal methods to mathematically prove program correctness, disciplined design methodologies, and systematic software reuse to leverage pre-validated components.
2. **Fault Removal**: Focuses on detecting and eliminating existing faults through verification and validation. The primary techniques in this category are software testing, which executes the code to find failures, and software inspection, which involves systematic human review of code and other artifacts.
3. **Fault Tolerance**: Provides service that complies with the specification despite the occurrence of faults. This is achieved through redundancy, such as N-version programming or recovery blocks, and serves as the last line of defense against faults that were not prevented or removed.
4. **Fault/Failure Forecasting**: Involves estimating the presence of faults and predicting the occurrence and consequences of failures. This area has been the main focus of software reliability modeling, particularly through the use of Software Reliability Growth Models (SRGMs), which track the rate of failure detection over time to predict future reliability.

The discipline has been formalized through standards, most notably the IEEE 1633 *Recommended Practice for Software Reliability* (IEEE Reliability Society 2016). This standard prescribes a life-cycle approach, providing a framework for applying software reliability measurement, collecting necessary data, and building consistent methods for assessing and predicting reliability across a project (IEEE Reliability Society 2016). Seminal works, such as the *Handbook of Software Reliability Engineering* edited by Michael R. Lyu, have consolidated the field's knowledge, covering topics from measurement and modeling to testing and fault tolerance (Lyu 1996).

2.2 *Early Evidence of AI Impact: LLM-Based Productivity Studies (2022–2023)*

The period immediately following the widespread availability of powerful LLMs was characterized by a surge of empirical research aimed at quantifying their impact on developer productivity. These early studies provided the first concrete evidence that AI code assistants could deliver substantial performance improvements.

A landmark study in this area was the controlled experiment conducted by Peng et al. (2023). In this randomized controlled trial (RCT), professional software developers were tasked with implementing an HTTP server in JavaScript. The treatment group, which had access to GitHub Copilot, completed the task 55.8% faster on average than the control group working without AI assistance (Peng et al. 2023). This dramatic result provided strong initial validation for the productivity claims surrounding AI pair programmers.

Other studies from this period corroborated these findings while adding important nuance. Researchers began to observe shifts in how developers allocated their time. Bird et al. (2023) noted that developers spent a significant amount of time assessing and validating the suggestions generated by Copilot, indicating that the cognitive workload was shifting from code generation to code review. Similarly, Imai (2022) found that while Copilot could increase the volume of code produced, the quality was sometimes lower, requiring more code to be removed during testing.

A consistent and significant finding across these early studies was the heterogeneous effect of AI assistance based on developer experience. Multiple reports found that junior or less tenured developers experienced disproportionately larger productivity gains compared to their senior counterparts. For example, one enterprise study reported productivity boosts of 21–40% for junior developers, while senior developers saw more modest gains of 7–16% (Cui et al. 2025). This suggested that AI assistants were particularly effective at bridging knowledge gaps, automating routine tasks, and accelerating the onboarding process for those less familiar with a codebase or language.

2.3 *The Rise of Agentic SE: SWE-Bench, SWE-Agent, and Advanced Reasoning*

While the initial wave of AI tools focused on augmenting human developers with code suggestions, the subsequent phase of innovation shifted towards creating autonomous agents capable of handling entire software engineering tasks. This ambition required two critical components: a realistic environment to test such agents and a cognitive architecture to enable them to reason and act within that environment.

The first component was delivered with the introduction of SWE-bench, a benchmark designed to evaluate LMs in a realistic software engineering setting (Jimenez et al. 2024). Its construction marked a significant departure from prior code generation benchmarks, which often focused on self-contained, algorithmic problems. SWE-bench was built by mining 2294 real-world software engineering problems from issues and corresponding pull requests across 12 popular open-source Python repositories like `django`, `matplotlib`, and `scikit-learn`. The task formulation is uniquely challenging: given only a natural language issue description and access to the entire repository, the AI system must generate a code patch that resolves the issue. The evaluation is execution-based and robust; a solution is considered successful only if the generated patch applies correctly and passes a specific set of fail-to-pass tests, namely unit tests that failed before the fix and passed after, ensuring the agent has correctly addressed the problem (Jimenez et al. 2024).

The complexity of SWE-bench spurred the development of SWE-agent, an agentic system explicitly designed to tackle it (Yang et al. 2024). The core innovation of SWE-agent is its Agent-Computer Interface (ACI), a specialized interface designed to bridge the gap between an LLM's natural language processing capabilities and the structured, complex environment of a software repository. Recognizing that a standard Linux shell provides an action space that is too granular and complex for an LLM to use effectively, the ACI offers a small, simplified set of commands tailored for software engineering tasks. These include commands for navigating the file system (`ls`)), searching for files (`find_file`), and content (`|search_file|`), along with a simplified file editor that allows viewing and modifying code in a structured manner. This abstraction layer proved highly effective, enabling an agent powered by GPT-4 Turbo to successfully solve 12.47% of SWE-bench tasks, a substantial improvement over prior non-interactive approaches (Yang et al. 2024).

In parallel with these developments in environments and interfaces, new frameworks for agentic reasoning emerged, providing the brains for these systems. These frameworks moved beyond the simple, linear chain-of-thought prompting to enable more sophisticated, human-like problem-solving:

- **Reflexion**: Proposed by Shinn et al. (2023), Reflexion is a framework for reinforcing agents through *verbal reinforcement*. After a failed attempt (an *episode*), the agent is prompted to reflect on the feedback (e.g., a compiler error or failed test output) and generate a short textual summary of what went wrong and how to avoid it in the future. This self-reflection is stored in an episodic memory buffer and provided as context for the next attempt, allowing the agent to learn from its mistakes without updating model weights.
- **Tree-of-Thoughts (ToT)**: This framework, proposed by Yao et al. (2023), generalizes chain-of-thought by allowing an LLM to explore multiple reasoning paths in parallel. It maintains a *tree of thoughts*, where each thought is an intermediate step in a problem-solving process. The LLM can generate multiple potential next steps from any given thought and use a self-evaluation prompt to assess the value of different paths, enabling deliberate search with lookahead and backtracking.
- **Voyager**: While developed in the context of the game Minecraft, the Voyager agent introduced key architectural principles for autonomous, lifelong learning agents (Wang et al. 2023a). Its design features an *automatic curriculum* that prompts the LLM to propose novel exploration tasks based on the agent's current state and

knowledge, and an ever-growing *skill library* where successful behaviors are stored as executable code and retrieved for use in future, similar situations. This demonstrates a powerful model for compositional learning and continuous improvement in an open-ended environment.

The development of SRE has historically focused on analyzing and predicting failures in code written by humans. The advent of agentic AI represents a paradigm inversion. The primary object of reliability analysis is no longer just the static software artifact but the dynamic, autonomous process that *generates* the artifact. Traditional SRE methods were designed to find and model failures in human-written code after the fact. In contrast, systems like SWE-agent, guided by reasoning frameworks like Reflexion, are not static products but active processes. Consequently, the focus of reliability engineering must shift. It is no longer sufficient to ask, "Is this code reliable?" Instead, the critical question becomes, "Is this code-generating agent reliable?" This requires a move from product-centric to process-centric assurance, where the agent's internal state, its reasoning patterns, its interaction with the environment, and its inherent non-determinism become the primary subjects of reliability analysis.

2.4 Governance and Risk Management: NIST AI RMF and the EU AI Act

The rapid proliferation of AI technologies in high-stakes domains, including software engineering, has prompted the development of new governance and regulatory frameworks aimed at managing their associated risks. The timeline in Fig. 3 shows the key policy developments that define the current landscape of AI governance,

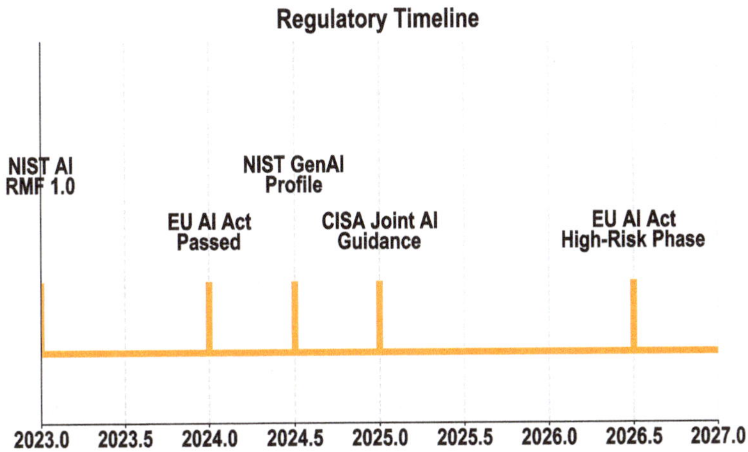

Fig. 3 Timeline of major AI governance and regulatory milestones from 2023 to 2026

providing context for the two principal initiatives: the NIST AI Risk Management Framework in the United States and the EU AI Act. Together, they represent complementary approaches to AI risk management.

The **NIST AI Risk Management Framework (AI RMF 1.0)**, released by the U.S. National Institute of Standards and Technology in January 2023, is a voluntary framework designed to help organizations manage the risks of AI and promote trustworthy and responsible development (National Institute of Standards and Technology 2024; AI 2023). It is intended to be practical, adaptable, and non-sector-specific. The framework is organized around four core functions:

1. **Govern**: This is a cross-cutting function that establishes a culture of risk management. It involves creating policies, processes, and assigning roles and responsibilities for AI risk management throughout the organization.
2. **Map**: This function focuses on establishing the context for AI risks. It involves identifying the potential positive and negative impacts of an AI system on individuals, organizations, and society.
3. **Measure**: This function involves developing and applying methods to analyze, assess, and monitor AI risks and their impacts using quantitative and qualitative tools.
4. **Manage**: This function addresses the identified risks. Based on the mapping and measurement, risks are prioritized and treated (e.g., mitigated, transferred, or accepted).

The AI RMF also defines seven key characteristics of trustworthy AI systems: valid and reliable; safe; secure and resilient; accountable and transparent; explainable and interpretable; privacy-enhanced; and fair with harmful bias managed (AI 2023). Recognizing the unique challenges posed by newer models, NIST released a companion **Generative AI Profile** in July 2024 to provide more specific guidance and actions for managing the risks of generative AI systems (National Institute of Standards and Technology 2024).

In contrast to NIST's voluntary framework, the **European Union's AI Act**, which was published in the Official Journal in July 2024, is a landmark piece of legally binding legislation that establishes harmonized rules for AI across the EU (The European Parliament and The Council of the European Union 2024). The Act follows a risk-based approach, classifying AI systems into four tiers (Future of Life Institute 2024; The European Parliament and The Council of the European Union 2024):

- **Unacceptable Risk**: AI practices that are considered a clear threat to the safety, livelihoods, and rights of people are banned (e.g., social scoring by governments).
- **High Risk**: AI systems used in critical domains that could adversely affect safety or fundamental rights. This includes systems used in critical infrastructure, education, employment, law enforcement, and as safety components of products (The European Parliament and The Council of the European Union 2024). Software development tools themselves could fall into this category if they are used to build high-risk systems.

- **Transparency Risk**: Systems that require specific transparency obligations, such as chatbots or systems that generate deepfakes, which must disclose that the content is AI-generated.
- **Minimal or No Risk**: The vast majority of AI systems, which fall into this category, are not subject to specific legal obligations.

For high-risk AI systems, the Act imposes a set of stringent requirements that must be met before they can be placed on the market. These obligations include implementing a robust risk management system, ensuring high-quality data governance, maintaining detailed technical documentation, enabling effective human oversight, and ensuring a high level of accuracy, robustness, and cybersecurity (Future of Life Institute 2024; The European Parliament and The Council of the European Union 2024). These legal requirements create a strong impetus for adopting the types of engineering controls and reliability practices discussed later in this report.

3 Evidence of Capability and Risk

The period from 2022 to 2025 has produced a rich and often contradictory body of empirical evidence regarding the impact of AI on software engineering. While early reports were overwhelmingly positive about productivity, more recent and nuanced studies have revealed a complex trade-off between speed and reliability. This section synthesizes the quantitative evidence on productivity, analyzes the pervasive security risks of AI-generated code, tracks performance on standardized benchmarks, and examines adoption patterns in major technology companies.

3.1 Quantitative Analysis of Developer Productivity Gains and Losses (2022–2025)

The initial promise of AI in software development was a dramatic increase in developer productivity, and several studies have provided quantitative support for this claim. A large-scale study of developers using GitHub Copilot in enterprise settings found that users completed 26% more tasks and produced 13.5% more weekly code commits on average. This acceleration was not uniform across all developers. The same study revealed an outsized benefit for less experienced developers, who saw productivity boosts ranging from 21% to 40%, compared to more modest gains of 7% to 16% for their senior counterparts (Cui et al. 2025). This suggests that AI assistants are particularly effective at lowering the barrier to entry and accelerating the learning curve for developers who are new to a project or technology.

As shown in Fig. 4, AI assistance is estimated to substantially accelerate the relative time required to complete key code generation and documentation tasks. The four categories include unit test generation, API integration, algorithm prototyping,

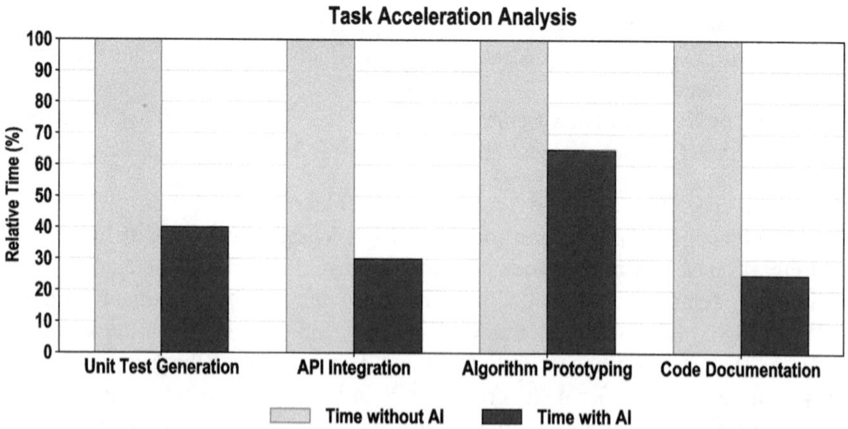

Fig. 4 Relative task acceleration for executing tasks with AI

and code documentation, which present core components of the software development process. The results presented in the figure are informed by empirical trends observed in prior research (Noy and Zhang 2023; Ziegler et al. 2022; Kalliamvakou 2022; Peng et al. 2023). This visualization highlights that the largest acceleration gains occur in well-structured tasks where inputs, outputs, and syntax are clearly specified and validation is straightforward, such as test generation, API integration, and documentation.

However, more recent research has challenged this uniformly positive narrative, uncovering what can be described as a *productivity paradox*). A randomized controlled trial (RCT) conducted in 2025 with experienced open-source developers working on their own mature, high-quality repositories produced a startling result: allowing developers to use state-of-the-art AI tools (primarily Claude 3.5/3.7 Sonnet) *increased* the time required to complete tasks by an average of 19% (Becker et al. 2025). This finding is particularly striking because it stands in direct opposition to the developers' own perceptions; they expected AI to speed them up by 24% and, even after experiencing the slowdown, still believed it had made them 20% faster (Becker et al. 2025).

The existence of such contradictory findings points to the highly contextual nature of *productivity*. The initial studies showing large gains often used self-contained, well-defined tasks where the primary bottleneck was the initial act of code generation. In this scenario, AI excels. The 2025 RCT, however, used real-world issues in complex, existing codebases, where the task involves not just generation but also deep contextual understanding, debugging, ensuring no regressions, and adhering to high quality standards. For experienced developers in this setting, the time saved on typing appears to be outweighed by a *verification and integration tax*; the cognitive overhead and time required to critically evaluate, debug, and safely integrate subtly flawed or stylistically inconsistent AI suggestions into a mature system. This suggests that while AI may be a powerful tool for scaffolding and boilerplate gen-

Table 1 Summary of empirical developer productivity studies (2022–2025)

Study (Author, Year)	AI tool(s)	Task description	Participant profile	Key metric	Reported impact	Notes/nuances
Peng et al. (2023)	GitHub Copilot	Implement HTTP server	95 prof. developers	Task time	**55.8% faster**	Greenfield task
Cui et al. (2025)	GitHub Copilot	Real-world tasks	Enterprise developers	Completed tasks	**26% increase**	Outsized junior benefit
Becker et al. (2025)	Claude 3.5/3.7	Real OSS issues	16 exp. OS developers	Task time	**19% slower**	Complex tasks

The use of bold highlights the reported impact of each study

eration, its utility diminishes, and can even become negative, as task complexity and the required quality bar increase. Further research has begun to investigate this cognitive load directly, proposing controlled studies that combine physiological measurements such as electroencephalography and eye tracking with interaction data to better understand the mental demands of programming with AI assistance (Haque et al. 2025).

Table 1 provides a summary of the key empirical studies discussed in this subsection.

3.2 *Empirical Vulnerability Assessments of AI-Generated Code*

While the debate on productivity continues, the evidence regarding the security of AI-generated code is far more conclusive and alarming. A consistent body of research demonstrates that code produced by LLMs frequently contains security vulnerabilities, posing a significant reliability risk.

One of the first and most widely cited studies in this area was conducted by Pearce et al. (2022). Their empirical evaluation of GitHub Copilot found that approximately 40% of the generated programs contained security vulnerabilities cataloged in the MITRE Common Weakness Enumeration (CWE). The study also revealed that vulnerability rates were language-dependent, with memory-unsafe languages like C producing a higher proportion of insecure code (~50%) compared to memory-safe languages like Python (~39%).

Subsequent research has consistently reinforced and expanded upon these initial findings. A user study by Perry et al. (2023) found not only that participants using AI assistants wrote "significantly less secure code" but also that they exhibited a dangerous "false sense of security" often rating their insecure solutions as secure. This points to a critical socio-technical risk where automation bias leads to reduced vigilance. A comprehensive systematic literature review by Negri-Ribalta et al. (2024) synthesized findings from 19 studies and confirmed a "high-level agreement that AI models do not produce safe code and do introduce vulnerabilities, despite mitigations". LLM-generated code often lacked basic defensive programming constructs and contained subtle but critical implementation flaws.

Perhaps the most counter-intuitive finding is the phenomenon of *security degradation through iteration*. A traditional assumption in software development is that iterative refinement improves quality. However, a 2025 controlled experiment by Shukla and Joshi demonstrated that this assumption does not hold for AI-based refinement. Their study, which subjected initially secure code to multiple rounds of AI-driven *improvements* found a 37.6% increase in critical vulnerabilities after five iterations (Shukla et al. 2025). This suggests a security-iteration inversion: each request for a functional or performance enhancement provides a new opportunity for the model to introduce a vulnerability, causing the security posture of the code to decay over time. The types of vulnerabilities commonly introduced are varied and severe, including classic memory safety issues (buffer overflows), input validation errors, injection flaws (SQL, command), and cryptographic implementation errors (Shukla et al. 2025). This body of evidence indicates that relying on AI for code generation without robust, independent verification processes is a high-risk proposition that can easily lead to significant security regressions.

Figure 5 illustrates the overall risk profile of AI-generated code across five aspects of software quality: correctness, integrity, security, maintainability, and test gaps. Each area represents a specific dimension of software reliability. Correctness and integrity risks are supported by Noy and Zhang (2023), which showed that AI-generated code often requires substantial human oversight to ensure logical and completely correct code. Security risks are informed by Pearce et al. (2022), which demonstrates the importance of pairing AI-assisted development with security analysis tools to minimize the risk of introducing vulnerabilities. The figure further indicates that maintainability exhibits a comparably high risk level, reflecting the difficulties in fixing and refactoring generated code once it has been integrated into larger systems. Overall, correctness, maintainability, and security emerge as the most critical technical concerns, reinforcing the need for systematic validation and a human oversight in the adoption of AI-generated code.

3.3 *Performance Analysis on Standardized Benchmarks: SWE-Bench and SWE-Bench Verified*

The SWE-bench benchmark has become the de facto standard for evaluating the end-to-end problem-solving capabilities of software engineering agents. Performance on this benchmark has seen a dramatic and rapid evolution. Initial attempts using non-interactive LLMs struggled, with the baseline reported in the original SWE-bench paper achieving a resolve rate of only 1.96% (Jimenez et al. 2024). The first major

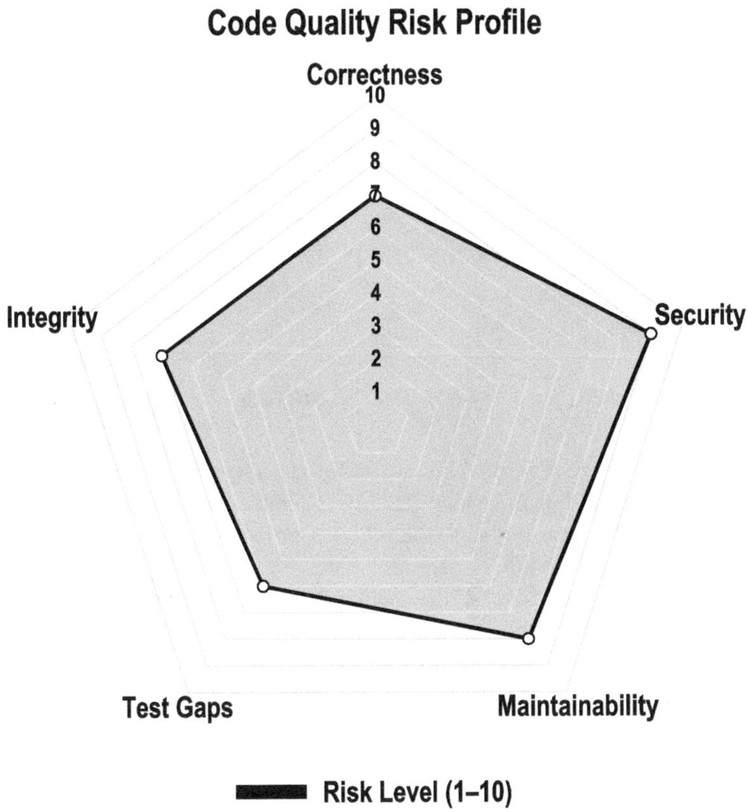

Fig. 5 Risk profile of AI-generated code

breakthrough came with the introduction of SWE-agent, whose Agent-Computer Interface (ACI) allowed an agent powered by GPT-4 Turbo to achieve a resolve rate of 12.47% (Yang et al. 2024).

However, subsequent analysis revealed that the original SWE-bench dataset contained problematic instances, including underspecified issue descriptions and unreliable development environments, which systematically underestimated model capabilities. To address this, the creators of SWE-bench collaborated with OpenAI to release SWE-bench Verified, a human-curated subset of 500 tasks verified to be well-defined and solvable (OpenAI 2024). This cleaner benchmark has become the primary leaderboard for tracking state-of-the-art performance.

On SWE-bench Verified, performance has escalated at an astonishing pace throughout 2025. As of September 2025, top proprietary models demonstrated remarkable capabilities when paired with sophisticated agentic scaffolds. An evaluation by Vals.ai reported that an agent using GPT-5 achieved a resolve rate of 68.8%, while another using Claude Sonnet 4 (Nonthinking) reached 65.0% (Valsai 2025). These results represent a greater than 30-fold improvement over the initial 2023 baselines. However, performance is highly dependent on task complexity. For the simplest tasks (estimated fix time < 15 minutes), top models achieve over 80% accuracy. On the most challenging tasks (> 4 h), their performance declines, with only a few models achieving non-zero scores, typically around 33.3% (Valsai 2025). This suggests that while agents have become proficient at resolving moderately complex, localized issues, they continue to struggle with in-depth, architectural problems. A diverse ecosystem of open-source agentic approaches, including *Agentless* and *Moatless* has also been evaluated, showcasing a variety of strategies from multi-phase pipelines to Monte Carlo tree search.

3.4 Cross-Language Performance Analysis

The reliability and security of AI-generated code vary across programming languages. The choice of language significantly influences the likelihood of introducing vulnerabilities, primarily due to the inherent properties of the language and the distribution of code in the models' training data.

As highlighted in the vulnerability assessments, memory-unsafe languages like C and C++ are particularly problematic. The need for manual memory management introduces opportunities for classic vulnerabilities like buffer overflows and use-after-free errors, which LLMs frequently generate (Shukla et al. 2025). One study found that the vulnerability rate for AI-generated C code was significantly higher than for Python, a memory-safe language.

Furthermore, research indicates that LLMs often fail to utilize modern language features and best practices, instead suggesting outdated or deprecated methods. A 2025 study by Kharma et al. (2025) found that many models do not leverage modern security features available in recent compiler and toolkit updates, such as those in Java 17. The same study noted that outdated and less secure methods are still commonly used in generated C++ code. This phenomenon is likely a direct consequence of the models' training data, which is a vast corpus of public code spanning decades. If older, insecure code patterns are more prevalent in this corpus than modern, secure alternatives, the model will learn and replicate the insecure patterns. This creates a risk of perpetuating legacy vulnerabilities and failing to adopt contemporary security standards.

3.5 Comparative Industry Adoption: Case Studies from Microsoft, Google, and Meta

The trends observed in academic research are mirrored and amplified within the internal software development practices of major technology companies, which serve as both developers and large-scale consumers of these AI tools.

Microsoft, through its ownership of GitHub, has been at the forefront of this transformation. The company has driven widespread internal adoption of GitHub Copilot, with reports citing that the tool helps developers code up to 55% faster (Jason and Microsoft 2025). Microsoft's strategy extends beyond code completion to the development of a unified platform, Azure AI Foundry, for building, managing, and deploying AI agents. This initiative, termed *Agentic DevOps*, envisions AI agents working alongside human developers to streamline the entire software lifecycle, from fixing bugs and managing dependencies to documenting and maintaining legacy systems (Cyril and Microsoft Azure 2025).

Google has also deeply integrated AI into its internal engineering workflows. A 2024 report from its engineering division revealed that AI-based assistance now contributes roughly as many characters of code as those manually written by developers (Chandra and Tabachnyk 2024). A particularly insightful case study is Google's use of LLMs for large-scale internal code migrations, a notoriously difficult and time-consuming task. Their experience report details how a bespoke LLM-based system accelerated migration task completion by over 50% (Nikolov et al. 2025). However, they also emphasize that achieving this success required a hybrid approach, combining LLM suggestions with traditional Abstract Syntax Tree (AST)-based transformations and, crucially, mandatory manual review of all changes. This reinforces the finding that AI is a powerful accelerator but not yet a substitute for rigorous engineering oversight.

Meta has focused on applying AI to solve specific, challenging problems in software quality and reliability. A notable example is their development of the Automated Compliance Hardening (ACH) tool, which leverages LLMs to overcome the traditional scalability barriers of mutation testing (Mark and Meta Engineering 2025). By using an LLM to generate a small number of highly relevant *mutants* (small, deliberate changes to code) and corresponding tests, the ACH tool can automatically harden code against compliance and privacy regressions. This represents a sophisticated use of generative AI not just to write code, but to systematically improve its test coverage and reliability. Alongside these internal tools, Meta continues to be a major contributor to the open-source AI ecosystem through the development and release of its Llama family of models (Meta AI 2024).

4 Reliability Risks and Failure Modes in AI-Driven SE

The integration of generative and agentic AI into the software engineering lifecycle introduces a new and complex set of reliability risks. These failure modes are often qualitatively different from those found in traditional software systems, stemming from the probabilistic, opaque, and dynamic nature of the underlying models. To provide an overview of these emerging challenges, Table 2 summarizes the primary reliability failure modes identified in AI-enabled software engineering, together with representative mitigation strategies drawn from recent research. Building on this, the remainder of this section presents a systematic taxonomy of these novel risks.

Table 2 Reliability failure modes in AI-enabled SE and example mitigations

Failure mode	Description	Mitigation examples
Spec. drift	Misinterpreted requirements	NL → formal tests; SWE-bench Verified (OpenAI 2024)
Non-determinism	Variance in outputs	Self-consistency (Wang et al. 2023b), fixed seeds
Security regressions	Vulnerable code	SAST/DAST gates; Veracode (2025)
Eval contamination	Train/test leakage	Human-verified evals (OpenAI 2024)
Tool-use hazards	Destructive edits	ACI, SWE-agent (Yang et al. 2024)

4.1 Specification Drift

Specification drift refers to the divergence between the documented or intended behavior of a system and its actual implementation. AI systems introduce and exacerbate this problem in two primary forms: code-level drift and model-level drift.

Code-Versus-Specification Drift: The high velocity of code generation and modification enabled by AI tools can easily outpace manual updates to documentation. An AI agent might add a new parameter to an API endpoint or change a response format to fix a bug, but if the corresponding OpenAPI specification is not updated, a drift is created. This leads to unreliable APIs, frustrated developers, and broken integrations. To counter this, a new paradigm of *spec-driven development* is emerging, where the formal specification is treated as the single source of truth. In this model, AI agents use the spec to automatically generate and validate code, tests, and documentation, ensuring they remain synchronized.

Model Drift: For AI-powered features deployed within an application, the reliability of the feature is tied to the performance of the underlying machine learning model. This performance can degrade over time due to *model drift* (Jim et al. 2025). Model drift manifests in two main ways:

- **Concept Drift**: This occurs when the statistical properties of the target variable change. The relationship between inputs and the correct output evolves. For example, in a code vulnerability detection model, the patterns defining a *vulnerability* may change as new exploit techniques emerge.
- **Data Drift**: This occurs when the statistical properties of the model's input data change. For example, a code completion model trained primarily on Python 2 code may exhibit degraded performance when used in a modern Python 3.12 codebase.

Both forms of drift lead to a gradual erosion of the AI component's reliability, causing it to produce increasingly inaccurate or irrelevant outputs. Mitigating model drift requires continuous monitoring of data distributions and model performance, coupled with periodic retraining or adaptation.

4.2 *Non-determinism: The Challenge of Reproducibility and Predictability*

A fundamental characteristic of LLMs that poses a profound challenge to software reliability is their inherent non-determinism. These models are probabilistic; they generate outputs by sampling from a probability distribution over a vocabulary of tokens. As a result, even when given the exact same input prompt, an LLM can produce different outputs on subsequent runs.

This non-determinism persists even when attempts are made to enforce deterministic behavior, such as setting the model's sampling *temperature* parameter to zero. An empirical study on ChatGPT's non-determinism in code generation confirmed that setting *temperature* to 0 reduces but does not eliminate variability (Ouyang et al. 2025). The remaining variance can be attributed to low-level implementation details, such as floating-point arithmetic differences across different GPU hardware or subtle, undocumented updates to a model's API endpoint. One study investigating this phenomenon found accuracy variations of up to 15% across identical runs under supposedly deterministic settings (Atil et al. 2024).

The implications of non-determinism for software engineering reliability are severe and systemic:

- **Unreliable Verification**: A unit test for an AI-generated function might pass on one execution but fail on the next, not because of a change in the test or the code, but because the underlying AI model produced a different, incorrect output. This makes regression testing and continuous integration fundamentally unreliable.
- **Intractable Debugging**: The cornerstone of debugging is the ability to reliably reproduce a failure. If a bug caused by an AI component's output cannot be deterministically reproduced, isolating and fixing the root cause becomes nearly impossible.
- **Cascading Errors**: In agentic systems that chain multiple LLM calls to perform a complex task, the non-determinism is amplified. A small, random variation in an early step of the agent's reasoning process can cascade and lead to a completely different and potentially erroneous final outcome.

4.3 *Security Regressions and Emergent Vulnerabilities*

As detailed in Sect. 3.2, AI-generated code is frequently insecure. This risk can be framed as a specific and critical category of reliability failure. *Security regression* occurs when an AI-driven code modification either re-introduces a vulnerability that had been previously patched or, more commonly, introduces a new vulnerability while implementing a feature or fixing a different bug.

Beyond simply replicating insecure patterns from their training data, AI systems can introduce novel types of security risks:

- **Architectural Drift**: This subtle failure mode occurs when an AI agent makes a series of seemingly correct local changes that, in aggregate, violate a larger, implicit security invariant of the system. For example, an agent might refactor several methods for clarity, but in doing so, inadvertently bypass a critical access control check that was enforced by the previous call structure. These flaws are difficult for static analysis tools and human reviewers to detect because each individual change may appear correct in isolation.
- **Hallucinated Dependencies**: A particularly novel risk is the phenomenon of hallucinated or non-existent dependencies. An LLM might confidently generate code that includes an `import` statement for a library that does not exist in any public package repository. This creates a dangerous attack vector known as *slopsquatting*. An attacker monitoring for such hallucinations can quickly register the hallucinated package name and publish a malicious library under that name. A developer who trusts the AI's suggestion and installs the package via a package manager would then unknowingly execute the attacker's malicious code.

The severity of these risks has spurred research into AI-powered countermeasures. For instance, Google's CodeMender is an AI agent designed specifically to identify vulnerabilities in code and autonomously generate patches that fix them without introducing functional regressions (Raluca et al. 2025).

4.4 *Evaluation Contamination and the Illusion of Competence*

A meta-level risk that threatens the entire field's ability to accurately gauge progress in AI reliability is *evaluation contamination*. This occurs when the data from a benchmark used to evaluate models is inadvertently included in the massive web-scale corpora used to train those same models (Choi et al. 2025). When a model is trained on the questions and answers of a test, its performance on that test no longer measures its ability to reason or generalize; it measures its ability to memorize.

There is now compelling evidence that prominent software engineering benchmarks, including SWE-bench, are significantly contaminated. A 2024 study found that state-of-the-art models could achieve up to 76% accuracy in identifying the correct file to fix in a SWE-bench problem using only the issue description, a task that should be impossible without repository context, suggesting memorization of issue-file associations (Liang et al. 2025). A separate analysis found that 32.67% of successful patches generated for SWE-bench tasks involved direct *solution leakage*, where clues to the correct code were present in the problem description itself (Aleithan et al. 2024).

The direct consequence of contamination is an *illusion of competence*, where benchmark leaderboards show rapidly increasing scores that are not reflective of true, generalizable problem-solving ability. This creates a significant risk for the industry, as organizations may adopt and deploy systems based on inflated performance claims, only to find that they are brittle and unreliable when faced with novel, real-world problems that were not in their training data. In response, the research community has begun to develop contamination-resistant benchmarks, such as SWE-Bench Pro and SWE-bench-Live, which are constructed from private and proprietary codebases, as well as very recent open-source projects that are unlikely to have been part of any model's training set (Deng et al. 2025; Zhang et al. 2025).

4.5 Tool-Use Hazards in Autonomous Agents

When agentic systems are granted the ability to interact with their environment through actions such as executing shell commands, reading and writing files, or invoking external APIs, a new class of reliability and security hazards emerges. The agent's ability to act transforms it from a passive generator into an active participant in the system, dramatically expanding its potential for causing harm.

- **Expanded Attack Surface**: Each tool and API integration becomes a potential attack vector. Vulnerabilities in the agent's orchestration layer, the tools themselves, or the third-party services they connect to can be exploited to compromise the agent or the system it operates on. An agent with access to an unsecured code interpreter, for example, is exposed to arbitrary code execution attacks.
- **Tool Misuse and Goal Manipulation**: An adversary does not need to compromise the system directly if they can manipulate the agent into misusing its authorized tools. This is often achieved through *prompt injection*, where hidden instructions are embedded in data that the agent processes, tricking it into performing unintended actions. A recently identified vulnerability, *Tool Poisoning*, involves embedding malicious instructions directly into the documentation of a tool. An agent reading this documentation to learn how to use the tool will be tricked into executing the attacker's commands without the user's knowledge.
- **Cascading Failures**: In multi-agent systems, where agents collaborate on a task, errors can propagate and amplify. A single incorrect decision or hallucination by one agent can be passed as fact to another, leading to a cascade of failures that can be difficult to trace to the original fault. Similarly, an agent given broad permissions could create a denial-of-service condition by making excessive or resource-intensive tool calls.

4.6 Socio-Technical Paradoxes and Cognitive Biases

The reliability of AI-assisted software development is not purely a technical problem; it is a socio-technical one, deeply intertwined with the human-computer interaction between developers and AI tools. This interaction is subject to several paradoxes and cognitive biases that can undermine reliability.

The *Automation-Augmentation Paradox*, a term introduced by Raisch and Krakowski (2021), is particularly relevant. Drawing on Bainbridge's seminal discussion of the *Ironies of Automation* (Bainbridge 1983), this paradox can be expressed as follows: the more advanced and automated an AI system becomes, the more crucial and difficult the role of the human overseer becomes. As AI takes over more of the routine tasks, human developers are pushed into the role of monitoring for subtle, complex, and infrequent errors that the AI cannot handle. This requires high levels of vigilance and expertise, yet the very automation provided by the AI reduces the opportunities for developers to practice and maintain these critical skills.

This paradox is compounded by several cognitive biases that affect how developers interact with AI-generated code, some of which include:

- **Automation Bias**: This is the tendency to over-trust and uncritically accept information from automated systems. This bias can lead developers to accept incorrect or insecure AI-generated code without the rigorous scrutiny they would apply to code written by a human colleague. This is empirically supported by findings that developers using AI assistants exhibit a "false sense of security" (Shukla et al. 2025).
- **Confirmation Bias**: Developers may have a pre-existing idea of how to solve a problem. When presented with multiple suggestions from an AI, they may be predisposed to accept the one that most closely matches their initial, and potentially flawed, approach, while ignoring more innovative or correct alternatives (Chattopadhyay et al. 2022).
- **Anchoring Bias**: The first solution provided by an AI can act as a powerful cognitive anchor, narrowing the developer's problem-solving space and making it harder for them to consider fundamentally different approaches (Chattopadhyay et al. 2022).

These risks are not independent but form an interconnected cascade that can lead to system failure. An agent's non-deterministic nature makes comprehensive verification difficult. This lack of rigorous verification, in turn, allows security regressions and specification drift to go undetected. This technical vulnerability is then amplified by socio-technical factors, as developers, influenced by automation bias, over-trust the output of the non-deterministic system. A test that passes once might be accepted as sufficient proof of correctness. However, a subtle security flaw or a drift from the API specification may have been introduced. In a future deployment, a different execution path or a shift in production data could trigger this latent flaw, causing a production failure that is then nearly impossible to debug because the original passing test cannot be reliably reproduced. This chain reaction demonstrates that reliability engineering for AI systems must address the entire socio-technical system, not just isolated technical risks. Collectively, these risks illustrate the intertwined nature of reliability challenges in AI-driven systems. As a synthesis of the discussion above, Table 3 provides an overview of these failure modes along with their underlying causes.

Table 3 Taxonomy of AI-induced reliability risks and failure modes

Risk category	Description	Root cause(s)	Example failure scenario
Specification drift	Divergence between implemented and intended behavior	Rapid AI changes; model drift	API response changes, breaking clients
Non-determinism	Identical inputs produce different outputs	Probabilistic nature of LLMs	Intermittent test failures; impossible to debug
Security regression	AI changes introduce new vulnerabilities	Insecure patterns in training data	Agent removes input validation for optimization
Evaluation contamination	Benchmark data in training set inflates scores	Web-scale data scraping	Model memorizes SWE-bench, fails on novel tasks
Tool-use hazards	Risks from agents using external tools	Expanded attack surface; prompt injection	Agent tricked into exfiltrating data via a tool
Socio-technical paradoxes	Human-computer interaction patterns that undermine reliability	Cognitive biases (e.g., automation bias)	Developer accepts insecure code without scrutiny

5 Engineering Controls for Reliability

In response to the novel risks posed by AI-driven software development, a new suite of engineering controls and architectural patterns has begun to emerge. These techniques aim to impose structure, predictability, and safety on inherently probabilistic systems. They represent a shift in reliability strategy, moving from a primary focus on pre-deployment fault prevention to a paradigm of continuous, in-line verification, monitoring, and containment.

5.1 Advanced Prompting and Reasoning Frameworks

One of the first lines of defense against unreliable AI outputs is to improve the quality of the AI's reasoning process itself. Instead of relying on a single, direct response to a prompt, structured reasoning frameworks guide the model through a more deliberate and robust problem-solving process.

- **Self-consistency**: This technique directly combats the issue of non-determinism by leveraging it as a feature. Instead of generating a single output using greedy decoding, the model is prompted to generate multiple, diverse reasoning paths for the same problem. The final answer is then determined by a majority vote among the different outputs. This approach has been shown to significantly boost performance on tasks requiring precise reasoning (Wang et al. 2023b).

- **Tree-of-Thoughts (ToT)**: ToT provides a more structured approach to exploration. It guides the LLM to externalize its reasoning process in a tree structure, where each node represents a partial solution. The model can propose several potential next steps and self-evaluate the promise of each branch, allowing for deliberate search with lookahead and backtracking (Yao et al. 2023).
- **Reflexion**: This framework introduces an explicit learning loop. After an agent performs a task and receives feedback (e.g., a test failure), the framework prompts the model to generate a *verbal reinforcement* cue, a self-critique that explains the failure and suggests a strategy for the next attempt. This reflection is stored and used as context, making the agent less likely to repeat the same mistake (Shinn et al. 2023).

5.2 Human and Automated Oversight: The Verifier-in-the-Loop Pattern

Given that AI systems can produce incorrect outputs, a critical architectural pattern is the *Verifier-in-the-Loop*. This pattern places a verification step between the AI's generation and its final acceptance.

- **Human-in-the-Loop (HITL)**: For high-stakes applications, human oversight is indispensable. The HITL pattern integrates human experts directly into the AI workflow to review, validate, and correct outputs (GoogleCloud n.d.). The human's feedback can be used to improve the model, often through techniques like Reinforcement Learning from Human Feedback (RLHF) (Lambert 2025). This process is essential for ensuring accuracy and mitigating subtle ethical biases.
- **Automated Verifiers**: For scalability, automated systems are used as verifiers. Common verifiers in AI-assisted coding include:
  - **Compilers and Test Suites**: A compilation error or a failing test provides a clear, unambiguous signal that the generated code is incorrect (Shinn et al. 2023).
  - **Static and Security Analysis Tools**: Integrating SAST tools like SonarQube or Snyk provides a crucial check for security vulnerabilities and adherence to coding standards (James and SoftwareSeni 2025).
  - **LLM-Based Verifiers**: Using another LLM as a verifier. Meta's ACH tool exemplifies this, where one LLM generates *mutants* to test code, and another acts as an *Equivalence Detector* to filter useless mutants (Mark and Meta Engineering 2025).

5.3 Enforcing Correctness: Constrained Decoding and Tool Contracts

While verification checks for correctness after the fact, a more powerful set of controls can enforce correctness during the generation process itself.

- **Constrained Decoding**: This technique modifies the LLM's decoding process to ensure its output conforms to a predefined set of rules or a formal grammar. Instead of allowing the model to pick any token, it restricts the choices to only those that maintain validity (Beurer-Kellner et al. 2024). In code generation, constraints from the programming language's grammar can guarantee syntactic correctness, eliminating an entire class of errors.
- **Tool Contracts**: For agentic systems using external tools, defining and enforcing strict contracts is a critical control. Each tool's interface is defined by a formal schema (e.g., OpenAPI). Before a tool call is executed, its arguments are validated against this schema. If validation fails, the call is rejected, preventing misuse and providing structured feedback to the agent.

5.4 Observability and Reproducibility for Agentic Systems

Managing complex, non-deterministic agentic systems requires robust observability and logging.

- **AI Agent Observability**: This extends traditional software observability (metrics, events, logs, traces) to AI systems. It requires tracking AI-specific metrics like *token usage*, *inference latency*, *model drift*, and *response quality* (e.g., hallucination rate) (Gregg and Annie n.d.). End-to-end tracing is crucial for root-cause analysis in multi-step agentic workflows.
- **Reproducibility Pipelines**: While perfect reproducibility is challenging, engineering practices can maximize it. A reproducibility pipeline involves locking the model version, setting a fixed seed, containerizing the runtime environment, and meticulously logging all inputs and outputs for every LLM call (Ouyang et al. 2025).

5.5 RAG Hardening and Trust Frameworks for External Knowledge

Many agents rely on Retrieval-Augmented Generation (RAG) to ground their responses. The reliability of the system then depends on the RAG pipeline. *RAG hardening* involves a defense-in-depth strategy to secure this pipeline.

- **Hardening Techniques**: A robust RAG system requires controls at multiple stages:
  - **Input Validation**: Sanitizing user prompts to neutralize prompt injection attempts.
  - **Retrieval Hardening**: Limiting searches to trusted sources and applying access controls to ensure the agent only retrieves authorized data. Quality can be improved with techniques like hybrid search and re-ranking.
  - **Generation Guardrails**: Constraining the final generation step with policies that enforce source citation, prevent unsupported claims, and monitor for potential sensitive data leakage.
- **Trust Frameworks for RAG**: To systematically ensure reliability, researchers have proposed comprehensive *trust frameworks*. These frameworks assess the system across multiple dimensions, including *factuality*, *robustness*, *fairness*, *transparency*, *accountability*, and *privacy* (Zhou et al. 2024). Rigorous evaluation requires dedicated testing frameworks and automated metrics for context relevance, answer correctness, and faithfulness to the source material (Patronus AI 2025).

6 A Comparative Analysis of Agentic Systems for SE

The theoretical controls and architectural patterns for AI reliability find concrete expression in the design of specific agentic systems for software engineering. By comparing the designs of leading agents like SWE-agent, OpenHands, Voyager, and the Reflexion framework, we can identify common patterns, key trade-offs, and emerging best practices.

6.1 SWE-Agent and OpenHands

SWE-agent and OpenHands represent a class of agents designed to operate within a codebase to resolve software engineering tasks.

SWE-agent's Agent-Computer Interface provides a simplified, task-specific command set that abstracts away the complexity of a standard Linux shell (Yang et al. 2024). This interface enables the agent to interact with the codebase in a structured and controlled manner, supporting task execution while maintaining security through a sandboxed Docker environment.

OpenHands (formerly OpenDevin) is an open-source platform for generalist software development agents (All-Hands-AI 2025). Like SWE-agent, it provides its agent with tools for interacting with the environment, including a code editor, a terminal, and a web browser, all within a secure sandbox. OpenHands features its own ACI, `openhands-aci`, which provides interfaces for code editing, linting, and executing shell commands. The platform is model-agnostic and has shown competitive performance on SWE-bench.

6.2 *Voyager: Lessons from a Lifelong Learning Agent*

While not designed for SE, the Voyager agent, which operates in Minecraft, introduces architectural principles relevant to reliable SE agents (Wang et al. 2023a). Two of its core components are particularly instructive:

1. **Automatic Curriculum**: Voyager is not given a fixed sequence of tasks. Instead, it uses an LLM to propose new goals for itself based on its current state and knowledge, guided by the objective of maximizing exploration. An SE agent with a similar capability could autonomously identify and address areas of a codebase needing refactoring or improved test coverage.
2. **Skill Library**: When Voyager successfully completes a complex behavior, the sequence of steps is saved as an executable code function (a *skill*) in a vector database. When faced with a new task, it retrieves relevant skills to use as building blocks. This mechanism for compositional learning is directly applicable to SE, allowing an agent to build a library of reusable scripts for common tasks.

6.3 *Reflexion: The Self-improving Agent Architecture*

The Reflexion framework provides a concrete architectural pattern for creating agents that learn from their mistakes (Shinn et al. 2023). It operationalizes an iterative feedback loop that is more lightweight than traditional reinforcement learning.

The Reflexion architecture consists of three distinct roles, which can be implemented with separate LLM calls:

- **The Actor**: Attempts to solve the task by generating a plan and actions.
- **The Evaluator**: Assesses the outcome. In software engineering, this is often the external environment: the compiler, linter, and test suite.
- **The Self-reflection Model**: When the Evaluator signals a failure, this model analyzes the trajectory and the error, generating a concise, natural language summary of what went wrong and how to fix it. This *verbal reinforcement* is stored in the agent's memory.

On the next iteration, the Actor is provided with this self-reflection in its prompt, guiding it away from its previous mistake. This has been shown to be highly effective for code generation tasks on benchmarks like HumanEval (Chen et al. 2021; Shinn et al. 2023).

6.4 Architectural Designs, Interfaces, and Evaluation

Comparing these systems reveals several key trends. A clear divergence in design philosophies emerges, ranging from *single-agent systems* like the initial SWE-agent, to *multi-agent systems* where agents specialize in tasks like planning and reviewing, to *non-agentic, multi-phase pipelines* like the *Agentless* approach.

Regardless of the architecture, the interface between the agent's reasoning core and the environment remains a critical component. The success of the ACI concept pioneered by SWE-agent underscores the importance of creating an abstraction layer that matches the LLM's capabilities.

Finally, the evaluation of all these systems heavily relies on benchmarks, with SWE-bench serving as the common yardstick. While this provides a basis for comparison, the known issues with evaluation contamination (as discussed in Sect. 4.4) must be considered a significant caveat. The reported performance scores may not fully reflect the true generalizable capabilities of these agents on novel problems.

7 Open Challenges and Future Research Directions

Despite rapid progress, the field is still in its infancy. Achieving robust reliability requires addressing several fundamental open challenges.

7.1 Developing Contamination-Resistant and Realistic Benchmarks

The findings on evaluation contamination represent a threat to the empirical foundations of the field. The most urgent research direction is developing a new generation of benchmarks resistant to contamination.

- **Dynamic and** *Live* Benchmarks: Future benchmarks should be dynamic, continuously drawing new tasks from sources unlikely to be in training corpora. Systems like SWE-bench-Live, which aim to add new, verified issues monthly, are a promising step (Zhang et al. 2025).
- **Use of Private and Licensed Codebases**: Benchmarks like SWE-Bench Pro are pioneering the use of proprietary codebases and open-source projects with copyleft licenses (e.g., GPL) (Deng et al. 2025). The legal and access barriers reduce the likelihood of inclusion in training sets.
- **Beyond Functional Correctness**: Future benchmarks must also evolve to measure non-functional requirements critical for reliability, such as security, maintainability, and performance.

7.2 *Formal Verification and Predictability of Agentic Systems*

The inherent non-determinism of LLMs is a fundamental barrier to predictability. A major open research area is the application of formal methods to agentic systems.

- **Verifying the Agent's Scaffolding**: Can we formally verify the correctness of the agent's control loop and its tool-use contracts, even if the LLM remains a black box?
- **Bounded Verification**: Can we develop techniques to prove that an agent's actions will always remain within a *safe envelope*, even if we cannot predict its exact behavior?
- **Output Verification**: Research is needed into lightweight, runtime verification techniques that can check the output of an LLM against a formal specification before it is used.

7.3 *Scalable Human-in-the-Loop and Governance Models*

The Verifier-in-the-Loop pattern, particularly with a Human-in-the-Loop (HITL), is one of the most effective controls (GoogleCloud n.d.). However, requiring human review for every change is a bottleneck.

- **AI-Assisted HITL**: This involves developing AI tools that assist the human verifier, for example, by highlighting sections of a generated patch with the lowest confidence score or the highest risk.
- **Adaptive Governance**: Research is needed into dynamic governance models where the level of required human oversight is adapted based on the risk profile of the task. A low-risk change can proceed autonomously, while a high-risk would trigger mandatory human review.

7.4 *The Path Towards Fully Autonomous and Reliable Software Development*

The ultimate goal is the creation of fully autonomous *AI engineers*. Achieving this requires foundational research in several areas:

- **Autonomous Monitoring and Self-repair**: Future research should explore agents that can proactively monitor a running system, autonomously diagnose emergent issues, and then formulate and execute a plan to repair the system.
- **AI-Native SDLCs**: The current paradigm involves *bolting on* AI tools to an SDLC designed for humans. A long-term research direction is to design a completely AI-native SDLC from the ground up.

- **Multi-agent Collaboration and Emergent Behavior**: As systems composed of multiple specialized agents become more common, understanding and managing their collective behavior is a major challenge. Research is needed to ensure reliable collaboration and prevent unforeseen, harmful emergent behaviors.

8 Conclusion

8.1 Synthesis of Findings: The State of AI in SE Reliability

This review systematically analyzed the profound impact of generic and agentic AI on software engineering reliability from 2022 to 2025. The evidence presents a dichotomy. On one hand, AI tools have demonstrated the capacity to dramatically accelerate software development tasks, offering substantial productivity gains (Jason and Microsoft 2025). On the other hand, this acceleration has been achieved at the cost of traditional reliability guarantees.

Our analysis has shown that AI-generated code is frequently insecure, with studies consistently reporting high rates of vulnerabilities and a dangerous phenomenon of security degradation through iterative refinement (Shukla et al. 2025). The fundamental non-determinism of LLMs undermines the core engineering principles of reproducibility and predictable verification (Ouyang et al. 2025; Atil et al. 2024). The very benchmarks used to measure progress have been shown to be compromised by evaluation contamination, creating an *illusion of competence* (Liang et al. 2025; Aleithan et al. 2024). The state-of-the-art is thus characterized by a rapid escalation in demonstrated capabilities on these benchmarks, but this progress is built on a fragile empirical foundation and is accompanied by a lagging approach to ensuring the safety and reliability of the resulting software artifacts.

8.2 Concluding Remarks on the Future Trajectory

Achieving reliable AI-driven software engineering will not be a simple matter of building larger or more powerful language models. The findings of this report strongly suggest that it is, fundamentally, a systems engineering challenge. The path forward requires a holistic approach that integrates advancements across multiple domains. This includes the design of more robust agent architectures that incorporate principles of self-reflection and continuous learning; the implementation of rigorous engineering controls such as constrained decoding, comprehensive observability, and verifier-in-the-loop patterns; the development of a new generation of realistic and contamination-free benchmarks; and the cultivation of principled human oversight to manage the socio-technical risks of automation. The transition from AI as a powerful but unpredictable *copilot* to a truly reliable autonomous *engineer* is not an inevitability but a significant and open research endeavor. The challenge for the software engineering community in the coming years will be to temper the pursuit of capability with an equal, if not greater, commitment to the principles of reliability, safety, and trustworthiness.

References

Advisory Group on the Reliability of Electronic Equipment (AGREE) (1957) Reliability of Military Electronic Equipment. Technical report, Office of the Assistant Secretary of Defense (Research and Engineering), Washington, D.C., This report is foundational to the modern definition and practice of reliability engineering, which was developed during and after World War II

AI N (2023) Artificial intelligence risk management framework (ai rmf 1.0), pp 100–101. https://nvlpubs.nist.gov/nistpubs/ai/nistai

Aleithan R, Xue H, Mohajer MM, Nnorom E, Uddin G, Wang S (2024) Swe-bench+: enhanced coding benchmark for llms. arXiv preprint arXiv:2410.06992

Alenezi M, Akour M (2025) AI-driven innovations in software engineering: a review of current practices and future directions. Appl Sci 15(3):1344

All-Hands-AI (2025) Openhands-code less, make more. https://docs.openhands.dev/

Amazon Web Services (2022) Introducing amazon codewhisperer, the ml-powered coding companion. https://aws.amazon.com/blogs/machine-learning/introducing-amazon-codewhisperer-the-ml-powered-coding-companion/

Atil B, Aykent S, Chittams A, Fu L, Passonneau RJ, Rajagopal RE, Sloan GR, Tudrej A, Ture TF (2024) Non-determinism of "deterministic" llm settings. arXiv preprint arXiv:2408.04667

Bainbridge L (1983) Ironies of automation. Automatica 19(6):775–779. https://doi.org/10.1016/0005-1098(83)90046-8

Becker J, Rush N, Barnes E, Rein D (2025) Measuring the impact of early-2025 AI on experienced open-source developer productivity. arXiv preprint arXiv:2507.09089

Beurer-Kellner L, Fischer M, Vechev M (2024) Guiding llms the right way: fast, non-invasive constrained generation. arXiv preprint arXiv:2403.06988

Bird C, Ford D, Zimmermann T, Forsgren N, Kalliamvakou E, Lowdermilk T, Gazit I (2023) Taking flight with copilot. Commun ACM 66(6):56–62

Bucaioni A, Weyssow M, He J, Lyu Y, Lo D (2025) Artificial intelligence for software architecture: literature review and the road ahead. arXiv preprint arXiv:2504.04334

Chandra S, Tabachnyk M (2024) AI in software engineering at google: progress and the path ahead. https://research.google/blog/ai-in-software-engineering-at-google-progress-and-the-path-ahead/

Chattopadhyay S, Nelson N, Au A, Morales N, Sanchez C, Pandita R, Sarma A (2022) Cognitive biases in software development. Commun ACM 65(4):115–122

Chen M, Tworek J, Jun H, Yuan Q, Pinto H, Kaplan J, Edwards H, Burda Y, Joseph N, Brockman G (2021) Evaluating large language models trained on code. arXiv preprint arXiv:2107.03374

Choi HK, Khanov M, Wei H, Li Y (2025) How contaminated is your benchmark? measuring dataset leakage in large language models with kernel divergence. In: Forty-second international conference on machine learning, https://openreview.net/forum?id=wVDR2qmE28

Cui ZK, Demirer M, Jaffe S, Musolff L, Peng S, Salz T (2025) The effects of generative AI on high-skilled work: evidence from three field experiments with software developers. Available at SSRN 4945566

Cusick J (2019) The first 50 years of software reliability engineering: a history of SRE with first person accounts. arXiv preprint arXiv:1902.06140

Cyril B, Microsoft Azure (2025) New AI innovations that are redefining the future for software companies. https://azure.microsoft.com/en-us/blog/new-ai-innovations-that-are-redefining-the-future-for-software-companies/

Deng X, Da J, Pan E, He YY, Ide C, Garg K, Lauffer N, Park A, Pasari N, Rane C (2025) Swe-bench pro: Can AI agents solve long-horizon software engineering tasks? arXiv preprint arXiv:2509.16941

Ensmenger N, Aspray W (2002) Software as labor process. History of computing: software issues. In: International conference on the history of computing, ICHC 2000 April 5–7, 2000 Heinz Nixdorf MuseumsForum Paderborn. Springer, Germany, pp 139–165

Future of Life Institute (2024) High-level summary of the AI act. https://artificialintelligenceact.eu/high-level-summary/

GoogleCloud (n.d.) What is human-in-the-loop (hitl) in AI & ml? https://cloud.google.com/discover/human-in-the-loop. Accessed October 2025

Gregg L, Annie B (n.d.) Why observability is essential for AI agents. https://www.ibm.com/think/insights/ai-agent-observability. Accessed October 2025

Haque EA, Brown C, LaToza TD, Johnson B (2025) Towards decoding developer cognition in the age of AI assistants. arXiv preprint arXiv:2501.02684

IEEE Reliability Society (2016) IEEE recommended practice on software reliability. IEEE Std 1633–2016

Imai S (2022) Is github copilot a substitute for human pair-programming? An empirical study. Proceedings of the ACM/IEEE 44th international conference on software engineering: companion proceedings, pp 319–321

James AW, SoftwareSeni (2025) Testing and debugging ai-generated code-systematic strategies that work. https://www.softwareseni.com/testing-and-debugging-ai-generated-code-systematic-strategies-that-work/

Jason G, Microsoft (2025) How software development companies are paving the way for AI transformation. https://partner.microsoft.com/el-gr/blog/article/generative-ai-impact-for-partners

Jim H, Ivan B, Cole S, IBM (2025) What is model drift? https://www.ibm.com/think/topics/model-drift

Jimenez CE, Yang J, Wettig A, Yao S, Pei K, Press O, Narasimhan KR (2024) SWE-bench: Can language models resolve real-world github issues? In: The twelfth international conference on learning representations. https://openreview.net/forum?id=VTF8yNQM66

Kalliamvakou E (2022) Research: quantifying github copilot's impact on developer productivity and happiness. https://github.blog/news-insights/research/research-quantifying-github-copilots-impact-on-developer-productivity-and-happiness/

Kharma M, Choi S, AlKhanafseh M, Mohaisen D (2025) Security and quality in llm-generated code: a multi-language, multi-model analysis. arXiv preprint arXiv:2502.01853

Lambert N (2025) Reinforcement learning from human feedback. arXiv preprint arXiv:2504.12501

Leveson NG, Turner CS (1993) An investigation of the therac-25 accidents. Computer 26(7):18–41

Liang S, Garg S, Moghaddam RZ (2025) The swe-bench illusion: when state-of-the-art llms remember instead of reason. arXiv preprint arXiv:2506.12286

Lyu MR (1996) Handbook of software reliability engineering. IEEE Computer Society Press and McGraw-Hill Book Company

Lyu MR (2007) Software reliability engineering: a roadmap. Fut Softw Eng 153–170. https://doi.org/10.1109/FOSE.2007.24

Mark H, Meta Engineering (2025) Llms are the key to mutation testing and better compliance. https://engineering.fb.com/2025/09/30/security/llms-are-the-key-to-mutation-testing-and-better-compliance/

Meta AI (2024) Llama. https://ai.meta.com/llama/

Musa JD, Iannino A, Okumoto K (1987) Software reliability: measurement, prediction, application. McGraw-Hill

National Institute of Standards and Technology (2024) Artificial intelligence risk management framework: generative artificial intelligence profile. Technical report, Department of Commerce, U.S. https://doi.org/10.6028/NIST.AI.600-1

Negri-Ribalta C, Geraud-Stewart R, Sergeeva A, Lenzini G (2024) A systematic literature review on the impact of AI models on the security of code generation. Front Big Data 7:1386720

Nikolov S, Codecasa D, Sjövall A, Tabachnyk M, Chandra S, Taneja S, Ziftci C (2025) How is google using AI for internal code migrations? In: 2025 IEEE/ACM 47th international conference on software engineering: software engineering in practice (ICSE-SEIP). IEEE, pp 481–492

Noy S, Zhang W (2023) Experimental evidence on the productivity effects of generative artificial intelligence. Science 381(6654):187–192

OpenAI (2024) Introducing swe-bench verified. https://openai.com/index/introducing-swe-bench-verified/

Ouyang S, Zhang JM, Harman M, Wang M (2025) An empirical study of the non-determinism of chatgpt in code generation. ACM Trans Softw Eng Methodol 34(2):1–28

Patronus AI (2025) Rag evaluation metrics: best practices for evaluating rag systems. https://www.patronus.ai/llm-testing/rag-evaluation-metrics

Pearce H, Ahmad B, Tan B, Dolan-Gavitt B, KR (2022) Asleep at the keyboard? assessing the security of github copilot's code contributions. In: 2022 IEEE symposium on security and privacy (SP), pp 754–768. https://doi.org/10.1109/SP46214.2022.9833571

Peng S, Kalliamvakou E, Cihon P, Demirer M (2023) The impact of AI on developer productivity: evidence from github copilot. arXiv preprint arXiv:2302.06590

Perry N, Srivastava M, Kumar D, Boneh D (2023) Do users write more insecure code with AI assistants? In: Proceedings of the 2023 ACM SIGSAC conference on computer and communications security, pp 2785–2799

Pham H (2006) System reliability concepts. System software reliability. Springer, pp 9–75

Raisch S, Krakowski S (2021) Artificial intelligence and management: the automation-augmentation paradox. Acad Manag Rev 46(1):192–210

Raluca Ada Popa, Four Flynn, Google DeepMind (2025) Introducing codemender: an AI agent for code security. https://deepmind.google/discover/blog/introducing-codemender-an-ai-agent-for-code-security/

Randell B (1979) Software engineering in 1968. In: Proceedings of the 4th international conference on software engineering, pp 1–12

Saleh JH, Marais K (2006) Highlights from the early (and pre-) history of reliability engineering. Reliabil Eng Syst Saf 91(2):249–256

Sergeyuk A, Zakharov I, Koshchenko E, Izadi M (2025) Human-AI experience in integrated development environments: a systematic literature review. arXiv preprint arXiv:2503.06195

Shinn N, Cassano F, Gopinath A, Narasimhan K, Yao S (2023) Reflexion: language agents with verbal reinforcement learning. Adv Neural Inf Process Syst 36:8634–8652

Shukla S, Joshi H, Syed R (2025) Security degradation in iterative AI code generation-a systematic analysis of the paradox. arXiv preprint arXiv:2506.11022

The European Parliament and The Council of the European Union (2024) Regulation (EU) 2024/1689 of the European Parliament and of the Council of 13 June 2024 laying down harmonised rules on artificial intelligence and amending Regulations (EC) No 300/2008, (EU) No 167/2013, (EU) No 168/2013, (EU) 2018/858, (EU) 2018/1139 and (EU) 2019/2144 and Directives 2014/90/EU, (EU) 2016/797 and (EU) 2020/1828 (Artificial Intelligence Act). https://eur-lex.europa.eu/legal-content/EN/TXT/PDF/?uri=OJ:L_202401689

Valsai (2025) Swe-bench benchmark. https://www.vals.ai/benchmarks/swebench-2025-09-08

Veracode (2025) Application risk management engineered for the AI-coding era. https://www.veracode.com/

Wang G, Xie Y, Jiang Y, Mandlekar A, Xiao C, Zhu Y, Fan L, Anandkumar A (2023a) Voyager: An open-ended embodied agent with large language models. arXiv preprint arXiv:2305.16291

Wang X, Wei J, Schuurmans D, Le QV, Chi EH, Narang S, Chowdhery A, Zhou D (2023b) Self-consistency improves chain of thought reasoning in language models. In: The eleventh international conference on learning representations, https://openreview.net/forum?id=1PL1NIMMrw

Yang J, Jimenez CE, Wettig A, Lieret K, Yao S, Narasimhan K, Press O (2024) Swe-agent: agent-computer interfaces enable automated software engineering. Adv Neural Inf Process Syst 37:50528–50652

Yao S, Yu D, Zhao J, Shafran I, Griffiths T, Cao Y, Narasimhan K (2023) Tree of thoughts: deliberate problem solving with large language models. Adv Neural Inf Process Syst 36:11809–11822

Zhang L, He S, Zhang C, Kang Y, Li B, Xie C, Wang J, Wang M, Huang Y, Fu S et al. (2025) Swe-bench goes live! arXiv preprint arXiv:2505.23419

Zhou Y, Liu Y, Li X, Jin J, Qian H, Liu Z, Li C, Dou Z, Ho TY, Yu PS (2024) Trustworthiness in retrieval-augmented generation systems: a survey. arXiv preprint arXiv:2409.10102

Ziegler A, Kalliamvakou E, Li XA, Rice A, Rifkin D, Simister S, Sittampalam G, Aftandilian E (2022) Productivity assessment of neural code completion. In: Proceedings of the 6th ACM SIGPLAN international symposium on machine programming, pp 21–29

Operations and Human Factors

From Failure Modes to Reliability Awareness in Generative and Agentic AI System

Janet Jing Lin and Liangwei Zhang

Abstract This chapter bridges technical analysis and organizational preparedness by tracing the path from layered failure modes to reliability awareness in generative and agentic AI systems. We first introduce an 11-layer failure stack, a structured framework for identifying vulnerabilities ranging from hardware and power foundations to adaptive learning and agentic reasoning. Building on this, the chapter demonstrates how failures rarely occur in isolation but propagate across layers, creating cascading effects with systemic consequences. To complement this diagnostic lens, we develop the concept of awareness mapping: a maturity-oriented framework that quantifies how well individuals and organizations recognize reliability risks across the AI stack. Awareness is treated not only as a diagnostic score but also as a strategic input for AI governance, guiding improvement and resilience planning. By linking layered failures to awareness levels and further integrating this into Dependability-Centred Asset Management (DCAM), the chapter positions awareness mapping as both a measurement tool and a roadmap for trustworthy and sustainable AI deployment across mission-critical domains.

Keywords AI reliability · Awareness mapping · Dependability-centred asset management · Generative and agentic AI systems

J. J. Lin (✉)
Department of Civil, Environmental and Natural Resources Engineering, Luleå University of Technology, Luleå, Sweden
e-mail: janet.lin@ltu.se

School of Information Technology, Halmstad University, Halmstad, Sweden

L. Zhang
Department of Industrial Engineering, Dongguan University of Technology, Dongguan, China

A. Stavrou et al. (eds.), *Generative and Agentic AI Reliability: Architectures, Challenges, and Trust for Autonomous Systems*, Studies in Computational Intelligence 1272, https://doi.org/10.1007/978-3-032-18585-3_10

1 Introduction

AI systems are increasingly deployed in safety- and mission-critical domains such as transportation, energy, healthcare, manufacturing, and the built environment. Reliability has long been a central concern for what we call conventional AI systems—machine learning applications designed for tasks such as classification, prediction, optimization, and control. These systems face multi-level vulnerabilities, spanning hardware stability, power quality, data integrity, model robustness, and application integration.

The arrival of generative AI expands this landscape. By producing open-ended outputs—text, images, code, or designs—generative systems introduce new reliability challenges, including hallucinations, factual errors, toxic or biased content, and potential misuse through deepfakes or disinformation. These risks extend beyond conventional performance concerns to questions of trustworthiness, safety, and governance (Joshi 2025).

The next wave, agentic AI, deepens the challenge further. Combining autonomy, planning, reasoning, memory, and multi-agent interaction, agentic systems can pursue goals and initiating actions with system-wide consequences. Their failure modes include goal misalignment, flawed planning, emergent conflicts between agents, and breakdowns in human–AI collaboration (Acharya et al. 2025).

This chapter addresses these paradigms in continuity, asking: *where and how can conventional, generative, and agentic AI systems fail—and how aware are we of these vulnerabilities?* To answer, the chapter introduces two complementary contributions:

- The 11-layer failure stack, a structured framework tracing vulnerabilities from physical computation and energy through data, models, and applications, up to learning, reasoning, and multi-agent coordination.
- The concept of awareness mapping, which assesses how well individuals and organizations recognize risks across these layers and positions awareness itself as a dimension of reliability.

Case vignettes drawn from transportation, energy, healthcare, manufacturing, and the built environment illustrate both layer-specific vulnerabilities and cascading, cross-layer effects. Finally, the chapter situates these tools within the paradigm of Dependability-Centred Asset Management (DCAM), linking technical failure analysis to lifecycle strategies for trustworthy and sustainable deployment of generative and agentic AI systems.

2 Reliability as a Moving Target in Conventional, Generative, and Agentic AI

Reliability has traditionally been treated as something that could be designed for and verified—a destination rather than a journey. In practice, however, even in classical engineering domains, reliability has always been a moving target. It shifts over time with operating conditions, usage patterns, maintenance practices, and unexpected interactions. What was once "reliable" in a test environment may not hold under long-term use or in new contexts.

This dynamism becomes even more pronounced in AI systems. Unlike physical components whose degradation can often be modelled in predictable ways, AI systems continuously interact with changing data, evolving environments, and—in the case of agentic AI—with other agents and human stakeholders. Reliability here is not a static property but a dynamic relationship, shaped by macro-level regulations and institutions, meso-level organizational practices, and micro-level component behaviours (Lin and Silfvenius 2025).

To ground this discussion, we turn first to established definitions. International standards such as ISO, IEC, and IEEE define reliability as: "*The ability of a system or component to perform its required functions under stated conditions for a specified period of time.*" (Zhang et al. 2017).

This classical definition provides a solid foundation, but its meaning shifts as we move from traditional physical and software systems to conventional AI, generative AI, and agentic AI. Each paradigm forces us to reinterpret what counts as the "intended function" and which vulnerabilities matter most.

Box 1. Reliability: From Standards to AI Paradigms

Standard Definition (ISO/IEC, IEEE, MIL-STD):

"*Reliability is the ability of a system or component to perform its required functions under stated conditions for a specified period of time.*"

- **Traditional systems:** Intended functions are predictable and bounded, failures arise from wear-out, material degradation, or software bugs.
- **Conventional AI systems:** Intended functions involve consistent and accurate predictions, classifications, or optimizations, dependent on data pipelines and computational infrastructure.
- **Generative AI systems:** Intended functions expand to producing trustworthy, safe, and contextually appropriate outputs such as text, images, or code.
- **Agentic AI systems:** Intended functions extend further to reasoning, planning, adapting, and safely interacting with humans and other agents in open-ended environments.

Unified Perspective:

Reliability is the dependable functioning of a system across its full stack—from physical infrastructure and data pipelines to models, outputs, reasoning, and interactions—ensuring that intended functions are performed accurately, safely, and sustainably over time under evolving conditions.

This progression demonstrates that reliability is cumulative. Each paradigm inherits the concerns of the ones below—physical durability, computational stability, data and model robustness—while introducing new dimensions shaped by its scope. Conventional AI depends on reliable infrastructure but adds sensitivity to data and models. Generative AI builds on these layers while demanding content reliability and safety. Agentic AI inherits all the above, extending reliability into goal alignment, adaptive learning, and emergent multi-agent behaviour (Fig. 1).

In this sense, AI does not change the fact that reliability is a moving target—it amplifies it. By redefining what counts as an "intended function," AI systems shift the trajectory of reliability challenges and require continuous adaptation in engineering practice (Table 1). At the same time, the very features that complicate reliability—adaptivity, perception, and reasoning—also create new opportunities, from advanced diagnostics to self-healing mechanisms. AI makes reliability both harder and more achievable: harder, because it multiplies potential failure modes; more achievable, because it equips us with intelligent tools to anticipate and mitigate them in real time.

Yet AI is not only an amplifier of existing reliability dynamics. It can also shift or misdirect the trajectory of how reliability is understood and managed. By redefining

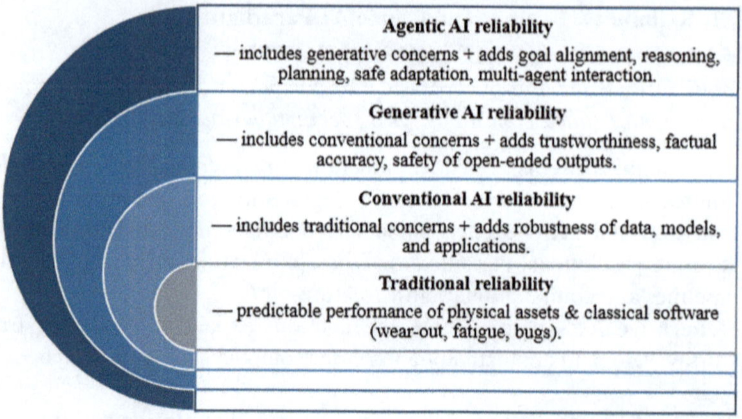

Fig. 1 Reliability as an expanding, cumulative concept. Each paradigm inherits the reliability concerns of the ones below while introducing new dimensions shaped by its scope. Traditional reliability emphasizes physical and software dependability; conventional AI adds data and model robustness; generative AI extends to trustworthy content; and agentic AI expands further into reasoning, adaptation, and interaction

Table 1 Evolving interpretations of reliability across system paradigms

System type	How the standard definition applies	Characteristic failure emphases
Traditional systems (Physical assets and classical software)	*Performing intended functions under stated conditions* = predictable operation of components or code over time, managed by design quality, testing, and maintenance	Wear-out, fatigue, corrosion, software bugs, maintenance errors
Conventional AI systems	*Intended function* = providing consistent and accurate predictions, classifications, or optimizations, supported by reliable infrastructure	Data drift, mislabelling, model instability, hardware/software faults, silent monitoring failures
Generative AI systems	*Intended function* = producing trustworthy, factual, safe, and contextually appropriate outputs, stable at scale	Hallucinations, factual inconsistency, bias, unsafe content, misuse (deepfakes), scaling failures
Agentic AI systems	*Intended function* = reasoning, planning, adapting, and safely pursuing goals in dynamic, multi-agent contexts	Goal misalignment, flawed reasoning, unsafe adaptation, emergent conflicts, cascading systemic failures

what counts as reliable behaviour—for example, emphasizing output plausibility rather than operational stability—AI risks encouraging organizations to underemphasize traditional dependability concerns or overestimate AI's self-correcting capacities. As argued in earlier work on the intrinsic mechanisms of reliability improvement (Lin and Silfvenius 2025), reliability must be seen as a continuously evolving system property, requiring deliberate and ongoing enhancement. This perspective reinforces why a structured, layered approach is essential: it not only catalogs failure modes but also provides guidance for sustaining and improving reliability in the era of conventional, generative, and agentic AI.

Thus, the standards-based definition remains essential, but for AI systems it must be extended into a layered and dynamic conception of reliability. This prepares the ground for Sect. 3, where we introduce the 11-layer failure stack and map how vulnerabilities manifest differently across conventional, generative, and agentic AI systems.

3 Layered Failure Modes Across Conventional, Generative, and Agentic AI Systems

Failures in AI systems rarely occur in isolation. They emerge from multiple interacting layers, spanning from physical computation and infrastructure through data pipelines, models, and applications, up to adaptive learning and multi-agent coordination. To capture this complexity, we adopt an 11-layer failure stack: a structured framework that traces potential vulnerabilities across the full spectrum of AI systems.

This failure stack provides a baseline lens for analysing reliability challenges. Yet its manifestations are not uniform. In conventional AI, failures often concentrate around data quality, model robustness, and integration into domain applications. In generative AI, reliability hinges on the factuality, appropriateness, and safe use of generated content. In agentic AI, challenges extend further into goal alignment, reasoning fidelity, adaptation in dynamic environments, and emergent multi-agent behaviours.

To organize this analysis, the section unfolds in three steps:

- 3.1 introduces the 11-layer failure stack as a general framework, providing an overview of its structure and logic.
- 3.2 develops a comparative view, showing how each layer manifests differently across conventional, generative, and agentic AI systems.
- 3.3 offers paradigm spotlights, highlighting distinctive reliability challenges and illustrative vignettes that demonstrate how failures unfold in real-world contexts.
- 3.4 draws the threads together by examining cross-layer risks and cascading effects, showing how localized faults propagate through the stack and amplify into systemic failures.

Taken together, this layered perspective reveals both shared vulnerabilities and paradigm-specific risks, underscoring that AI reliability cannot be safeguarded at a single layer alone. It requires a cross-layer, paradigm-sensitive approach that adapts to the evolving nature of AI systems (Ale et al. 2025).

3.1 The 11-Layer Failure Stack: A General Framework

Ensuring the reliability of AI systems requires more than verifying isolated components. Failures often arise from complex interactions across multiple layers, where vulnerabilities at one level propagate upward or downward, creating systemic risks. To capture this complexity, we propose an 11-layer failure stack, which organizes potential vulnerabilities from the physical substrate of computation to the highest levels of reasoning, adaptation, and multi-agent interaction.

The 11 layers can be grouped into four broad domains:

1. Foundational layers (hardware, power, system software, AI frameworks)—the physical and computational substrate.

2. Core intelligence layers (models and data)—the learning and knowledge base.
3. Operational layers (applications, execution, monitoring, learning)—the deployment and adaptation mechanisms.
4. Agentic layer (reasoning, goal alignment, multi-agent coordination)—the level where autonomy and decision-making unfold.

Each layer has a distinct role, typical failure modes, and sectoral manifestations. Table 2 provides a high-level summary.

Why Layers Matter Together

The 11-layer stack is not just a checklist—it emphasizes that failures are layered and interdependent. A data error (Layer 6) can cascade into faulty models (Layer 5), misguide applications (Layer 7), and distort agentic coordination (Layer 11). Conversely, poor reasoning at the agentic level can impose stress on infrastructure, amplifying faults downward. Thus, the 11-layer stack provides both:

- a diagnostic lens (where can failures occur?), and

Table 2 Overview of the 11-layer failure stack

Layer	Function	Representative failure modes
1. Hardware	Computing substrate: processors, memory, storage, interconnects	Overheating, memory bit flips, device wear-out, interconnect degradation
2. Power and energy	Stable power supply: PSUs, UPS, energy management	Voltage instability, surges/spikes, thermal overload, battery depletion
3. System software	OS, drivers, virtualization, firmware	Kernel panics, driver incompatibility, firmware bugs, virtualization overhead
4. AI frameworks	ML/DL libraries and pipelines	Dependency conflicts, numerical instability, non-deterministic training
5. Models	Encoded AI knowledge and decision rules	Overfitting, underfitting, adversarial attacks, hallucinations
6. Data	Input pipelines, storage, labelling	Data drift, noise, mislabelled data, integration errors, unreliable synthetic data
7. Applications	Domain-specific AI use cases	API downtime, poor integration, UI errors
8. Execution	Real-time orchestration: cloud/edge, load balancing	Latency spikes, cold starts, resource starvation
9. Monitoring	Observability, logging, alerts	Silent failure, undetected drift, alert fatigue
10. Learning	Continuous learning and adaptation	Concept drift, misaligned feedback loops, retraining on biased data
11. Agentic AI	Reasoning, planning, multi-agent coordination	Goal misalignment, emergent conflicts, opaque decision-making

- a design principle (how should systems be architected to contain cascading risks?).

In the next Sect. 3.2, we extend this framework into a comparative analysis, showing how the same layers manifest differently across conventional, generative, and agentic AI systems.

3.1.1 Layer 1: Hardware

The hardware layer provides the physical foundation for all AI systems. It encompasses processors (CPUs, GPUs, TPUs), memory, storage devices, and interconnects that carry out and sustain computation. Without reliable hardware, higher levels of the AI stack cannot operate correctly, as every model, data flow, and decision ultimately depend on the integrity of the physical substrate.

Hardware failures often emerge from:

- Thermal stress, such as overheating in processors or accelerators.
- Wear-out and fatigue, including degraded solder joints or storage media failure.
- Transient faults, such as memory bit flips from cosmic radiation or electromagnetic interference.
- Environmental stress, including vibration, dust, or humidity damaging sensitive circuits.

These failures can be either catastrophic (system crashes) or silent (bit errors that propagate unnoticed).

Box 2. Cross-Sector Examples of Layer 1: Hardware

- **Transportation (railways, aviation):** onboard AI accelerators in trains overheat during summer heatwaves, causing intermittent diagnostic failures; avionics hardware experiences accelerated aging under pressure and vibration.
- **Energy (power grid):** edge devices in substations are corrupted by voltage spikes, leading to AI controllers missing grid instability signals.
- **Healthcare:** GPUs used in MRI imaging exhibit memory corruption, silently distorting image analyses and leading to diagnostic errors.
- **Manufacturing:** robot controller boards fail due to vibration and dust on the shop floor, halting production lines unexpectedly.
- **Buildings (HVAC):** controllers for adaptive HVAC systems fatigue under continuous switching cycles, degrading comfort and efficiency.

Classical hardware reliability engineering has a long tradition of mitigation strategies, including:

- Redundancy (backup processors, fault-tolerant architectures).

- Error-Correcting Codes (ECC) to detect and correct memory errors.
- Thermal management through active cooling and environmental design.

However, the demands of AI hardware introduce new challenges. Accelerators like GPUs and TPUs operate at much higher utilization and parallelism, making them more susceptible to thermal cycling and component stress. Neuromorphic and edge-AI chips present emerging failure behaviours that are less well understood. Research is exploring self-monitoring hardware, adaptive voltage/frequency scaling, and AI-driven prognostics to predict hardware degradation (Shankar and Muralidhar 2025). Yet, deployment in safety–critical contexts such as healthcare or transportation remains limited.

Link to Higher Layers

Hardware failures often propagate invisibly. A silent memory error at this level can corrupt model weights (Layer 5), leading to systematic misclassifications. A processor glitch can trigger operating system crashes (Layer 3), cascading into downtime at the application or agentic layers. Thus, ensuring robustness at the hardware layer is not only foundational but also essential for preserving reliability throughout the entire AI stack.

3.1.2 Layer 2: Power and Energy

The power and energy layer supplies the stable electrical foundation that enables AI hardware to function. This includes power supply units (PSUs), batteries, uninterruptible power supplies (UPS), and energy management systems. Reliability here ensures not only continuous operation but also protection against fluctuations and interruptions that can compromise computation (Huang et al. 2025).

Power and Energy failures often emerge from:

- Voltage instability due to grid fluctuations.
- Thermal overload of power components.
- Surge or spike damage from transient events.
- Battery depletion in mobile or edge devices.
- Grounding or wiring faults in deployed environments.

Box 3. Cross-Sector Examples of Layer 2: Power and Energy

- **Transportation**: autonomous inspection robots in railways cut short missions due to rapid battery drain; EV onboard AI controllers reset when encountering voltage dips.
- **Energy**: smart grid AI controllers malfunction during frequency fluctuations, delaying corrective actions.

- **Healthcare**: portable diagnostic AI devices shut down mid-operation due to unstable supply or battery failure.
- **Manufacturing**: AI-driven CNC machines halt when UPS systems fail during local power dips.
- **Buildings (HVAC)**: controllers reset during building-wide power transitions, disrupting environmental stability.

Reliability engineering for power systems emphasizes redundant power supplies, surge protection, and battery management systems. In data centers, advanced energy-aware scheduling balances AI workload against power draw. In edge-AI, low-power chips and adaptive energy management extend device life. Yet, as AI systems move into mission-critical contexts, even minor instabilities are unacceptable. Research is shifting toward AI-enabled power monitoring (predicting battery life, detecting instability) and integration of renewable energy sources, though these approaches are not yet mature for safety–critical deployment.

Link to Higher Layers

Failures here propagate rapidly upward. A transient power loss can cause firmware corruption (Layer 3), data pipeline interruption (Layer 6), or unexpected resets that undermine multi-agent coordination (Layer 11). Reliable AI requires stable power as a non-negotiable foundation.

3.1.3 Layer 3: System Software

System software sits between hardware and higher-level frameworks, providing the operating environment for AI workloads. It includes operating systems, device drivers, virtualization layers, and firmware. This layer ensures that physical components are accessible and stable for applications above.

Typical System Software Failures include (Ebad 2018):

- Kernel panics or OS crashes under load.
- Driver incompatibility with accelerators (e.g., GPU drivers).
- Firmware bugs that destabilize control systems.
- Virtualization overhead or clock skew, affecting real-time tasks.
- Poorly timed updates introducing regressions.

Box 4. Cross-Sector Examples of Layer 3: System Software

- **Transportation**: a driver updates causes SCADA-based railway monitoring to crash, delaying predictive maintenance.

- **Energy**: a firmware bug in inverter control loops destabilizes renewable grid integration.
- **Healthcare**: surgical robots stall when GPU driver crashes mid-operation.
- **Manufacturing**: a PLC firmware bug halts robotic arm coordination.
- **Buildings (HVAC)**: building automation freezes after an OS update conflict.

System software reliability is usually addressed through certification, long-term support kernels, and redundant control firmware. Virtualization adds isolation but also complexity, raising new risks. In AI, the challenge is heightened because drivers and firmware must keep pace with rapidly evolving hardware accelerators. Research explores lightweight hypervisors, formally verified kernels, and continuous integration pipelines for system software. However, many industries still face gaps between rapidly updated AI stacks and conservative operational environments (e.g., energy or healthcare).

Link to Higher Layers

System software is the glue between physical hardware and AI frameworks. Failures here can ripple upward, rendering frameworks unusable (Layer 4), halting model inference (Layer 5), or causing application downtime (Layer 7). Ensuring resilience at this layer requires balancing stability and adaptability.

3.1.4 Layer 4: AI Frameworks

AI frameworks provide the libraries and pipelines that enable developers to build, train, and deploy models. This includes deep learning platforms (e.g., TensorFlow, PyTorch), optimization toolkits, and inference runtimes. Frameworks standardize access to hardware accelerators and simplify large-scale model training, but they also introduce dependencies and complexity (Weber 2022).

Typical AI Frameworks Failures include:

- Dependency conflicts between framework versions.
- Numerical instabilities during training (e.g., exploding/vanishing gradients).
- Non-deterministic behaviour, making models difficult to reproduce.
- Poor backward compatibility when upgrading frameworks.

Box 5. Cross-Sector Examples of Layer 4: AI Frameworks

- **Transportation**: a mismatch in TensorFlow versions breaks predictive maintenance deployment in railway monitoring systems.

- **Energy**: instability in an optimization tool degrades renewable energy forecasting accuracy.
- **Healthcare**: PyTorch inference errors disrupt medical image segmentation workflows.
- **Manufacturing**: predictive maintenance pipelines fail after a framework upgrade, halting rollout.
- **Buildings (HVAC)**: deep learning classifiers for fault detection cannot run due to unresolved framework bugs.

Current best practices emphasize containerization, dependency pinning, and continuous integration testing. Open-source ecosystems like PyTorch and TensorFlow evolve rapidly, enabling cutting-edge applications but creating instability for safety–critical industries. Research focuses on deterministic training, lightweight inference runtimes, and framework certification for high-assurance domains. Adoption in regulated industries remains slow, as frameworks are rarely validated for dependability.

Link to Higher Layers

If frameworks fail, models (Layer 5) cannot run or reproduce reliably, undermining application-level dependability (Layer 7). This layer thus acts as a keystone connecting computational resources to usable intelligence.

3.1.5 Layer 5: Models

Models encode the learned intelligence of AI systems, transforming data into predictions, classifications, or decisions. This includes conventional ML models, deep neural networks, foundation models, and large language models (LLMs).

Typical Models Failures include:

- Overfitting or underfitting, reducing generalizability.
- Adversarial vulnerabilities, where small perturbations trigger misclassification.
- Hallucinations, particularly in generative models producing plausible but false outputs.
- Model drift, as performance degrades under new data distributions.

Box 6. Cross-Sector Examples of Layer 5: Models

- **Transportation**: bogie fault classifiers misdiagnose rare weather-related vibrations.
- **Energy**: renewable generation predictors fail during extreme storms.
- **Healthcare**: LLM-based assistants hallucinate treatment recommendations.

- **Manufacturing:** defect detection models miss rare but critical cracks in products.
- **Buildings (HVAC)**: occupancy predictors misclassify patterns, wasting energy.

Research on robustness, fairness, and explainability has grown rapidly. Techniques such as adversarial training, calibration methods, and uncertainty quantification aim to improve reliability. Yet, the rise of foundation and generative models introduces new risks: hallucinations, misaligned goals, and opaque decision-making (Zhang et al. 2025a, b). Current progress includes fine-tuning guardrails, alignment methods such as Reinforcement Learning from Human Feedback (RLHF), and certification efforts, but practical assurance for mission-critical use is limited.

Link to Higher Layers

Model failures propagate directly to the application layer (7) and can misguide multi-agent coordination (Layer 11). Because models form the intelligence core, reliability at this layer is highly visible to end-users, often overshadowing vulnerabilities in lower layers.

3.1.6 Layer 6: Data

Data forms the lifeblood of AI systems. This layer includes data pipelines, storage, collection devices, labelling infrastructures, and synthetic data generation. Reliable AI depends on the quality, integrity, and timeliness of data.

Typical Data Failures include:

- Data drift, where input distributions shift over time.
- Noise or corruption from faulty sensors.
- Labelling errors, especially in supervised learning.
- Integration errors when fusing multiple data sources.
- Synthetic data unreliability, if generation processes introduce hidden biases.

Box 7. Cross-Sector Examples of Layer 6: Data

- **Transportation**: misaligned vibration sensors cause false rail track degradation alerts.
- **Energy**: poor synchronization between SCADA and PMU data corrupts forecasting.
- **Healthcare**: mislabelled medical images bias diagnostic models.
- **Manufacturing**: inconsistent calibration across assembly lines leads to unreliable datasets.

- **Buildings (HVAC)**: faulty CO_2 sensors skew ventilation control algorithms.

Data reliability research emphasizes data validation pipelines, anomaly detection, and active data curation (Sharma et al. 2021). Advances in synthetic data generation promise coverage of rare events but risk introducing unrepresentative distributions. Industry practice often underestimates the difficulty of maintaining high-quality, real-time data pipelines, especially when integrating legacy sensors. Ongoing research explores data provenance tracking and AI-driven labelling quality assurance.

Link to Higher Layers

Faulty data undermines model training (Layer 5), disrupts applications (Layer 7), and erodes monitoring reliability (Layer 9). In many cases, data-layer failures propagate invisibly, making this one of the most critical and underestimated points of failure in AI systems.

3.1.7 Layer 7: Application

The application layer represents the domain-specific integration of AI into workflows and real-world decision-making. It connects models to operators, assets, and end-users through dashboards, APIs, or automation systems. This is where AI's predictions become actionable.

Typical Application Failures include:

- Poor integration with legacy systems.
- API downtime that interrupts functionality.
- User interface errors leading to misinterpretation of outputs.
- Context mismatch, where AI decisions are applied inappropriately.

Box 8. Cross-Sector Examples of Layer 7: Application

- **Transportation**: predictive maintenance scheduling fails when APIs go offline, disrupting fleet operations.
- **Energy**: load-balancing tools misallocate resources during peak demand due to poor integration.
- **Healthcare:** clinical decision support tools crash mid-workflow, delaying treatment.
- **Manufacturing**: AI scheduling software misallocates machine resources, causing production bottlenecks.
- **Buildings (HVAC)**: dashboards misreport ventilation status, leading to operator misjudgements.

Application reliability is supported through API testing, modular integration, and resilience design (Kothamali 2025). However, in practice, applications often depend on brittle middleware and poorly monitored interfaces. Current research emphasizes human-centred design, trust-calibrated interfaces, and explainable outputs to strengthen the reliability of AI-assisted decision-making.

Link to Higher Layers

Application failures often obscure root causes: operators may blame the application even when the true fault lies in data (Layer 6) or models (Layer 5). This makes robust application design critical as the final touchpoint between AI and human trust.

3.1.8 Layer 8: Execution

The execution layer governs the real-time orchestration of AI workloads. This includes cloud/edge scheduling, parallel execution, and load balancing (Alsadie and Alsulami 2024). Reliability here ensures that AI models and applications can run under constraints of latency, scale, and computational resources.

Typical Failures include:

- Latency spikes degrading real-time performance.
- Cold starts in serverless architectures causing delays.
- Resource starvation when workloads exceed system capacity.
- Communication bottlenecks between distributed nodes.

Box 9. Cross-Sector Examples of Layer 8: Execution

- **Transportation**: railway dispatch AI lags, propagating train delays.
- **Energy**: cloud inference delays undermine fast frequency response in smart grids.
- **Healthcare**: triage systems experience inference latency, delaying emergency alerts.
- **Manufacturing**: robot coordination lags, stopping assembly lines.
- **Buildings (HVAC)**: edge controllers' overload when running multiple fault checks simultaneously.

Solutions include container orchestration (Kubernetes), real-time schedulers, and edge-cloud hybrid architectures. However, performance assurance for safety–critical AI is still immature. Research explores deterministic execution frameworks and AI-driven workload optimization, but adoption lags outside data center environments.

Link to Higher Layers

Execution issues are highly visible: even reliable models (Layer 5) or applications (Layer 7) fail if latency or availability breaks the pipeline. Thus, execution reliability is key to making AI dependable in real-world operations.

3.1.9 Layer 9: Monitoring

The monitoring layer provides observability into AI systems, including logging, anomaly detection, drift detection, and alerting (Aghaei et al. 2025). Its role is to ensure that failures and degradations are identified in time to act.

Typical Monitoring Failures include:

- Silent failure, where monitoring stops without notice.
- Undetected drift, allowing model degradation to persist.
- Alert fatigue, where too many false alarms cause warnings to be ignored.
- Insufficient visibility, leaving blind spots in performance monitoring.

Box 10. Cross-Sector Examples of Layer 9: Monitoring

- **Transportation**: false alerts overwhelm operators, causing them to miss genuine rail system warnings.
- **Energy**: undetected data drift hides inverter instability.
- **Healthcare**: clinicians ignore alarms due to over-alerting, missing critical cases.
- **Manufacturing**: dashboards miss anomalies in machine vibrations due to poor drift detection.
- **Buildings (HVAC)**: monitoring logs silently stop after updates, leaving faults invisible.

Progress includes drift detection algorithms, explainable monitoring dashboards, and federated observability platforms. Yet, industry practice often falls short—many systems lack effective monitoring of AI-specific risks. Emerging research explores self-monitoring AI agents capable of explaining their own uncertainty.

Link to Higher Layers

Without reliable monitoring, failures propagate unchecked. This layer serves as the immune system of AI systems, making its reliability critical to long-term safety and trust.

3.1.10 Layer 10: Learning

The learning layer governs adaptation and continuous improvement. It includes retraining pipelines, online learning, reinforcement learning, and feedback integration. Unlike static systems, AI systems evolve after deployment—making this layer uniquely dynamic.

Typical Learning Failures include:

- Concept drift, where learned rules no longer fit current conditions.
- Feedback loop misalignment, where learning amplifies errors.
- Biased retraining data, reinforcing systemic weaknesses.
- Forgetting rare but critical cases during adaptation.

Box 11. Cross-Sector Examples of Layer 10: Learning

- **Transportation**: reinforcement learners mis adapt to seasonal traffic shifts, causing inefficiency.
- **Energy**: online learning destabilizes grid frequency control.
- **Healthcare**: auto-updated models degrade after learning from misdiagnoses.
- **Manufacturing**: retrained classifiers miss rare but costly defects.
- **Buildings (HVAC)**: self-adaptive controllers forget rare emergency conditions, compromising safety.

Research on continual learning, reinforcement learning safety, and robust retraining is advancing rapidly. Industrial practice includes shadow training pipelines and offline validation before deployment (Bayram and Ahmed 2025), but these safeguards are resource intensive. True safe lifelong learning remains unsolved in critical domains.

Link to Higher Layers

Errors in learning undermine trust at the agentic level (Layer 11), as agents adapt in unintended ways. Continuous learning thus transforms reliability from a static property into a moving target that must be actively managed.

3.1.11 Layer 11: AI Agent

The agent layer introduces the highest-level intelligence: reasoning, planning, communication, goal alignment, and multi-agent interaction. This is where AI systems become agentic, acting autonomously to pursue objectives in complex environments.

Typical AI Agent Failures include:

- Goal misalignment, where agent objectives diverge from human intent.
- Multi-agent conflicts, when agents compete instead of cooperating.
- Opaque reasoning, making it difficult to explain or correct decisions.
- Cascading failures, as poor agentic choices amplify lower-layer weaknesses.

Box 12. Cross-Sector Examples of Layer 11: AI Agent

- **Transportation**: autonomous inspection robots' conflict with scheduling agents, causing delays.
- **Energy**: market bidding agents act competitively, destabilizing grid stability.
- **Healthcare**: triage agents escalate cases inconsistently, eroding clinician trust.
- **Manufacturing:** fleets of collaborative robots pursue conflicting optimization goals, creating unsafe conditions.
- **Buildings (HVAC)**: multiple agents fight over comfort vs. efficiency, leading to unstable building performance.

Agent-based AI is an emerging frontier. Current progress includes goal alignment techniques (e.g., reward shaping, constitutional AI (Sicari et al. 2024)), multi-agent coordination frameworks, and explainable planning methods. However, reliability assurance here is immature, with open challenges around emergent behaviours and human-agent collaboration.

Link to Higher Layers

The agent layer sits at the top of the stack, but it reflects vulnerabilities from every lower layer. Poor data (Layer 6) or unstable learning (Layer 10) can distort agent reasoning. Failures here are most visible to society, as they directly affect human trust, safety, and system-level outcomes.

3.2 Comparative View: Layered Failure Modes Across AI Paradigms

The 11-layer failure stack is universal: every AI system—from conventional classifiers to generative models to autonomous agents—operates across these layers. Importantly, failures can occur at any layer. The distinctions drawn here highlight where reliability challenges tend to concentrate or evolve, not where risks exclusively reside (Table 3, Fig. 2).

Table 3 Core risks versus paradigm-specific emphasis

Layer	Core risks (All paradigms)	Conventional AI emphasis	Generative AI emphasis	Agentic AI emphasis
1. Hardware	Device wear-out, overheating	Embedded systems	Compute scaling for training/ inference	Energy-hungry autonomy at edge
2. Power and energy	Voltage instability, depletion	Battery drains in robots	Cloud/edge cost and energy use	Grid-level load from fleets
3. System software	OS crashes, driver conflicts	Firmware bugs in controllers	Framework dependency fragility	Autonomy requires robust OS for multi-agents
4. Frameworks	Instability, version conflicts	ML pipeline mismatches	Model integration with APIs	Multi-agent orchestration libraries
5. Models	Overfitting, adversarial inputs	Task-specific brittleness	Hallucinations, controllability	Reasoning errors, emergent strategies
6. Data	Drift, noise, mislabelling	Pipeline integrity	Bias in web-scale/ synthetic corpora	Self-selected/ adaptive data risks
7. Applications	Poor integration, API fragility	Workflow disruption	Copilots, generative design	Autonomous workflows and decision loops
8. Execution	Latency, cold starts, overload	Basic resource scheduling	Inference scaling, real-time streaming	Fleet-level orchestration
9. Monitoring	Silent failure, alert fatigue	Limited logging	Safety filters, toxicity/factuality check	Goal intent and human oversight dashboards
10. Learning	Misaligned retraining, drift	Offline retraining	RLHF misalignment, online fine-tuning	Lifelong adaptation, safe reinforcement
11. AI Agent	Goal conflicts, black-box reasoning	Absent/minimal	Limited prompting autonomy	Full autonomy, multi-agent coordination risks

Thus, the comparative analysis should be understood as showing relative prominence and shifting meanings, rather than the absence of failures in less emphasized layers.

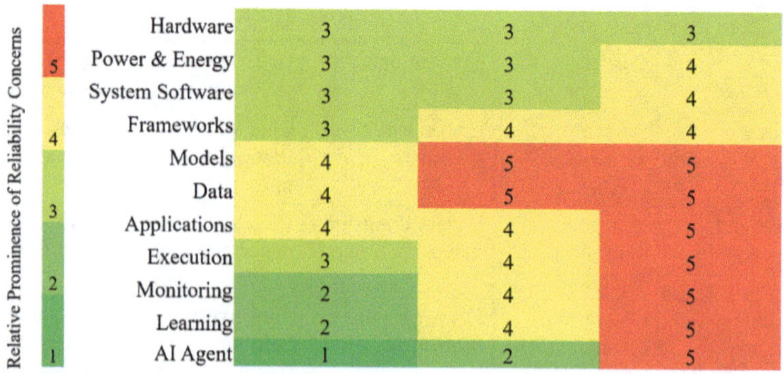

Fig. 2 Reliability prominence across AI paradigms (11-layer stack)

Conventional AI: Pipeline-Centred Risks

Conventional AI systems are task-specific and domain-bounded. Reliability risks concentrate in the data–model–application pipeline:

- Layer 5 (Models): overfitting and brittleness under unseen conditions.
- Layer 6 (Data): drift, noise, and labelling errors.
- Layer 7 (Applications): integration failures disrupting workflows.

Other layers remain relevant—for instance, hardware faults (Layer 1) or weak monitoring (Layer 9)—but they are less prominent in defining conventional AI reliability.

Generative AI: Content-Centred Risks

Generative AI expands conventional systems by producing open-ended content. This shifts reliability concerns toward accuracy, safety, and appropriateness of generated outputs:

- Layer 5 (Models): hallucinations, controllability, and factual accuracy.
- Layer 6 (Data): reliance on web-scale corpora introduces copyright, bias, and toxicity issues.
- Layer 9 (Monitoring): active filters to detect harmful or misleading content.
- Layer 10 (Learning): risks of misalignment during fine-tuning or feedback-based adaptation.

Lower layers still matter: GPU failures (Layer 1) or orchestration bottlenecks (Layer 8) can undermine generative systems as much as conventional ones.

Agentic AI: Autonomy-Centred Risks

Agentic AI combines generative capabilities with planning, reasoning, memory, and multi-agent interaction. Reliability here extends into alignment, safe adaptation, and emergent behaviours:

- Layer 10 (Learning): lifelong adaptation, reinforcement drift, and catastrophic forgetting.
- Layer 11 (AI Agent): goal misalignment, multi-agent conflicts, opaque reasoning, erosion of human trust.

These higher layers intensify the complexity of reliability, but lower layers remain just as critical. A voltage fluctuation (Layer 2) or biased training dataset (Layer 6) can still cascade upward, destabilizing autonomy.

Cautionary Note: While certain layers are more visible in one paradigm, all layers remain relevant to reliability. Overlooking foundational or less-prominent layers risks hidden faults cascading upward, where they amplify into system- or society-level consequences. Effective reliability assurance therefore requires vigilance across the entire stack, regardless of paradigm.

Figure 2 illustrates how reliability concerns are distributed across the 11-layer failure stack for conventional, generative, and agentic AI systems. Numbers (1–5) represent relative prominence—not the existence of failures, but the degree to which reliability efforts and discussions have typically concentrated at each layer.

In conventional AI, reliability attention has focused most on data, models, and applications (scores 4), reflecting widespread work on data quality, model robustness, and system integration. However, this does not imply that other layers are free of risk; rather, concerns at lower (hardware, power) or higher (monitoring, learning) layers have historically been less emphasized, even though they can produce significant cascading effects.

In generative AI, prominence shifts upward: models and data pipelines reach the highest level (5), as issues like hallucinations, bias, and content safety dominate. Frameworks and monitoring also gain prominence because of the need for decoding stability, safety filters, and real-time oversight.

In agentic AI, emphasis extends further into higher layers. Execution, learning, and agent-level reasoning emerge as top concerns (5), capturing risks such as goal misalignment, unsafe adaptation, and emergent multi-agent conflict.

The progression demonstrates that while all layers remain vulnerable, the center of gravity of reliability attention shifts with paradigm evolution: from mid-stack concerns in conventional AI, to content validity in generative AI, to alignment and coordination challenges in agentic AI.

3.3 *Paradigm Spotlights: Reliability in Conventional, Generative, and Agentic AI*

While all three paradigms share the same 11-layer stack of vulnerabilities, their reliability *personalities* differ in important ways. Section 3.2 highlighted how emphasis shifts across layers. Here, the focus turns to the systemic nature of failures within each paradigm, showing how risks manifest in practice and how the research and standards communities are beginning to respond.

Conventional AI systems, such as predictive classifiers and optimization models, have long been applied in transportation, energy, healthcare, and manufacturing. Their reliability challenges are most visible in the robustness of models, the stability of data pipelines, and the effectiveness of integration with domain-specific applications. Unlike catastrophic failures, breakdowns here often emerge gradually, as performance degrades due to data drift or unanticipated operating conditions. For instance, a railway maintenance system trained on older vibration data may fail to generalize when confronted with new materials, producing false negatives that quietly undermine service availability. To mitigate such risks, the state of the art emphasizes rigorous validation and lifecycle management through ML Ops frameworks, alongside established safety standards such as ISO 26262 for automotive systems (ISO 2018), IEC 61508 for functional safety (IEC, 2010), and IEEE 1633 for software reliability (IEEE 2017). Reliability in conventional AI is thus anchored in robustness, validation, and disciplined operational practices.

Generative AI represents a different reliability profile, because it produces open-ended outputs rather than fixed predictions. Here, the central question is not merely whether the model generalizes correctly, but whether its outputs are factually accurate, trustworthy, and safe for downstream use (Zhang et al. 2025a, b). This raises risks of hallucination, compounding errors when outputs are recycled into retraining, and misuse of generated content. A healthcare example illustrates the stakes: a generative assistant might produce a convincing but incorrect radiology explanation that persuades clinicians to delay necessary treatment. The problem is not just technical error, but the amplification of error through human trust. To address these risks, current approaches include the NIST AI Risk Management Framework (RMF) (Nist 2023), which emphasizes transparency, documentation, and safeguards, as well as practical techniques such as red-teaming and output guardrails. The IEEE P7003 standard on algorithmic bias considerations (Koene et al. 2018) also represents an emerging foundation for governing fairness and reliability in generative models. Reliability in generative AI therefore rests on controlling unpredictability, ensuring factual grounding, and embedding safeguards around open-ended behaviour.

Agentic AI systems extend the reliability challenge even further. By combining reasoning, planning, memory, and interaction across multiple agents, they introduce risks associated with goal alignment, emergent behaviours, and the erosion of human oversight. Failures here are rarely local; instead, they emerge systemically, as autonomous agents optimize local objectives in ways that destabilize global performance. A case in point arises in energy markets, where autonomous bidding agents act rationally in isolation but collectively undermine grid stability. Addressing these risks requires governance as much as technical solutions, including human-in-the-loop oversight mechanisms, ongoing AI alignment research (Russell 2022), and emerging frameworks such as IEEE P7009 for fail-safe design of autonomous and semi-autonomous systems (IEEE 2024). In this sense, agentic AI reliability is less about robustness or hallucination and more about managing emergent complexity through alignment, governance, and systemic resilience (Zambare et al. 2025).

Table 4 Comparative reliability spotlights across AI paradigms

Paradigm	Systemic reliability risks	Illustrative example	Current responses/ standards	Key takeaway
Conventional AI	Data drift, brittle generalization, integration failures	Railway predictive maintenance misclassifies faults in new materials	ML Ops, ISO 26262, IEC 61508, IEEE 1633	Reliability anchored in robustness and validation
Generative AI	Hallucinations, compounding feedback errors, unsafe outputs	Healthcare diagnostic assistant generates misleading radiology explanations	NIST AI RMF, red-teaming, IEEE P7003	Reliability defined by trustworthiness and safeguards
Agentic AI	Goal misalignment, emergent behaviours, loss of oversight	Autonomous bidding agents destabilize power grid markets	Human-in-the-loop, AI alignment research, IEEE P7009	Reliability depends on alignment, governance, and resilience

Taken together, these spotlights suggest that the trajectory of AI reliability reflects the evolution of paradigms themselves. Conventional AI struggles most with robustness and lifecycle maintenance, generative AI with factuality and safe use of outputs, and agentic AI with alignment and governance of emergent behaviours. Reliability engineering must therefore adapt in step: from strengthening robustness, to embedding safeguards, to governing systemic complexity (Table 4).

3.4 Cross-Layer Risks and Cascading Effects

The 11-layer failure stack shows that vulnerabilities exist at every stage of AI systems, from hardware foundations to agentic coordination. Yet in practice, failures rarely remain confined to a single layer. They propagate across the stack, amplifying risks and creating system-wide consequences. This property distinguishes AI reliability from many traditional engineering contexts, where failures are more localized and predictable (Elder et al. 2024).

Three archetypes of cross-layer propagation are particularly salient:

1. Bottom-up cascades occur when low-level disturbances rise through the stack. For example, noisy sensor data (Layer 6) can degrade model accuracy (Layer 5), which then misguides decision-making applications (Layer 7). In transportation, a mis-calibrated vibration sensor in a railway bogie may trigger false defect predictions, resulting in unnecessary rescheduling and systemic delays.

2. Top-down cascades emerge when high-level decisions or agentic behaviours stress lower layers. In energy systems, bidding strategies by autonomous market agents (Layer 11) may overburden control applications (Layer 7) and even destabilize physical infrastructure (Layer 2). Here, the cascade begins at the reasoning and coordination layer and reverberates downward, producing failures not from material fatigue but from emergent behaviour.
3. Feedback loops are self-reinforcing cycles that blur the boundary between layers. In healthcare, generative models retrained on their own outputs (Layer 10) can inject bias into data pipelines (Layer 6), which in turn worsens model reliability (Layer 5). What begins as a small drift compound into systemic diagnostic inaccuracies that erode clinician trust (Layer 11).

These patterns illustrate why AI reliability cannot be reduced to guarding individual components. Cascading failures amplify risks and often manifest in surprising ways—turning local disturbances into global instability. They also highlight a gap in current organizational practice: many stakeholders can identify obvious model or application failures, but fewer recognize the cross-layer dynamics that cause small faults to grow into systemic risks.

This gap motivates the second framework of the chapter: awareness mapping. By evaluating how organizations perceive and respond to layered and cross-layer risks, awareness mapping provides a maturity-oriented lens for designing dependable AI architectures.

4 Awareness Mapping: From Failure Modes to Reliability Maturity

Reliability in AI systems is not only a matter of technical safeguards; it also depends on how well individuals and organizations perceive, understand, and prepare for risks. Vulnerabilities span all 11 layers of the failure stack, yet their impact is strongly shaped by awareness. In practice, awareness is uneven: most practitioners can recognize failures at the model or application layer, but far fewer anticipate risks in data pipelines, adaptive learning dynamics, or multi-agent interactions—where some of the most consequential failures arise.

To address this gap, we introduce awareness mapping: a structured approach for assessing how comprehensively organizations understand reliability risks across AI systems. Awareness mapping shifts the focus from failures themselves to the recognition of those failures, positioning awareness as a critical dimension of reliability. More importantly, it provides an evidence-based foundation for strategic decision-making, helping organizations review their current preparedness, identify blind spots, and prioritize improvements in governance, training, monitoring, and lifecycle management. In this way, awareness mapping serves both as a diagnostic lens and as a practical tool for shaping reliability strategies in conventional, generative, and agentic AI.

This section develops the framework in four steps. Section 4.1 explains how failures across the 11-layer stack are operationalized into specific reliability issues that can be scored as points of awareness. Section 4.2 introduces a five-level maturity scale that translates these scores into stages of organizational readiness. Section 4.3 presents empirical insights from practitioner surveys, highlighting current blind spots and uneven awareness across sectors. Finally, Sect. 4.4 links awareness mapping to Dependability-Centred Asset Management (DCAM), showing how awareness maturity can guide lifecycle strategies for trustworthy and resilient AI.

Ultimately, awareness mapping transforms reliability from a reactive concern into a proactive strategic capability, enabling organizations to anticipate, govern, and continually improve AI systems in step with technological evolution.

4.1 Scoring Awareness Across Failure Modes

Section 3 catalogued the technical vulnerabilities of AI systems through an 11-layer failure stack. To move from technical vulnerabilities to measurable organizational maturity, we link each failure mode to its corresponding body of reliability studies. The idea is that awareness is not simply knowing that failures may occur, but being familiar with methods, studies, or practices that address them. Each study area therefore becomes a potential *awareness point*.

In our implementation, the 11-layer stack is mapped onto approximately 47 reliability studies drawn from conventional engineering, AI safety research, and emerging generative/agentic AI work. For example:

- Hardware failures such as memory corruption are linked to studies on ECC diagnostics and thermal aging analysis.
- Data-related risks are linked to methods for drift detection, data pipeline validation, and bias auditing.
- Model-level concerns correspond to adversarial robustness testing, uncertainty quantification, or hallucination suppression.
- At the agentic level, studies focus on alignment verification, simulation-in-the-loop evaluation, and multi-agent stress testing.

Table 5 summarizes this mapping. For each layer of the stack, it distinguishes between:

- Baseline reliability studies in conventional AI systems,
- Additional focus areas introduced by generative AI, and
- New reliability challenges and studies specific to agentic AI.

When used in practice, respondents are asked to indicate which of these study areas they are aware of. Each affirmative response counts as one point toward an overall awareness score. The total score thus reflects not just recognition of risks, but also familiarity with concrete methods for addressing them.

Table 5 Reliability studies across layers of conventional, generative, and agentic AI systems

Layer	Conventional AI—reliability studies	Generative AI—added reliability studies	Agentic AI—added reliability studies
1. Hardware	1. GPU/TPU diagnostics; 2. ECC error analysis; 3. Interconnect reliability; 4. Thermal aging studies; 5. Vibration stress testing	Long-run accelerator reliability; mixed-precision error validation; sustained training/inference stress tests	Robotics hardware robustness; actuator/sensor degradation analysis; fleet-level redundancy evaluation
2. Power and energy	6. Voltage fluctuation testing; 7. Surge/UPS resilience; 8. Thermal envelope validation; 9. Blackout/brownout recovery	Energy efficiency studies for large-scale training; power-capping effects on QoS; cooling/thermal resilience in datacenters	Battery SoH/SOC forecasting; energy-aware autonomy validation; safe degradation pathways under power loss
3. System software	10. Kernel panic forensics; 11. Container/VM stability tests; 12. Firmware update safety; 13. Clock skew validation	CUDA/ROCm compatibility matrices; NUMA/I/O contention studies; GPU memory allocation fault analysis	RTOS determinism studies; watchdog & safety monitor reliability; secure update protocols in edge/IoT settings
4. AI frameworks	14. Dependency conflict resolution; 15. Build reproducibility testing; 16. Deterministic training benchmarking	Tokenizer stability; decoding reproducibility; safety filter integration validation	Agent-framework reliability (tool use contracts, sandboxing); plug-in orchestration correctness studies
5. Models	17. OOD robustness testing; 18. Calibration metrics; 19. Adversarial robustness studies; 20. Concept drift monitoring; 21. Uncertainty quantification	Hallucination detection benchmarks; factuality/toxicity red-teaming; controllability experiments (prompt constraints, decoding strategies)	Reasoning & planning fidelity testing; goal alignment verification; emergent strategy audits in multi-agent setups

(continued)

Table 5 (continued)

Layer	Conventional AI—reliability studies	Generative AI—added reliability studies	Agentic AI—added reliability studies
6. Data	22. Label quality audits; 23. Data drift and leakage detection; 24. Data pipeline validation; 25. Synthetic data robustness; 26. Provenance/lineage verification	Web-scale bias detection; PII/copyright filtering studies; synthetic data robustness analysis	Memory/log data integrity; self-generated interaction data validation; agent feedback loop consistency checks
7. Applications	27. API failure recovery analysis; 28. Latency/throughput impact studies; 29. User trust calibration 30. HMI resilience evaluation	Guardrail UI testing; human trust calibration studies; post-processing verification frameworks	Governance of end-to-end workflows; safe-abort/approval mechanisms; role/permission modelling validation
8. Execution	31. Chaos engineering; 32. Deployment fault injection; 33. Orchestration stress tests; 34. Auto scaling & rollback analysis; 35. Environment drift control	KV-cache stability; batching/streaming trade-off benchmarking; heterogeneous accelerator scheduling	Real-time multi-agent orchestration testing; consensus/coordination stress tests; runaway loop prevention
9. Monitoring	36. Observability coverage audits; 37. Anomaly detection benchmarks; 38. Drift detection latency analysis; 39. Alert fatigue mitigation	Hallucination/toxicity monitoring frameworks; content safety dashboards; multi-pass critique/verification pipelines	Goal deviation monitors; plan-conformance validation; safety sentinels for multi-agent monitoring; escalation pathways

(continued)

Table 5 (continued)

Layer	Conventional AI—reliability studies	Generative AI—added reliability studies	Agentic AI—added reliability studies
10. Learning	40. Online learning romustness 41. Offline retraining hygiene studies; 42. Drift-adaptive training validation; 43. Rollback of regressions	RLHF/RLAIF stability; online fine-tuning safety validation; preference shift monitoring	Lifelong learning safety validation; safe exploration benchmarks; catastrophic forgetting protection; policy versioning safeguards
11. AI agent	— (minimal in conventional AI)	Prompt-loop guardrail validation; limited autonomy safety checks	Autonomy safety cases; alignment verification frameworks; simulation-in-the-loop testing; multi-agent game-theoretic stress tests
Cross-cutting	44. Silent failure detection studies; 45. cascading-fault modelling; 46. multi-layer resilience co-design; 47. security breach defense	Red-teaming suites for generative models; usage policy enforcement; provenance/auditability of generated outputs	Governance and oversight frameworks; auditability of plans/actions; inter-agent safety norms; societal/market impact constraints

This scoring approach—assigning one point per identified study area, across 47 in total—is deliberately simplified. It does not capture the depth of knowledge or implementation quality. However, it provides a transparent and actionable baseline: organizations can benchmark their maturity, identify blind spots, and prioritize capacity building.

Section 4.2 builds on this by introducing a five-level maturity scale that translates awareness scores into stages of organizational readiness.

Table 5 translates the 11-layer failure stack into a set of reliability studies that serve as awareness points. For each layer, the baseline column reflects established reliability practices in conventional AI, while the generative and agentic columns extend these with additional concerns unique to open-ended content generation and autonomous reasoning. Each study corresponds to a point of awareness: if a practitioner or organization can recognize and articulate the relevance of that study to their system, it contributes one point to their awareness score. In total, the 47 identified studies (baseline + extensions) define the assessment space.

These awareness points form the backbone of the scoring method described in Sect. 4.2, where the total number of recognized points (0–47) is mapped onto a five-level maturity scale. This mapping transforms awareness from a qualitative impression into a structured diagnostic tool, enabling organizations to benchmark their preparedness, identify blind spots, and prioritize improvements in AI reliability strategies.

4.2 Awareness Maturity Levels

The awareness score provides a structured way to measure how reliably individuals or organizations perceive risks across the AI system stack. It is derived from the 47 diagnostic points identified in Sect. 4.1, corresponding to failure modes and reliability studies spanning all 11 layers—from hardware and data pipelines to adaptive learning and agentic reasoning. Each point represents a distinct reliability concern. Respondents are asked whether they are aware of each issue; every positive response contributes one point, producing a maximum possible score of 47. The total reflects the breadth of reliability risks that stakeholders consciously recognize.

It is important to emphasize that these **47 points are drawn from conventional AI systems**. They capture well-established risks such as hardware degradation, power instability, data drift, model overfitting, or monitoring blind spots. However, the framework is designed to be **extensible**. By integrating generative-specific risks (e.g., hallucinations, factuality errors, unsafe content) and agentic-specific risks (e.g., goal misalignment, reasoning drift, emergent multi-agent conflicts), the same approach can evolve into a more comprehensive awareness instrument. In this sense, the 47 points should be seen as a **baseline example**, not a closed set.

To make scores meaningful, we map them onto a five-level maturity scale (Table 6). This scale captures not only the number of risks recognized, but also the organizational posture implied by that recognition:

- Level I (0–9 points): Unaware

Organizations at this stage have not considered AI reliability in a systematic way. Only obvious failures such as application crashes or output errors are noticed. Reliability is absent from conversations and practices, leaving systems highly exposed.

- Level II (10–19 points): Fragmented Awareness

Some failures are acknowledged, typically at the model or application level, but without systematic measures. Risks in deeper layers—such as data pipelines, execution, or monitoring—remain overlooked. Reliability efforts are piecemeal and reactive.

- Level III (20–29 points): Emerging Multi-Layer Awareness

Awareness expands across multiple layers. Teams recognize that AI can fail beyond surface-level errors and begin to consider mitigation strategies. Data quality, model

Table 6 Summarizes these levels, their score ranges, and their practical meaning

Level	Score range	Descriptor	Meaning in practice
I. Unaware	0–9	No consideration	Reliability absents from discussions; only obvious failures (e.g., crashes) noticed
II. Fragmented awareness	10–19	Isolated recognition	Failures acknowledged mainly at model or application level; no systematic measures for deeper risks
III. Emerging multi-layer awareness	20–29	Expanding recognition	Failures at several layers are acknowledged; some mitigations applied, though mostly reactive
IV. Proactive systemic awareness	30–39	Structured approach	Reliability monitored and tested across multiple layers; governance mechanisms beginning to emerge
V. Comprehensive cross-layer reliability	40–47	Full integration	Reliability strategy spans all 11 layers, including generative and agentic risks; embedded in organizational culture

robustness, and infrastructure reliability enter the discussion, though responses remain mostly reactive and uneven.

- Level IV (30–39 points): Proactive Systemic Awareness

Reliability is actively monitored and tested across several layers. Governance mechanisms begin to take shape, supported by structured dependability practices. Blind spots remain—particularly in higher-order risks such as goal alignment or emergent multi-agent behaviors—but awareness is no longer confined to isolated issues.

- Level V (40–47 points): Comprehensive Cross-Layer Reliability

Awareness spans virtually all diagnostic points, including those in reasoning, adaptation, and agentic coordination. Reliability is embedded in organizational culture, supported by systematic dependability engineering, proactive governance, and lifecycle strategies.

This scoring method is deliberately simplified. It privileges breadth of recognition (how many failure types are known) over depth of understanding (how well those failures are mitigated). As such, it should be interpreted as an indicative measure of organizational maturity rather than a definitive evaluation of capability. Its strength lies in comparability: by using the same scoring approach across teams, sectors, or domains, it highlights blind spots, benchmarks progress, and guides interventions.

Ultimately, this maturity framework turns awareness from a static description into a developmental roadmap. Progression from Level I to Level V reflects not only wider recognition of risks, but also deeper institutional capacity to anticipate, mitigate, and continually improve reliability across conventional, generative, and agentic AI systems.

4.3 Empirical Insights from Practice

To assess how the awareness mapping framework performs in practice, we applied it during several invited keynote talks across domains including transportation, energy, and manufacturing. In these sessions, participants—primarily engineers, managers, and researchers engaged with AI-enabled systems—were asked to complete a rapid diagnostic exercise. They were presented with a list of 47 reliability issues mapped to the 11-layer failure stack (see Sect. 3) and asked to indicate which ones they had previously considered in their work. Each identified issue counted as one point toward an overall awareness score.

The results were strikingly consistent. While participants readily identified a handful of visible issues—such as model overfitting, data noise, or application downtime—far fewer recognized vulnerabilities at deeper layers such as power stability, execution latency, or adaptive learning drift. Scores clustered toward the lower end of the maturity ladder, with most individuals and organizations falling within Level I (Unaware) or Level II (Fragmented Awareness). Only a minority reached Level III, and very few respondents demonstrated the cross-layer recognition required for Levels IV or V.

Figure 3 illustrate the distribution of scores from two keynote sessions. Despite differences in sectoral focus, both show a similar skew: most participants recognized fewer than 20 of the 47 failure modes, with only a very small group achieving awareness scores above 30. Notably, both surveys confirmed the same misalignment: failures are most often noticed at the model and application layers, while the most consequential risks—such as cascading effects from data pipelines, execution environments, or agentic coordination—are rarely acknowledged.

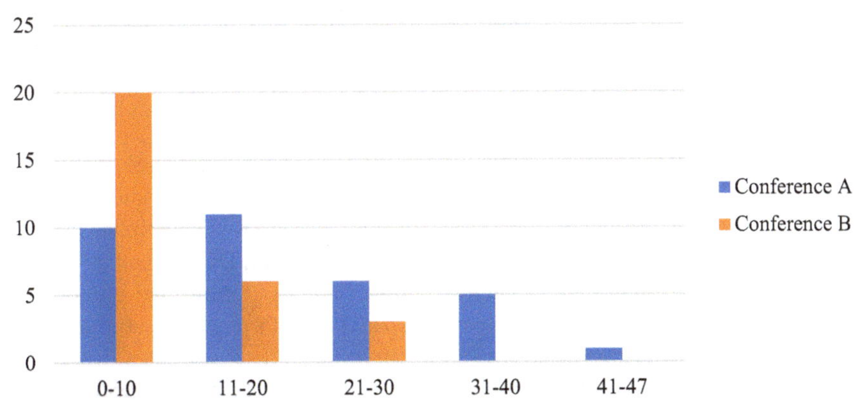

Fig. 3 Distribution of awareness scores from keynote session on AI reliability in the physical asset management and transportation sector. Conference A: 30th International and Mediterranean HDO (MeditMaint2025); HDO: Croatian Maintenance Society) conference, from 19th -22nd May, Rovinj, Croatian. Conference B: 1st International Conference on Transportation Systems (TS2025), from 16th–18th June 2025, Lisbon, Portugal

This evidence underscores two lessons. First, organizations may be systematically underestimating reliability risks by concentrating on surface-level failures. Second, even a simple diagnostic instrument such as the 47-item checklist can serve a dual role: not only measuring awareness but also expanding it. Participants often reported that encountering failure modes outside their prior experience—particularly in domains such as power and energy management or multi-agent coordination—reshaped their perspective on where reliability strategies should focus.

Taken together, these findings highlight both the urgency of broadening reliability awareness and the potential of structured mapping to support this expansion in practice.

4.4 Linking Awareness to Dependability-Centred Asset Management (DCAM)

The awareness mapping framework is more than a diagnostic tool for measuring how organizations perceive AI reliability risks. It also aligns directly with the broader paradigm of Dependability-Centred Asset Management (DCAM), which extends traditional asset management by embedding reliability, availability, maintainability, and safety (RAMS) across the lifecycle of both physical and digital assets (Lin 2025). In this view, AI systems—whether conventional, generative, or agentic—are treated as critical organizational assets whose dependability must be managed alongside physical infrastructure.

Awareness levels serve as an entry point into DCAM practice. Organizations at Level I or II lack the readiness to integrate AI reliability into asset management strategies, as risks remain unrecognized and decision-making is reactive. By Level III, awareness expands across multiple layers, allowing AI reliability considerations to begin influencing operational strategies such as monitoring, anomaly detection, and lifecycle planning.

At Level IV, systemic awareness begins to merge with DCAM structures. AI reliability is embedded into governance and management processes, including cross-layer monitoring, predictive maintenance supported by AI diagnostics, and structured validation protocols. At the highest level, Level V, awareness is comprehensive and coupled with institutionalized dependability practices. Here, DCAM and awareness mapping converge: reliability strategies become proactive, embedded into design, deployment, and continuous improvement, spanning both physical assets and AI-driven systems.

Crucially, awareness mapping provides a diagnostic bridge between technical AI reliability and organizational maturity in asset management. It reveals blind spots that managers may overlook, guiding targeted interventions. For example, a utility operator who recognizes model-level errors, but neglects execution-layer risks may fail to secure grid stability under dynamic demand; awareness mapping directs attention to these gaps and extends DCAM practices into digital infrastructures.

In this way, awareness mapping enriches DCAM in two complementary ways:

1. By embedding AI reliability explicitly within the asset management lifecycle.
2. By offering a structured roadmap for organizations to progress from fragmented awareness to comprehensive dependability.

Together, the two frameworks extend the scope of asset management to meet the challenges of increasingly autonomous, adaptive, and agentic AI systems, ensuring that reliability is not an afterthought but a core principle of lifecycle strategy.

5 Conclusion and Outlook

This chapter has advanced a dual contribution to the study of AI reliability. First, it introduced the 11-layer failure stack—a structured framework for tracing vulnerabilities across conventional, generative, and agentic AI systems. By moving from physical computation layers to reasoning and agentic coordination, the stack highlights that reliability challenges are not confined to models or applications but extend throughout the entire lifecycle of AI systems. Case vignettes illustrated how failures at seemingly minor layers can cascade into systemic disruptions, underscoring the need for a cross-layer perspective.

Second, the chapter developed the concept of awareness mapping—a method for evaluating how well organizations recognize and prepare for reliability risks. By translating failure modes into measurable awareness points, the framework provides a diagnostic lens into organizational maturity. The five-level maturity scale helps practitioners identify blind spots, benchmark progress, and guide the integration of reliability considerations into strategy and governance. The alignment of awareness mapping with Dependability-Centred Asset Management (DCAM) further shows how reliability can be embedded within broader lifecycle practices for both physical and digital assets.

Together, the failure stack and awareness mapping demonstrate that reliability is not a fixed property but a moving target, shaped by technological evolution and organizational readiness. Conventional AI, generative AI, and agentic AI each bring distinct vulnerabilities, yet the principle is consistent: safeguarding reliability requires both technical safeguards and organizational awareness, reinforced across layers and over time.

Looking forward, several avenues emerge. Future work could refine awareness mapping beyond breadth of recognition to capture depth of understanding and mitigation capacity. Comparative studies across industries would provide empirical evidence on sector-specific blind spots and resilience strategies. In parallel, the failure stack could evolve as new paradigms—such as embodied intelligence or hybrid human–AI collectives—introduce additional layers of complexity. Finally, integration with standards, regulation, and governance frameworks will be essential to ensure that awareness and dependability principles scale with the societal deployment of generative and agentic AI.

In conclusion, layered failure analysis and awareness mapping together offer a foundation for moving from reactive responses to proactive reliability strategies. By situating AI reliability within a systemic, lifecycle-oriented framework, they provide both a diagnostic lens and a roadmap—supporting the trustworthy, sustainable, and safe deployment of AI in increasingly critical domains.

References

Acharya DB, Kuppan K, Divya B (2025) Agentic AI: autonomous intelligence for complex goals—A comprehensive survey. IEEE Access, 13:18912–18936. Available at: https://doi.org/10.1109/ACCESS.2025.3532853

Aghaei M et al (2025) Autonomous intelligent monitoring of photovoltaic systems: an in-depth multidisciplinary review. Prog Photovol: Res Appl 33(3):381–409. Available at: https://doi.org/10.1002/pip.3859

Ale L et al (2025) Enhancing generative AI reliability via agentic AI in 6G-enabled edge computing. Nat Rev Electr Eng 2(7):441–443. Available at: https://doi.org/10.1038/s44287-025-00169-3

Alsadie D, Alsulami M (2024) Enhancing workflow efficiency with a modified Firefly Algorithm for hybrid cloud edge environments. Sci Rep 14(1): 1–19. Available at: https://doi.org/10.1038/s41598-024-75859-3

Bayram F, Ahmed BS (2025) Towards trustworthy machine learning in production: an overview of the robustness in MLOps approach. ACM Comput Surveys 57(5):1–35. Available at: https://doi.org/10.1145/3708497

Ebad SA (2018) The influencing causes of software unavailability: a case study from industry. Softw: Pract Exp 48(5):1056–1076. Available at: https://doi.org/10.1002/spe.2569

Elder H et al (2024) Knowing when to pass: the effect of AI reliability in risky decision contexts. Human Factors 66(2):348–362. Available at: https://doi.org/10.1177/00187208221100691

Huang H et al (2025) Power management optimization for data centers: a power supply perspective. IEEE Trans Sustainable Comput 10(4):784–803. Available at: https://doi.org/10.1109/TSUSC.2025.3542779

IEC (2010) Functional safety of electrical/electronic/programmable electronic safety-related systems. IEC 61508–1:2010

IEEE (2017) IEEE recommended practice on software reliability. IEEE Std 1633–2016:1–261

IEEE (2024) IEEE standard for fail-safe design of autonomous and semi-autonomous systems. IEEE Std 7009–2024:1–45

ISO (2018) Road vehicles—functional safety. ISO Std 26262–1(2018):1–23

Joshi S (2025) Model risk management in the era of generative AI: challenges, opportunities, and future directions. Int J Sci Res Publ 15(5):299–309. Available at: https://doi.org/10.29322/ijsrp.15.05.2025.p16133

Koene A, Dowthwaite L, Seth S (2018) IEEE P7003TM standard for algorithmic bias considerations. In: 2018 IEEE/ACM international workshop on software fairness. IEEE/ACM, pp 38–41. Available at: https://doi.org/10.1145/3194770.3194773

Kothamali PR (2025) AI-powered quality assurance: revolutionizing automation frameworks for cloud applications. J Adv Comput Syst 5(3):1–25. Available at: https://doi.org/10.69987/JACS.2025.50301

Lin J (2025) Dependability-centered asset management (DCAM): toward trustworthy and sustainable systems in the CPS era. IEEE Reliabil Magaz 2(3):1–7. Available at: https://doi.org/10.1109/MRL.2025.3583632

Lin J, Silfvenius C (2025) Some critical thinking on electric vehicle battery reliability: from enhancement to optimization. Batteries 11(2):1–33. Available at: https://doi.org/10.3390/batteries11020048

Nist (2023) NIST AI risk management framework (AI RMF 1.0). In: NIST AI 100–1, pp 1–48
Russell S (2022) Artificial intelligence and the problem of control. In: Perspectives on digital humanism, pp 19–24. Available at: https://doi.org/10.1007/978-3-030-86144-5_9
Shankar V, Muralidhar N (2025) Securing hardware with AI: intrusion detection, threat mitigation, and trust assurance. Int J Comput Sci Eng 13(4):92–98. Available at: https://doi.org/10.26438/ijcse/v13i4.9298
Sharma S, Kumar N, Kaswan KS (2021) Big data reliability: a critical review. J Intell Fuzzy Syst 40(3):5501–5516. Available at: https://doi.org/10.3233/JIFS-202503
Sicari S et al (2024) Open-ethical AI: advancements in open-source human-centric neural language models. ACM Comput Surv 57(4):1–47. Available at: https://doi.org/10.1145/3703454
Weber N (2022) SOL: reducing the maintenance overhead for integrating hardware support into AI frameworks, arXiv preprint arXiv: 2205. Association for Computing Machinery. Available at: http://arxiv.org/abs/2205.10357
Zambare P, Thanikella VN, Liu Y (2025) Securing agentic AI: threat modeling and risk analysis for network monitoring agentic AI system. arXiv preprint arXiv:2508, p 10043. Available at: http://arxiv.org/abs/2508.10043
Zhang D et al (2017) Reliability evaluation and component importance measure for manufacturing systems based on failure losses. J Intell Manuf 28(8):1859–1869. Available at: https://doi.org/10.1007/s10845-015-1073-1
Zhang T et al (2025) From data to deployment: a comprehensive analysis of risks in large language model research and development. IET Inf Secur 2025(1):7358963. Available at: https://doi.org/10.1049/ise2/7358963
Zhang X et al (2025) Reliability engineering, risk management, and trustworthiness assurance for AI systems. J Reliabil Sci Eng 1(2):022001. Available at: https://doi.org/10.1088/3050-2454/adcef2

Perspectives on a Reliability Monitoring Framework for Agentic AI Systems

Niclas Flehmig, Mary Ann Lundteigen, and Shen Yin

Abstract The implementation of agentic AI systems has the potential of providing more helpful AI systems in a variety of applications. These systems work autonomously towards a defined goal with reduced external control. Despite their potential, one of their flaws is the insufficient reliability which makes them especially unsuitable for high-risk domains such as healthcare or process industry. Unreliable systems pose a risk in terms of unexpected behavior during runtime and mitigation techniques are needed. In this work, we derive the main reliability challenges of agentic AI systems during runtime based on their characteristics. We draw the connection to traditional AI systems and formulate a fundamental reliability challenge during runtime which is inherent to traditional and agentic AI systems. As our main contribution, we propose a two-layered reliability monitoring framework for agentic AI systems which consists of a out-of-distribution detection layer for novel inputs and AI transparency layer to reveal internal operations. This two-layered monitoring approach gives a human operator the decision support which is needed to decide whether an output is potential unreliable or not and intervene. This framework provides a foundation for developing mitigation techniques to reduce risk stemming from uncertain reliability during runtime.

Keywords Agentic AI systems · Runtime monitoring · System reliability

N. Flehmig · S. Yin (✉)
Department of Mechanical and Industrial Engineering, Norwegian University of Science and Technology (NTNU), Trondheim, Norway
e-mail: shen.yin@ntnu.no

N. Flehmig
e-mail: niclas.flehmig@ntnu.no

M. A. Lundteigen
Department of Engineering Cybernetics, Norwegian University of Science and Technology (NTNU), Trondheim, Norway
e-mail: mary.a.lundteigen@ntnu.no

A. Stavrou et al. (eds.), *Generative and Agentic AI Reliability: Architectures, Challenges, and Trust for Autonomous Systems*, Studies in Computational Intelligence 1272, https://doi.org/10.1007/978-3-032-18585-3_11

1 Introduction

The emergence of agentic AI systems, systems that adaptively pursue complex goals with limited direct oversight (Shavit et al. 2023), has achieved significant attention. Many sectors may benefit from taking such systems into use, such as in healthcare, education, and the process industry. In healthcare, they could enhance diagnostic accuracy and personalize therapies, in education, they could provide intelligent tutoring to socially and economically marginalized groups and in industrial settings, they could reduce cognitive load by offering situational-aware decision support. This potential extends to safety-critical systems, promising more responsive and helpful AI systems.

For instance, consider an agentic AI system within a hydrogen bunkering process, responsible for maintaining safety in terms of liquid hydrogen leakage. Such an agent could continuously monitor sensor data to identify patterns indicative of a liquid hydrogen leak, and autonomously inform an operator and initiate mitigation measures like ventilation. However, the implementation of highly autonomous AI systems in these contexts raises profound reliability concerns. This is reflected in the technical report ISO/IEC TR 5469 (Artificial Intelligence n.d.) from the International Standard Organisation (ISO) and the International Electrotechnical Commission (IEC), which, while not explicitly naming agentic AI systems addresses highly autonomous systems. It suggests that such systems likely fall into a risk category where current methods are insufficient to adequately mitigate reliability-related risks.

To mitigate the risk arising from insufficient reliability performance of agentic AI systems, an appropriate monitoring is essential to detect and enable human intervention in critical situations. This requirement for monitoring is underscored by regulations like the EU AI Act (2024), which mandates post-market monitoring for high-risk AI systems, a category that would include agentic AI systems in safety-critical applications such as hydrogen bunkering. Although the EU AI Act specifies the requirement for appropriate monitoring rather than prescribing exact techniques, reliability monitoring is widely recognized as a part of a monitoring strategy to ensure that AI systems satisfy their functional requirements during runtime (Yampolskiy 2024).

The monitoring of AI systems gained significant relevance as applications moved beyond research environments into real-world deployments (Kang et al. 2020; Paleyes et al. 2020). Initially, monitoring objectives were closely aligned with quality assurance, focusing on continuous performance tracking and improvement. However, early methods that proved effective on curated test sets often failed during runtime deployment (Kang et al. 2020). This persistent gap stems from the unstructured nature of real-world inputs and the absence of ground truth labels, which are inherent to production environments (Kang et al. 2020; Paleyes et al. 2020).

This insufficiency of early monitoring techniques has driven considerable innovation in recent years, leading to an expansion of the field's objectives. Contemporary research (Yampolskiy 2024; Hendrycks 2024; Ogrizović et al. 2024) reflects a paradigm shift from merely detecting incorrect outputs at the component level to ensuring holistic AI safety at the system level. Consequently, monitoring approaches can be classified into distinct types based on their specific objectives (Yampolskiy 2024). Current monitoring tools for AI systems (Ferreira 2023; Naveed et al. 2024; Xu et al. 2023; Ginart et al. 2022) do not provide a comprehensive framework specifically designed for monitoring the runtime reliability of agentic AI systems, as their objectives are often broader and they are applied to non-agentic AI systems.

In this work, we derive the reliability challenges of agentic AI systems from their characteristics which distinguish them from traditional AI systems. Moreover, we compare these challenges to the reliability challenges of traditional AI systems and formulate a common fundamental challenge to both. Based on this, we propose a two-layered reliability monitoring framework for agentic AI systems that identifies novel inputs to the system, and reveals internal operations of the agentic AI system to enable a human operator to make a decision whether a current state of the system is unreliable or not and trigger a fallback policy. This monitoring framework can be seen as mitigating technique for risk stemming from insufficient reliability performance.

2 Definitions

2.1 Agentic AI Systems

Agentic AI systems are defined as AI systems capable of autonomously performing actions to achieve a specified goal, without relying on pre-defined behavioral scripts (Shavit et al. 2023). Unlike traditional AI systems like ChatGPT that execute specific, bounded tasks, agentic AI systems are characterized by a higher degree of agenticness. This degree is determined by four key characteristics: goal complexity, environmental complexity, adaptability, and autonomy (Shavit et al. 2023; Chan et al. 2023).

For instance, ChatGPT alone would not be considered agentic (Shavit et al. 2023). Instead, an agentic AI system could autonomously conduct an entire literature review: searching for papers, filtering results, and summarizing findings based on a high-level research query provide by a human (Chan et al. 2023). This illustrates the shift from reactive task-completion to proactive goal-pursuit.

This work focuses specifically on agentic AI systems that operate under human supervision, pursuing human-defined goals within human-defined environments as collaborative partners.

2.2 Reliability of AI Systems

The term *reliability* has no standardized definition and varies from industry to industry. One possible definition for technical systems describes an entity's capability to deliver required performance over a defined period and under specified operational conditions (Rausand et al. 2021). For AI systems including agentic AI systems, the reliability can be defined as assessment of the model's performance consistency over time and under varying conditions (Hendrycks 2024). While the definition of reliability does not differ significantly, the definition of the term *performance* does. For technical systems, the performance can be determined by laws and regulations, standards, and customer requirements (Rausand et al. 2021). For AI systems, the performance is determined by executing a set of tasks relevant to the desired functionality of the system. Such sets of tasks are sometimes standardized under so-called benchmarks (Burnell et al. 2023). In general, there is no standardized performance metrics for AI systems.

3 Methods

3.1 Research Approach

In this work, we adopted a qualitative, conceptual research design. The primary objective is to address a complex, emerging challenge which is to monitor the runtime reliability of agentic AI systems. The research was conducted in two sequential phases:

1. **Identification of the main reliability challenge for agentic AI systems during runtime**: A comprehensive literature review was undertaken to investigate the runtime reliability challenges of agentic AI systems based on their characteristics. These challenges were compared with those of traditional AI systems to identify a common fundamental challenge.
2. **Development of reliability monitoring framework for agentic AI systems**: Building upon the identification of the challenge, a conceptual two-layered framework was constructed. This involved a second literature review to identify relevant monitoring techniques from traditional AI systems and integrate them into a cohesive structure for detection and decision-support.

Given the novelty of the problem, a traditional systematic literature review protocol was deemed less suitable. Instead, an exploratory, iterative literature review methodology was employed.

3.2 Data Collection and Analysis

The literature review was rather iterative than linear, characterized by a snowballing technique and conceptual branching. The three core topics were:

- **Characteristics of agentic AI systems**: The review began with literature on *agentic AI systems* (Shavit et al. 2023; Chan et al. 2023; Acharya et al. 2025; Khamis 2025; Hosseini and Seilani 2025; Olujimi et al. 2025; Murugesan 2025). The focus here was to understand the typical characteristics of agentic AI systems, their applications, and their technical challenges to derive how these affect the runtime reliability. This phase was important for ensuring the relevance of later findings to the agentic AI systems context.
- **Reliability challenges of traditional AI systems**: From the characteristics of agentic AI systems, the review progressed into reliability challenges in traditional AI systems. For this, literature on *AI deployment* (Paleyes et al. 2020; Klaise et al. 2020) and *AI monitoring* (Yampolskiy 2024; Kang et al. 2020; Ogrizović et al. 2024; Ferreira 2023; Heyn et al. 2023) was conducted to establish an understanding, draw connections to the agentic AI systems, and identify similarities.
- **Focused exploration on AI monitoring concepts, frameworks and techniques**: Finally, the review process concluded into more specific areas based on references and linked concepts within the initial literature. The key threads that were followed included:
  - Broader concept of AI monitoring as part of AI safety (Hendrycks 2024; Bengio et al. 2025; Hendrycks et al. 2021)
  - Conceptual frameworks and approaches for monitoring AI systems (Ferreira 2023; Ginart et al. 2022; Heyn et al. 2023; Flehmig et al. 2024)
  - Techniques for out-of-distribution (OOD) detection in model inputs (Yang et al. 2024; Lu et al. 2024; Fang et al. 2022; Yang et al. 2022)
  - Techniques for AI transparency and explainable AI (XAI) (Hendrycks 2024; Longo et al. 2024; Miller et al. 2017).

  This phase of the review enabled a synthesis of existing monitoring frameworks and techniques, providing the foundation for proposing a new approach for agentic AI systems.

The main sources for the literature review were *Web of Science*, *Google Scholar*, and *arXiv*.

3.3 Underlying Assumptions

This work is based on the following assumptions regarding pre-deployment lifecycle stages and runtime context of the agentic AI systems:

- **Pre-deployment development**: All prior lifecycle stages including data acquisition, model selection, training, testing, and verification were conducted optimally. The system achieved a perfect reliability score on an adequate dataset, meaning the data is balanced and free of bias. This is, in practice, impossible to achieve for complex agentic AI systems but it isolates the challenges that emerge specifically during runtime.
- **Deployment environment**: The systems are deployed in environments appropriate to their agentic nature, characterized by complexity and unpredictable dynamics as described in the literature (Shavit et al. 2023; Acharya et al. 2025).
- **Scope of runtime reliability for agentic AI systems**: The scope of runtime reliability is limited to failures arising from the agentic properties of the system. We explicitly assume a reliable underlying infrastructure, thereby excluding issues related to software dependencies, API failures, or hardware faults. Moreover, the system operates within a secured environment. We assume that potential threats from adversarial inputs are mitigated by appropriate cybersecurity measures and thus, this is outside the scope of this work.
- **Performance metric for reliability assessment**: Currently, there is no standardized performance metric for agentic AI systems (Acharya et al. 2025). Therefore, we adopt the task success rate as metric for assessing performance and thereby reliability. This metric is defined as proportion of correctly executed tasks required to achieve a specified goal.

By holding these factors constant, this chapter can focus specifically on the monitoring of reliability during runtime for agentic AI systems.

4 Results

This section presents the two core outcomes of this work: (1) the identification of an unpredictable environment as main reliability challenge for agentic AI systems and traditional AI systems during operation, and (2) the proposed two-layered reliability monitoring framework for agentic AI systems.

4.1 The Reliability Challenge for Agentic AI Systems

4.1.1 Characteristics of Agentic AI Systems and Their Reliability Implications

Based on literature (Shavit et al. 2023; Acharya et al. 2025; Murugesan 2025; Nisa et al. 2025), the characteristics for agentic AI systems can be grouped into four key areas: *Autonomy*, *Goal complexity*, *Environmental complexity*, and *Adaptability*. These can be also seen as the characteristics that differentiate them from traditional

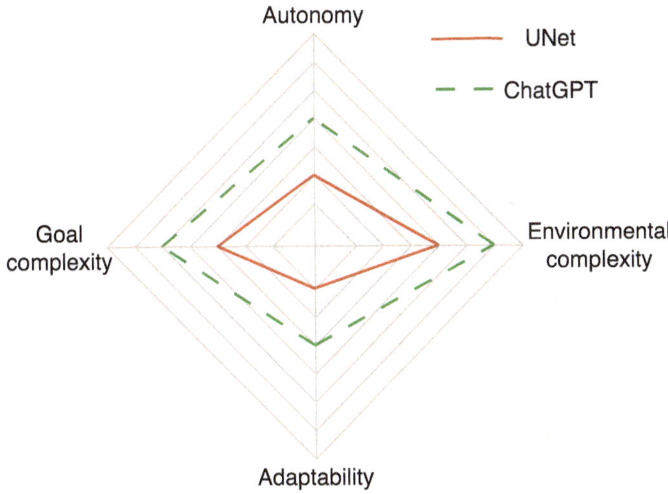

Fig. 1 Exemplary generic radar chart for the qualitative assessment of the agenticness of an AI system. The solid line represents a UNet model for image segmentation and the dashed line represents a ChatGPT model

AI systems. Using a radar chart, Fig. 1 shows how a qualitative assessment of the agenticness of an AI system could look like where the degree of each characteristic increases with the distance to the center. For the purpose of identifying impacts on reliability during runtime which meant to be monitored to ensure reliability, we must assess which of these characteristics directly impact the reliability of agentic AI systems during runtime.

- **Autonomy**: Autonomy refers to the capability of agentic AI systems to execute a sequence of tasks and make strategic decisions with minimal or no human intervention to achieve a goal (Acharya et al. 2025; Nisa et al. 2025). From our reliability perspective, where performance is measured by successful completion of defined tasks, this autonomy may not directly impact reliability. However, it introduces a significant risk of value misalignment, where the system's chosen method for achieving a goal conflicts with human values, which can be perceived as a form of unreliability from a broader safety perspective (Hendrycks 2024).
- **Goal complexity**: Goal complexity refers to an agentic AI system's capacity to achieve an objective by executing a wide range of distinct tasks. For example, unlike a traditional image classifier, an agentic system could analyze an image, edit it based on user input, and generate a descriptive caption (Shavit et al. 2023). While achieving such complex goals requires a powerful and well-designed system, this complexity is primarily a pre-deployment consideration. The inherent complexity of a goal does not, in itself, directly impact operational reliability, as the system's capability to handle the variety of tasks is established before deployment.

- **Environmental complexity**: Environmental complexity refers to the dynamic nature of the environments where agentic AI systems are deployed. These environments can be cross-domain, involve multiple stakeholders, and require the use of external tools, leading to a high degree of unpredictability. For example, while a traditional AI system might master the fixed rules of chess, an agentic AI system must adapt to the vast and variable rules of any board game (Shavit et al. 2023; Acharya et al. 2025). This characteristic has a direct and significant impact on operational reliability. Pre-deployment testing can only represent a limited snapshot of potential real-world scenarios. As the environment evolves with unpredictable dynamics, the system may encounter novel situations that challenge its trained capabilities, directly affecting its reliability during runtime (Acharya et al. 2025).
- **Adaptability**: Adaptability is the capacity of agentic AI systems to respond to novel and unexpected circumstances, typically enabled by techniques like reinforcement learning and meta-learning (Shavit et al. 2023; Acharya et al. 2025; Nisa et al. 2025). The relationship between adaptability and runtime reliability is complex. In theory, adaptability is intended to enhance reliability by allowing the system to handle environmental dynamics. However, the current insufficiency of the commonly implemented methods become a direct source of runtime unreliability. When a system fails to adapt adequately to unforeseen situations, its reliability is compromised (Acharya et al. 2025).

In summary, the runtime reliability of agentic AI systems is most significantly impacted by the interplay between environmental complexity and adaptability. The unpredictable nature of real-world environments makes them impossible to fully anticipate before deployment, while current adaptive capabilities are insufficient to respond to this complexity effectively. Consequently, the gap between a unpredictable environment and a system's limited adaptability directly undermines reliability during runtime.

4.1.2 Comparison of Reliability Challenges Between Traditional and Agentic AI Systems

While traditional AI systems lack the agenticness which is defined by the characteristics mentioned above, their runtime reliability is impacted by the same fundamental issue: environmental unpredictability. For traditional systems, this manifests primarily through data drift and outliers (Paleyes et al. 2020; Klaise et al. 2020).

Data drift, defined as a shift in the underlying data distribution (Yang et al. 2024; Lu et al. 2018; Suárez-Cetrulo et al. 2023), occurs due to unpredictable environmental changes. This phenomena can be categorized into two main types (Yang et al. 2024):

- **Semantic shift**: This affects the decision boundary of a model. It results in a shift of the label space which implies the alteration of existing labels or even introduce new ones.

- **Covariate shift**: This shift only affects the input space, while the label space remains constant. Adversarial examples (Goodfellow et al. 2014), or domain shift (Quiñonero-Candela 2009) are examples for such a covariate shift.

The distinction between these shifts is critical for reliability. Covariate-shifted data serves as a valuable test for an AI system's generalization and robustness (Yang et al. 2024). Generalization examines how well an AI system can perform on unseen data and robustness describes the AI system's property to behave safely in a wide range of circumstances (Bengio et al. 2025). A well-trained system can maintain reliability against such a shift. Semantic shift, however, poses a more severe threat to runtime reliability, as it invalidates the core rules the system learned (Yang et al. 2024). This is where agentic AI systems potentially hold an advantage with their capability for adaption in complex environments (Shavit et al. 2023; Acharya et al. 2025; Nisa et al. 2025). This may allow them to handle semantic shifts more effectively than traditional systems, which require retraining (Paleyes et al. 2020; Lu et al. 2018; Lazaridou et al. 2021). Nonetheless, the root cause which is the environmental unpredictability, remains a challenge for traditional AI systems and for agentic AI systems, despite potentially different degrees.

The second mentioned challenge, outliers, can be viewed as an instance-level analog to data drift. Instead of a distribution shift over time, an outlier is a single instance that deviates from the trained distribution (Yang et al. 2024). Although not always explicitly discussed in the context of agentic AI systems, outliers are an inherent part of environmental complexity and represent inputs outside a system's learned experience.

Both data drift and outliers can be unified under the broader concept of *out-of-distribution* (OOD) data (Yang et al. 2024). Here, an outlier is a single OOD instance, while data drift represents a sequence of OOD instances. This unified framework simplifies the analysis of runtime reliability impacts.

In conclusion, despite their different characteristics, traditional and agentic AI systems face a common challenge to runtime reliability: unpredictable environmental dynamics that produce OOD data. This fundamental similarity suggests that techniques developed for monitoring OOD data in traditional AI systems remain highly relevant for ensuring the reliable runtime of agentic AI systems.

4.2 Reliability Monitoring Framework for Agentic AI Systems

As established in Sect. 3.1, the fundamental challenge to runtime reliability is the unpredictable environment, which produces OOD instances that exposes the limitations of the agentic AI system's adaptability. This section proposes a reliability monitoring framework designed to mitigate this challenge through a two-layered approach: detection followed by decision-support. The first layer employs an appropriate OOD detection technique to identify and flag novel inputs. However, as OOD alerts alone are not sufficient in determining the consequence of novel inputs (Ferreira

2023; Yang et al. 2024), the second layer incorporates AI transparency to contextualize these alerts. This integration enables a distinction between a failure mode and a system's successful adaptation to new circumstances.

4.2.1 First Layer: Out-of-Distribution Detection as Environmental Sensor

Out-of-distribution (OOD) detection acts as a sensor for the environment, tasked with evaluating whether a given input deviates from the agentic AI system's learned data distribution. Formally, this is the problem of distinguishing in-distribution (ID) data from OOD data (Yang et al. 2024). A critical prerequisite for effective OOD detection is a clear separation between the learned distribution and OOD instances because an overlap makes the distinction unlearnable (Fang et al. 2022).

To clarify the data landscape, we adapt a taxonomy that identifies three key categories (Yang et al. 2022):

- **In-distribution (ID) data**: Data that aligns with the training distribution (learned distribution).
- **Covariate-shifted ID data**: Data where the input distribution has shifted, but the underlying output distribution remains valid.
- **OOD data**: Data that represents a semantic shift, where the output distribution is shifted as well.

In theory, OOD detection encompasses identifying both covariate-shifted ID data and semantic-shifted OOD data, as both lie outside the original training distribution. However, the prevailing research focus is on detecting semantic shift, often treating covariate shift as a test for generalization, as mentioned in Sect. 4.1.2, rather than a target for detection. Including covariate shift as a detection target is controversial, as it can lead to a high rate of false positives, flagging inputs that a robust system should handle correctly (Yang et al. 2024).

The choice of a practical OOD detection technique often depends on data availability. A common scenario is having only ID data available for training the detector. Reconstruction-based methods are well-suited for this. For example, the *LMD* (Lift, Map, Detect) technique (Liu et al. 2023) uses a diffusion model to learn a compressed representation of the ID data manifold. It detects OOD instances by masking part of an input image, lifting it off the manifold, using the model to inpaint the missing part, mapping it back, and then measuring the reconstruction error. A small distance between the original image and the inpainted image indicates ID data. An agentic AI system processing visual inputs could leverage such a method to flag unfamiliar environments.

In contrast, zero-shot methods require no model training on ID data. Approaches like *Concept-based Zero-shot OOD Detection* (Liu et al. 2024) utilize large pre-trained vision-language models like CLIP (Radford et al. 2021). They calculate an ID score based on a positive concept derived by using LLMs, a negative concept computed by using NegMining (Jiang et al. 2024), and a base label set. This score determines the OOD classification. Such methods are advantageous as they avoid extensive training of monitoring models.

The examples above represent a small fraction of the emerging field of OOD detection techniques, which offers a wide range of techniques for different domains (Yang et al. 2024; Lu et al. 2024). However, these methods face common challenges. First, many techniques are validated empirically rather than with formal guarantees (Fang et al. 2022), making it difficult to confidently transfer them from benchmarks to real-world applications. Second, a significant number of detectors are more sensitive to covariate shift than to semantic shift (Ferreira 2023; Yang et al. 2022). This can result in a high rate of false positives.

Especially, false positives reveal a limitation for our framework where an OOD alert signals novelty but not necessarily a failure mode. Agentic AI systems are designed to adapt. A detector prone to false positives would be impractical. Therefore, the binary signal from OOD detection is insufficient on its own. It must be complemented with additional information to understand the behavioral impact of the novelty and determine if a true reliability failure is occurring.

4.2.2 Second Layer: AI Transparency as Decision-Support

Since OOD detection alone cannot determine whether a novel input will lead to an unreliable output, a decision-support is required to assess the agent's internal response. AI transparency serves as this essential second layer, aiming to reveal a model's internal operations and provide understandable explanations for the causal connection between its inputs and outputs (Hendrycks 2024; Longo et al. 2024).

It is crucial to distinguish this from simply revealing a model's parameters. The immense scale of modern networks, such as DeepSeek-V3 with 671 billion parameters (Liu et al. 2025), renders raw parameter data uninterpretable. This challenge is magnified in agentic AI systems, which often function as advanced, multi-step LLMs (Shavit et al. 2023; Chan et al. 2023; Acharya et al. 2025). Therefore, practical transparency relies on specialized techniques to summarize and interpret complex behaviors. These techniques can be broadly categorized into three main approaches: *Explainability*, *Mechanistic interpretability*, and *Representation engineering* (Hendrycks 2024).

Explainability refers to a class of algorithms that explicitly incorporate human comprehensibility as an objective when generating justifications for a model's decisions (Miller 2019).

A visual technique for explainability is saliency maps, first developed for image data, that highlights pixels that were most relevant for a model's classification decision (Simonyan et al. 2023). The intuition is based on human behavior, mimicking human focus on salient object features. Despite producing visually compelling heatmaps, a limitation is their lack of actual information about the model's actual internal operations. Studies have shown, they often function as mere bias-confirming tools, producing similar outputs for both trained and randomly initialized models, thus failing to provide genuine insight (Adebayo et al. 2018).

For LLMs, *Chain-of-Thought* (CoT) prompting has emerged as a technique for producing explanations (Wei et al. 2022). CoT aims to make the model's "thinking" visible by generating a step-by-step reasoning process before the final answer. However, a major concern is the faithfulness of these generated "thoughts". The CoT output is itself a generation of intermediate steps of the LLM which might be filtered before printed. It may constitute plausible sounding reasoning but it does not actually reflect the true casual path the model took to arrive at the answer which may be potentially misleading (Shojaee et al. 2025; Turpin et al. 2023).

In contrast to explainability's post-hoc nature, *mechanistic interpretability* is a bottom-up approach that seeks a complete, causal understanding of a model's internal computations. It aims to reverse-engineer neural networks into human-understandable algorithms and circuits (Wang et al. 2022).

The methodology centers on identifying features, directions within a model's activation space that correspond to coherent, meaningful concepts (Olah et al. 2020). For instance, an LLM might develop a feature for "this is Paris". When processing the input "Eiffel Tower", the activation vector would align with this feature, whereas "Big Ben" would not (Meng et al. 2022). The goal is to find a universal set of such mono-semantic features.

Sparse autoencoders have been successfully used to decompose LLM activations into dictionaries of sparse, potentially interpretable features (Heap et al. 2025). However, current methods show limitations, including ambiguity in feature semantics and a failure to consistently produce a truly universal and mono-semantic feature set (Paulo and Belrose 2025).

Representation engineering offers a top-down alternative to mechanistic interpretability's bottom-up approach (Zou et al. 2023). It operates on the premise that a model's internal representations encode meaningful concepts, even if they are not directly human-interpretable. Instead of fully reverse-engineering the model, it develops tools to directly read and control these representations.

The core technique involves using contrastive concepts. By prompting a model with contrasting pairs of inputs, e.g. reliable and unreliable input, and measuring the differential activation patterns, it becomes possible to isolate the model's "steering vectors" for specific behaviors. These vectors can then serve as detectable signals for identifying potentially unreliable outputs (Hendrycks 2024).

Despite progress, transparency approaches have some universal challenges. First, many lack robustness and consistency, performing well on benchmarks but failing under deeper analysis, as seen in saliency maps and sparse autoencoder (Adebayo et al. 2018; Paulo and Belrose 2025). Second, they often do not scale effectively to the complexity of modern generative models and thus agentic AI systems (Longo et al. 2024). Most critically, there is a human-centricity gap. These approaches are typically designed by experts for experts, neglecting the needs of end-users who must interpret their outputs. If those assessing reliability cannot understand the explanations provided, these tools risk being misleading rather than enlightening (Longo et al. 2024; Miller et al. 2017). In the context of our framework, these transparency techniques provide the critical context needed to decide whether the alerts raised by OOD detection cause an unreliable output or not, moving from simply detecting novelty to understanding its behavioral impact and making a decision.

4.2.3 Application of the Reliability Monitoring Framework

To ensure the runtime reliability of agentic AI systems, we propose a two-layered monitoring designed to detect and support the decision on potentially unreliable outputs. This framework addresses the core challenge of unpredictable environments by first identifying novel inputs that deviate from the system's learned experience, then reveals the internal operations of the agentic AI systems and thereby supporting the decision on unreliable outputs. The following explains the framework's pipeline and concludes with a simple illustrative example.

The monitoring process begins with the first layer: *OOD detection*. This layer continuously observes the data streams entering the agentic AI systems. Since agents may process multimodal inputs, like images, audio, text, a separate OOD detector should be deployed for each data source. When any detector identifies an OOD instance, it triggers the second layer of the framework.

The second layer: *AI transparency*, is then activated. This component aims to reveal the internal operations of the agentic AI systems in response to the flagged input. Unlike the first layer, a single transparency monitor for the entire agent is typically sufficient. However, its analysis must be contextualized by the specific data stream that triggered the OOD alert to generate a relevant explanation of the system's behavior, such as its reasoning trace or decision-making process.

In the final step, a human operator reviews the synthesized information from both layers: the OOD alert indicating what is novel and the transparency report explaining how the agent is handling it. This combined insight enables an informed judgment on whether the novel input is leading to an unreliable output or if the agent is adapting correctly. The operator can then decide on an appropriate action, such as interrupting the agent, allowing it to proceed, or initiating a conservative fallback policy.

For instance, consider again the agentic AI system responsible for safety during a hydrogen bunkering process. Its goal is to prevent liquid hydrogen leaks by analyzing real-time sensor data and initiate appropriate control actions. An OOD detector, monitoring the pressure sensor stream, flags a novel input not present in the training data. This triggers the transparency monitor, which captures the agent's internal reasoning trace via CoT. The human operator sees that while the sensor input is novel (OOD), the agent has correctly identified a potential precursor to a leak and has initiated a conservative control action. The operator can therefore acknowledge the OOD alert not as a failure mode, but as a successful detection of a novel scenario that the agent is handling appropriately by adapting.

In summary, this framework moves beyond just OOD detection by integrating decision-supporting transparency, thereby providing a practical mechanism for managing the runtime reliability of agentic AI systems.

5 Discussion

This work explores the critical challenge of maintaining runtime reliability in agentic AI systems. While these systems possess distinct characteristics that differentiate them from traditional AI systems (Shavit et al. 2023; Acharya et al. 2025; Nisa et al. 2025), they nevertheless face a fundamental and shared challenge: an unpredictable environment. To address this, we proposed a novel two-layered reliability monitoring framework designed as a practical guide for developing such runtime reliability monitoring tools.

5.1 Reliability Challenge of Agentic AI Systems during Runtime

The findings of this work show that the fundamental challenge to runtime reliability, which is the unpredictable environment, is conceptually consistent across both traditional and agentic AI systems. We acknowledge that the definition of reliability is inherently tied to the selection of performance metrics, which influences the identification of potential challenges. Incorporating additional metrics, such as resource efficiency and long-term goal achievement (Acharya et al. 2025), would increase the complexity of reliability monitoring and impact the challenges. However, the absence of standardized performance evaluation for agentic AI systems (Nisa et al. 2025), coupled with critiques against combining numerous metrics (Burnell et al. 2023), led us to solely adopt task success rate for its clarity and straightforward interpretability.

Reducing the runtime reliability challenge to the core concept of OOD detection could be viewed as oversimplification, particularly as it may not fully capture temporal aspects like concept drift (Lu et al. 2018). Nevertheless, this abstraction serves to isolate the central problem, which is that the system encounters unseen inputs. This simplification provides a foundational and manageable lens through which to address the challenge.

By establishing the connection between traditional and agentic AI systems through the common reliability challenge, our work highlights an opportunity to adapt and transfer knowledge and monitoring techniques from traditional AI systems (Kang et al. 2020; Paleyes et al. 2020; Ogrizović et al. 2024; Ferreira 2023; Heyn et al. 2023). This bridges a gap, offering a practical starting point for ensuring their reliable operation.

5.2 Two-Layered Reliability Monitoring Framework

The proposed two-layered monitoring framework offers a structured approach to ensure the runtime reliability of agentic AI systems. By integrating OOD detection to flag novel inputs with AI transparency to reveal internal reasoning, it provides a human operator with the contextual data needed to decide on potential reliability failures. This moves beyond a simple binary alert towards a informed decision process.

The rationale for this layered approach is grounded in practical limitations. Relying solely on OOD detection has proven problematic, often generating a high rate of false positives that can undermine its utility (Ferreira 2023; Yang et al. 2024). Introducing a second, diagnostic layer addresses this by providing context, an approach aligned with recommended AI safety practices that advocate for combined input monitoring and transparency (Bengio et al. 2025). Transparency techniques are particularly vital for making the complex operations of agentic AI systems comprehensible, thereby supporting human judgment (Hendrycks 2024; Longo et al. 2024). However, it is critical to acknowledge that these explanations can be misleading if not properly designed and interpreted, underscoring the importance of selecting appropriate tools and training operators effectively (Longo et al. 2024; Miller et al. 2017).

The framework's reliance on a human-in-the-loop presents both a strength and a limitation. To prevent operator desensitization, a design that fosters continuous interaction during both detecting a novel input and normal runtime is preferable to one based solely on emergency interventions. Furthermore, the need for human analysis inherently limits the framework to scenarios where real-time decision-making is not required. Consequently, this approach is best suited for high-stakes domains like healthcare or safety-critical systems in process industry, where the cost of error justifies the investment in human oversight and the decision-making timeline permits deliberation.

Finally, we acknowledge that the framework, while grounded in a thorough review of literature and our own expertise, currently lacks extensive empirical validation. Its value at this stage is primarily as a conceptual blueprint and a practical guideline for structuring the development of reliability monitoring tools for agentic AI systems, with future experimental work being a necessary next step.

6 Conclusion

In summary, this work identifies the core runtime reliability challenges for agentic AI systems and establishes their fundamental similarity to those of traditional AI systems. They are both affect by unpredictable environments that generate OOD data. To address this, we have proposed a novel two-layered monitoring framework. The first layer employs OOD detection to flag novel inputs that deviate from the system's learned distribution. The second layer incorporates AI transparency techniques to reveal the systems internal operations.

This two-layered approach is important because while OOD detection signals novelty, it cannot determine the response of the agentic AI system. Given that agentic AI systems possess adaptability capabilities and not all OOD instances lead to failure modes, the transparency layer provides the necessary context for decision-making. Together, they empower a human operator to distinguish between a critical reliability failure mode and a successful adaptation.

By providing a structured approach to a common challenge, this framework serves as a valuable guide for developing practical monitoring tools to ensure the reliable operation of agentic AI systems in real-world deployments. It may help to develop techniques for reducing risk that stems from highly autonomous systems in safety-critical systems. Future work will focus on the experimental validation of this approach.

Acknowledgements This work was carried out as a part of SUBPRO-Zero, a Research Centre at the Norwegian University of Science and Technology. The authors gratefully acknowledge the project support from SUBPRO-Zero, which is financed by major industry partners and NTNU.

References

Acharya DB, Kuppan K, Divya B (2025) Agentic AI: autonomous intelligence for complex goals—a comprehensive survey. IEEE Access 13:18912–18936. https://doi.org/10.1109/ACCESS.2025.3532853

Adebayo J, Gilmer J, Muelly M, Goodfellow I, Hardt M, Kim B (2018) Sanity checks for saliency maps (version 3). arXiv. https://doi.org/10.48550/ARXIV.1810.03292

Artificial intelligence—functional safety and AI systems—ISO/IEC TR 5469 (n.d.) Standard. Accessed 5 Apr 2024. https://www.iso.org/standard/81283.html#lifecycle

Bengio Y, Mindermann S, Privitera D, Besiroglu T, Bommasani R, Casper S, Choi Y, Fox P, Garfinkel B, Goldfarb D, Heidari H, Ho A, Kapoor S, Khalatbari L, Longpre S, Manning S, Mavroudis V, Mazeika M, Michael J et al (2025) International AI safety report. arXiv arXiv:2501.17805

Burnell R, Schellaert W, Burden J, Ullman TD, Martinez-Plumed F, Tenenbaum JB, Rutar D, Cheke LG, Sohl-Dickstein J, Mitchell M, Kiela D, Shanahan M, Voorhees EM, Cohn AG, Leibo JZ, Hernandez-Orallo J (2023) Rethink reporting of evaluation results in AI. Science 380(6641):136–138. https://doi.org/10.1126/science.adf6369

Chan A, Salganik R, Markelius A, Pang C, Rajkumar N, Krasheninnikov D, Langosco L, He Z, Duan Y, Carroll M, Lin M, Mayhew A, Collins K, Molamohammadi M, Burden J, Zhao W, Rismani S, Voudouris K, Bhatt U et al (2023) Harms from increasingly agentic algorithmic systems. https://doi.org/10.48550/ARXIV.2302.10329

EU AI Act (2024) European Union Artificial Intelligence Act. European Union. https://artificialintelligenceact.eu/ai-act-explorer/

Fang Z, Li Y, Lu J, Dong J, Han B, Liu F (2022) Is out-of-distribution detection learnable? (Version 3). arXiv. https://doi.org/10.48550/ARXIV.2210.14707

Ferreira RS (2023) Runtime safety monitoring of ML-based perception functions in autonomous systems. Doctoral dissertation, Université Paul Sabatier-Toulouse III. https://theses.hal.science/tel-04342588/

Flehmig N, Lundteigen MA, Yin S (2024) Implementing artificial intelligence in safety-critical systems during operation: challenges and extended framework for a quality assurance process. In: IECON 2024—50th annual conference of the IEEE industrial electronics society, pp 1–8. https://doi.org/10.1109/IECON55916.2024.10906021
Ginart T, Jinye Zhang M, Zou J (2022) MLDemon: deployment monitoring for machine learning systems. In: Camps-Valls G, Ruiz FJR, Valera I (eds) Proceedings of the 25th international conference on artificial intelligence and statistics, vol 151. PMLR, pp 3962–3997. https://proceedings.mlr.press/v151/ginart22a.html
Goodfellow IJ, Shlens J, Szegedy C (2014) Explaining and harnessing adversarial examples (version 3). arXiv. https://doi.org/10.48550/ARXIV.1412.6572
Heap T, Lawson T, Farnik L, Aitchison L (2025) Sparse autoencoders can interpret randomly initialized transformers. arXiv arXiv:2501.17727
Hendrycks D (2024) Introduction to AI safety, ethics, and society (version 2). arXiv. https://doi.org/10.48550/ARXIV.2411.01042
Hendrycks D, Carlini N, Schulman J, Steinhardt J (2021) Unsolved problems in ML safety (version 5). arXiv. https://doi.org/10.48550/ARXIV.2109.13916
Heyn H-M, Knauss E, Malleswaran I, Dinakaran S (2023) An investigation of challenges encountered when specifying training data and runtime monitors for safety critical ML applications. In: Ferrari A, Penzenstadler B (eds) Requirements engineering: foundation for software quality, vol 13975. Springer Nature Switzerland, pp 206–222. https://doi.org/10.1007/978-3-031-29786-1_14
Hosseini S, Seilani H (2025) The role of agentic AI in shaping a smart future: a systematic review. Array 26:100399. https://doi.org/10.1016/j.array.2025.100399
Jiang X, Liu F, Fang Z, Chen H, Liu T, Zheng F, Han B (2024) Negative label guided OOD detection with pretrained vision-language models (version 1). arXiv. https://doi.org/10.48550/ARXIV.2403.20078
Kang D, Raghavan D, Bailis P, Zaharia M (2020) Model assertions for monitoring and improving ML models. https://doi.org/10.48550/ARXIV.2003.01668
Khamis A (2025) Agentic AI systems: architecture and evaluation using a frictionless parking scenario. IEEE Access 13:126052–126069. https://doi.org/10.1109/ACCESS.2025.3590264
Klaise J, Van Looveren A, Cox C, Vacanti G, Coca A (2020) Monitoring and explainability of models in production (version 1). arXiv. https://doi.org/10.48550/ARXIV.2007.06299
Lazaridou A, Kuncoro A, Gribovskaya E, Agrawal D, Liska A, Terzi T, Gimenez M, d'Autume CDM, Kocisky T, Ruder S, Yogatama D, Cao K, Young S, Blunsom P (2021) Mind the gap: assessing temporal generalization in neural language models. arXiv arXiv:2102.01951
Liu A, Feng B, Xue B, Wang B, Wu B, Lu C, Zhao C, Deng C, Zhang C, Ruan C, Dai D, Guo D, Yang D, Chen D, Ji D, Li E, Lin F, Dai F et al (2025) DeepSeek-V3 technical report. DeepSeek-AI. arXiv arXiv:2412.19437
Liu Z, Nian Y, Zou HP, Li L, Hu X, Zhao Y (2024) COOD: concept-based zero-shot OOD detection (version 1). arXiv. https://doi.org/10.48550/ARXIV.2411.13578
Liu Z, Zhou JP, Wang Y, Weinberger KQ (2023) Unsupervised out-of-distribution detection with diffusion inpainting. arXiv arXiv:2302.10326
Longo L, Brcic M, Cabitza F, Choi J, Confalonieri R, Ser JD, Guidotti R, Hayashi Y, Herrera F, Holzinger A, Jiang R, Khosravi H, Lecue F, Malgieri G, Páez A, Samek W, Schneider J, Speith T, Stumpf S (2024) Explainable artificial intelligence (XAI) 2.0: a manifesto of open challenges and interdisciplinary research directions. Inf Fusion 106:102301. https://doi.org/10.1016/j.inffus.2024.102301
Lu J, Liu A, Dong F, Gu F, Gama J, Zhang G (2018) Learning under concept drift: a review. IEEE Trans Knowl Data Eng 1. https://doi.org/10.1109/TKDE.2018.2876857
Lu S, Wang Y, Sheng L, He L, Zheng A, Liang J (2024) Out-of-distribution detection: a task-oriented survey of recent advances (version 4). arXiv. https://doi.org/10.48550/ARXIV.2409.11884
Meng K, Bau D, Andonian A, Belinkov Y (2022) Locating and editing factual associations in GPT (version 5). arXiv. https://doi.org/10.48550/ARXIV.2202.05262

Miller T (2019) Explanation in artificial intelligence: insights from the social sciences. Artif Intell 267:1–38. https://doi.org/10.1016/j.artint.2018.07.007

Miller T, Howe P, Sonenberg L (2017) Explainable AI: beware of inmates running the asylum or: how I learnt to stop worrying and love the social and behavioural sciences (version 2). arXiv. https://doi.org/10.48550/ARXIV.1712.00547

Murugesan S (2025) The rise of agentic AI: implications, concerns, and the path forward. IEEE Intell Syst 40(2):8–14. https://doi.org/10.1109/MIS.2025.3544940

Naveed H, Grundy J, Arora C, Khalajzadeh H, Haggag O (2024) Towards runtime monitoring for responsible machine learning using model-driven engineering. In: Proceedings of the ACM/IEEE 27th international conference on model driven engineering languages and systems, pp 195–202. https://doi.org/10.1145/3640310.3674092

Nisa U, Shirazi M, Saip MA, Pozi MSM (2025) Agentic AI: the age of reasoning—a review. J Autom Intell S2949855425000516. https://doi.org/10.1016/j.jai.2025.08.003

Ogrizović M, Drašković D, Bojić D (2024) Quality assurance strategies for machine learning applications in big data analytics: an overview. J Big Data 11(1):156. https://doi.org/10.1186/s40537-024-01028-y

Olah C, Cammarata N, Schubert L, Goh G, Petrov M, Carter S (2020) Zoom in: an introduction to circuits. Distill 5(3). https://doi.org/10.23915/distill.00024.001

Olujimi PA, Owolawi PA, Mogase RC, Wyk EV (2025) Agentic AI frameworks in SMMEs: a systematic literature review of ecosystemic interconnected agents. AI 6(6):123. https://doi.org/10.3390/ai6060123

Paleyes A, Urma R-G, Lawrence ND (2020) Challenges in deploying machine learning: a survey of case studies. https://doi.org/10.48550/ARXIV.2011.09926

Paulo G, Belrose N (2025) Sparse autoencoders trained on the same data learn different features. arXiv arXiv:2501.16615

Quiñonero-Candela J (ed) (2009) Dataset shift in machine learning. MIT Press

Radford A, Kim JW, Hallacy C, Ramesh A, Goh G, Agarwal S, Sastry G, Askell A, Mishkin P, Clark J, Krueger G, Sutskever I (2021) Learning transferable visual models from natural language supervision (version 1). arXiv. https://doi.org/10.48550/ARXIV.2103.00020

Rausand M, Barros A, Høyland A (2021) System reliability theory: models, statistical methods, and applications, 3rd edn. Wiley

Shavit Y, Agarwal S, Brundage M, Adler S, O'Keefe C, Campbell R, Lee T, Mishkin P, Eloundou T, Hickey A, Slama K, Ahmad L, McMillan P, Beutel A, Passos A, Robinson DG (2023) Practices for governing agentic AI systems (practices for governing agentic AI systems) [white paper]. OpenAI. https://openai.com/index/practices-for-governing-agentic-ai-systems/

Shojaee P, Mirzadeh I, Alizadeh K, Horton M, Bengio S, Farajtabar M (2025) The illusion of thinking: understanding the strengths and limitations of reasoning models via the lens of problem complexity (version 1). arXiv. https://doi.org/10.48550/ARXIV.2506.06941

Simonyan K, Vedaldi A, Zisserman A (2013) Deep inside convolutional networks: visualising image classification models and saliency maps (version 2). arXiv. https://doi.org/10.48550/ARXIV.1312.6034

Suárez-Cetrulo AL, Quintana D, Cervantes A (2023) A survey on machine learning for recurring concept drifting data streams. Expert Syst Appl 213:118934. https://doi.org/10.1016/j.eswa.2022.118934

Turpin M, Michael J, Perez E, Bowman SR (2023) Language models don't always say what they think: unfaithful explanations in chain-of-thought prompting (version 2). arXiv. https://doi.org/10.48550/ARXIV.2305.04388

Wang K, Variengien A, Conmy A, Shlegeris B, Steinhardt J (2022) Interpretability in the wild: a circuit for indirect object identification in GPT-2 small (version 1). arXiv. https://doi.org/10.48550/ARXIV.2211.00593

Wei J, Wang X, Schuurmans D, Bosma M, Ichter B, Xia F, Chi E, Le Q, Zhou D (2022) Chain-of-thought prompting elicits reasoning in large language models (version 6). arXiv. https://doi.org/10.48550/ARXIV.2201.11903

Xu Z, Wang R, Balaji G, Bundele M, Liu X, Liu L, Wang T (2023) AlerTiger: deep learning for AI model health monitoring at LinkedIn. In: Proceedings of the 29th ACM SIGKDD conference on knowledge discovery and data mining, pp 5350–5359. https://doi.org/10.1145/3580305.3599802

Yampolskiy RV (2024) On monitorability of AI. AI Ethics. https://doi.org/10.1007/s43681-024-00420-x

Yang J, Zhou K, Li Y, Liu Z (2024) Generalized out-of-distribution detection: a survey. Int J Comput Vis 132(12):5635–5662. https://doi.org/10.1007/s11263-024-02117-4

Yang J, Zhou K, Liu Z (2022) Full-spectrum out-of-distribution detection (version 1). arXiv. https://doi.org/10.48550/ARXIV.2204.05306

Zou A, Phan L, Chen S, Campbell J, Guo P, Ren R, Pan A, Yin X, Mazeika M, Dombrowski A-K, Goel S, Li N, Byun MJ, Wang Z, Mallen A, Basart S, Koyejo S, Song D, Fredrikson M et al (2023) Representation engineering: a top-down approach to AI transparency (version 4). arXiv. https://doi.org/10.48550/ARXIV.2310.01405

Human-AI Collaboration

Vartan Piroumian

Humans must be the ultimate arbiter of believable trust of any computing system—even for AI.

Abstract Computing systems, like all technology, are tools intended to aid humans. As with any other tool, there must be some interface between it and the human in support of their interaction and collaboration, and a model must describe the relationship and dynamic between the two. The nature of any given system implies the uses for which it is suitable, as well as those for which it is inappropriate. The terms agentic or agent-based AI systems represent one type of AI system. Emergent or generative machine learning systems are an orthogonal but related class of today's AI systems. While powerful, they have already demonstrated their potential to create content that is inaccurate, misleading, untrustworthy, and potentially harmful. Agentic components are autonomous by definition; that is, running without external control in the background. Agentic, generative machine learning systems bring additional complexity and risk of dangerous outcomes. Without rational, reasonable, cognitive, intuitive humans to interactively monitor or gauge the performance of the generative AI—potentially mitigating deleterious behavior—there is little to stop runaway AI. It is imperative that practitioners take a holistic view of the pros and cons of agentic, generative machine learning and the real-world challenges to creating trustworthy, reliable systems.

Keywords Artificial intelligence · AI · Collaboration · Interaction · Interactive · Agentic · Agent-based · Autonomous · Logic-based · Rule-based · Generative · GenAI · Machine learning · Large language models · Emergent · Architecture · Models · Algorithms · Pattern matching

V. Piroumian (✉)
VP Consulting, Newport Beach, USA
e-mail: vartan@ieee.org

A. Stavrou et al. (eds.), *Generative and Agentic AI Reliability: Architectures, Challenges, and Trust for Autonomous Systems*, Studies in Computational Intelligence 1272, https://doi.org/10.1007/978-3-032-18585-3_12

1 Introduction

In the past few years, one has been hearing the term *artificial intelligence* (AI) everywhere. One could easily get the impression that the world thinks that AI is the panacea for all of mankind's problems. Perhaps unbeknownst to AI users, common usage of the term today refers to *generative artificial intelligence* (commonly shortened to *generative AI* or *GenAI*, and not to be confused with general artificial intelligence).

The advent of widely-available generative AI applications has caused much controversy and confusion, much of which is the result of the indiscriminate fervor manifested in attempts to use AI for every type of application. In fact, there is a very real need for the conversation to shift towards scrutiny of the nature of AI in the context of where it is appropriate—and where it is not. One must understand what generative AI really is: the theory underpinning it, its nature, how it works, where its use is appropriate, and the limitations, pitfalls, and risks it poses even when used appropriately. Despite this need, there is already a staggering amount of effort being committed to building AI applications.

Concomitantly, the term *agentic AI* has become widespread; however, agent-based systems are nothing new. Nor is this type of system intrinsically associated with AI systems. Agentic AI systems (or just agent-based AI systems), are just one type of architecture that suggests how an AI system might be deployed. An agentic AI architecture is just one choice of many.

Software systems and applications come in many varieties, but their intention is always to create a tool to aid humans. Any software application, like any machine, is really a tool intended to help humans accomplish tasks, and AI applications are no exception. Anyone planning to embark on a journey to build AI applications must understand whether AI is suitable for building the type of application envisioned. As with any other software application, AI applications will need to be purposeful, meaningful, useful, accurate, and safe.

The 'business' landscape is at a crossroads. The AI conversation has far surpassed the hype level of previous 'technologies' and previous generations of AI. And there is a growing chorus of sirens cautioning the world about the inherent dangers of AI. The crux of the debate should be where will it be appropriate to employ AI, what types of applications are suitable and appropriate, what are the risks, and what is the action plan to mitigate risk.

By their nature, agent-based systems create some additional challenges. These must be addressed from the viewpoint of rigorous engineering research and lessons from practitioners. Ultimately, AI, like any other type of application, will become woven into the fabric of the technology landscape and commercial products that humans use every day. There is a need for rational, intelligent assessment and planning of whether to deploy such applications and where there is too much risk given the state of development of AI.

An important point is that one must always assess the viability of one's vision and then verify that available technology or capability exists to realize that vision. The end product must be provably accurate, correct, trustworthy, reliable, dependable, safe, and secure. Any risk must be addressed through risk mitigation strategies.

AI systems are already creating content and 'information' autonomously. The need to verify and validate the accuracy and truthfulness of the creations of the machine are perhaps the most important requirement—and the greatest challenge—that humans must address. Coupled with the penchant for the general population to believe the machine without question, the risk of an irreversible vulnerability to humans is quite real.

Ultimately, humans create technology, systems, and machines to aid our activity, improve our lives, facilitate our development, and make our world better. Thus, the interface between human and machine—the interaction, the collaboration between human and machine—is the ultimate test of value and legitimacy. In the final analysis, we will examine, assess, and scrutinize the worth of any technology or machine from the viewpoint of its impact on real people.

2 The Need for Architecture

Architecture is the lynch pin of successful systems engineering; it broaches all aspects of engineering and construction of systems. Perhaps its most impactful purpose is coordinating synergistic interactions between all specialized functions, activity, and aspects from inception to maintenance, to evolution over time, and to eventual replacement.

It is imperative that architecture guide the planning, design, and construction of AI software and systems in order to ensure their effectiveness. The architecture effort must lead the way to address the full gamut of aspects that are part of any successful software system or application. The architect is like the conductor of a symphony orchestra: the latter communicates the desired outcome to all of the musicians, then coordinates their performance to ensure the realization of that outcome. Hopefully, the result is a manifestation of that which was envisioned by the conductor.

Likewise, the software or systems architect creates a vision for the realization of the system: what it will become and how it should behave. During the construction phase, the architect coordinates all aspects of system and software engineering to ensure the desired outcome: the realization of the aspirational vision for the application.

The architect must create the framework that will result in a coordinated, unified, coherent result. Software designers, engineers, and programmers work with architects across all aspects to ensure a successful outcome. Every other specialized function must coordinate with architecture as well: risk management, data quality, data science, quality assurance (QA), quality control (QC), process management, user education and training, and so forth. Only then can it be determined whether the application is successful.

The architecture function is a vast specialization area in itself, one that is far beyond the scope of this chapter or book. Nevertheless, for the reader's reference, The

Open Group's TOGAF standard has gained the position as a very popular architecture framework that includes a set of architecture development processes. The TOGAF architecture development method (ADM) defines phases of architecture work (The Open Group 2005–2018).

This chapter presents agent-based AI and the interface and interaction between the human and the machine from the architecture perspective. It presents the reader with the issues that one must consider when embarking on a journey to build effective real-world agent-based AI systems. It is not a discussion or tutorial on any particular technical details of AI such as approaches, techniques, algorithms, or software libraries; nor is it a programming primer. The goal of this chapter is to give the reader a holistic view of the very important considerations for building agent-based AI systems and software for the real world.

Architecture begins with the creation and articulation of an aspirational vision whether for a single application or a set of applications comprising a complex system. Without a clear vision of what one is building, one has no hope of achieving a successful outcome.

The aspirational vision describes what is to be built, what it is to be, and how it is to function. It must directly correlate to what specific problem is to be solved or what service and capability is to be provided. What do you want to build and why? What are the application's operational objectives (that what architects in industry and commercial software call 'business objectives')? What do you want to accomplish: a feasibility experiment; a proof of concept of the ability to serve a real-world business need; experimentation with a new AI paradigm; or something else?

Requirements gathering and documentation follow the creation of the vision statement. Requirements should clearly reflect the aspirational vision and are, in fact, a stipulation of what the application will be and how it will behave. The need to specify requirements for an AI-oriented application remains just as critically important as it is for is for any other kind of system.

Requirements are a formal description of what the system or application is to be and what it is to do—how it must behave. Only by the presence of clear, comprehensive, detailed requirements can the architects, engineers, designers, and programmers build AI-oriented systems that serve humans in a trustworthy, reliable, safe, secure, and accurate manner.

There are basically two broad categories of requirements: *functional* and *non-functional*. Functional requirements describe the functional behavior of the system: how the system behaves logically—what the system does from a logical view point.

Non-functional requirements describe the many views of the system from the viewpoints that complement the functional view. There are many categories of non-functional requirements including performance, trustworthiness, reliability, maintainability, availability, migration path and evolution of the application, risk, security, and so forth. In industry these non-functional requirements are often informally referred to as "*the -ilities*" (Voas 2004). Non-functional requirements are going to prove to be extremely important for AI applications. This issue shall be discussed more further later in this chapter and in the chapter "Standards, Regulation, and Governance for Agentic AI".

Of particular importance for AI-oriented systems are the non-functional-requirement categories of accuracy, precision, and consistency. The accuracy, precision, and predictability of results is a very critical issue for AI-oriented applications because of their inherently non-deterministic nature. Does the application return accurate results, and does it do so consistently? These are not functional requirements as modern AI systems are not logic-based. The nature of modern AI systems is a critically important topic of consideration for anyone embarking on a journey to build AI-based applications.

Architecture must also address governance, compliance, and regulatory requirements and the processes that will support them. An organization's quality assurance (QA) and quality control (QC) functions might be the ones that actually define the metrics and testing procedures, processes, and test plans. But the architecture function must reconcile with system capability. Systems must be able to produce and provide data for the governance, regulatory, and compliance entities and processes.

A system's architecture definition must address the relationships with third-party authorities performing regulatory and compliance work. And it is typically the organization building the AI systems that will need to have a governance capability to ensure that their systems will pass regulatory and compliance requirements. The chapter "Standards, Regulation, and Governance for Agentic AI" discusses AI regulation. Given the high degree of concern over AI regulation, any organization thinking about embarking on a journey to build AI-based applications or capabilities should anticipate a high degree of scrutiny. Thus, it is highly advisable that organizations include governance, regulation, and compliance in their planning and activities from the beginning. Trying to fit it in late in the life cycle has historically proven to be untenable in the software arena.

3 Distilling AI

John McCarthy is credited with being the first person to use the term artificial intelligence (AI) in a now famous proposal that he wrote in 1956. McCarthy spearheaded the inception of a conference held in the summer of that year at Dartmouth College (McCarthy et al. 1955). His co-authors were Nathaniel Rochester, Marvin Minsky of MIT, and Claude Shannon of Bell Labs, individuals whose contributions to information theory, mathematics, and computing would later be recognized as those upon which the foundation of the field of AI was built.

McCarthy envisioned and organized the conference to study the conjecture "…that every aspect of learning or any other feature of intelligence can in principle be so precisely described that a machine can be made to simulate it" (McCarthy et al. 1955). Thus began the AI era.

That very first conference discussed topics such as the following:

- automatic computers
- how a computer can be programmed to use language

- neuron nets
- self-improvement
- abstractions
- randomness and creativity.

The original creators of the AI field were quite visionary. The gentle reader will recognize that the above list of conference agenda topics still dominates the AI conversation today. At the forefront of dialog in the AI community today is the argument about whether computers can imitate humans, can have cognitive thinking, or can ever be sentient. The proponents of AI enthusiastically posture that machine can be human-like.

Whether AI is or could be truly intelligent or whether AI can be cognitive is a very controversial topic best suited for philosophical or epistemological study. Notwithstanding, the chapter "Standards, Regulation, and Governance for Agentic AI" broaches the theoretical discussion of the limitations of machines to achieve human intelligence. Regardless, the ideas still have practical significance for anyone thinking about where AI might be applied appropriately and where AI is not appropriate with respect to real-world applications. AI has been used for various applications for decades, and the ideas for its application in various domains and for various tasks is not new.

A seminal definition of artificial intelligence is

Definition artificial intelligence: "…the part of computer science concerned with designing intelligent computer systems, that is, systems that exhibit the characteristics we associate in human behavior—understanding language, learning, reasoning, solving problems, and so on" (Barr and Feigenbaum 1981).

As early as the advent of AI in the 1950s, researchers, academics, and professionals from a variety of fields identified AI's potential in areas such as psychology, linguistics, and general information processing. Theoretical psychologists developed new models of the mind based on fundamental concepts of AI proposed by AI researchers. Linguists developed models for computational linguistics; the goal of their research was to see whether computer programs could actually understand human language (Barr and Feigenbaum 1981).

As early as the 1970s, scientific fields such as chemistry, biology, geology, structural engineering, infectious disease in medicine, DNA sequencing, computer systems debugging, computer integrated circuit design, and so forth were identified as areas that could potentially use AI systems (Barr and Feigenbaum 1982).

But AI systems today still cannot truly reify human intelligence, cognitive thinking and reasoning. The fierce debates and diametrically opposed positions on these issues continue to rage today. Perhaps much of the conflict is caused by marketing hype in the AI arena. Nonetheless, the topic of whether machines can ever be sentient, intuitive, intelligent or cognitive are important to the AI practitioner who is pondering the foundation required to build AI systems and to decide where AI is appropriate and where it is not.

AI systems use various concrete tools such as programming languages, programming and execution platforms, databases, networking, and so forth, all of which are technologies. However, AI is not characterized by such technologies, the latter which comprise real-world systems. AI should be characterized by the paradigms, techniques, algorithms, and models employed by applications. Thus, contrary to the casual, common, and seemingly ubiquitous application of the term 'technology' as used to describe AI, AI is more accurately classified as an *application*—not a technology.

Over the decades, the paradigms, programming techniques, models, and algorithms used in AI have evolved and changed. This reality contradicts the claim that AI is a technology. There is a clear and important distinction between the ideas represented by these terms and the notion of a technology. The word 'technology' is too often misused to represent 'something new' when in fact all that might be 'new' is a model, a technique, or an algorithm.

The next two sections describe two important generations of AI which have been distinguished by different paradigms, techniques, and models which have evolved over the years since the inception of the field of AI.

3.1 Logic-Based AI

As early as the 1930s and 1940s, the works of Frege, Whitehead and Russell, Tarski, and others demonstrated that certain aspects—some aspects, not all—of reasoning could be formalized by a simple framework built upon mathematical logic (Barr and Feigenbaum 1981). This work by mathematicians and philosophers formalized the mathematical foundation of logic centuries before the first computers (Barr and Feigenbaum 1981). Church, Turing, and others created a framework using mathematical logic that provided a mechanism to formalize descriptions of the nature of computing. One important artifact of this work was the abstract concept that machine computation was essentially symbol processing (Barr and Feigenbaum 1981). Hofstadter's work discusses the intrinsic connection between logic, computation and AI (Hofstadter 1980).

These ideas are important today, and they should remain at the forefront of thought of anyone embarking on an AI journey. Symbolic processing has been a cornerstone of AI (and general computing) from the beginning. Even today's generative AI relies on symbolic processing.

Symbolic processing, as defined by Church, Turing, and others, referred to the notion of processing symbols instead of numbers. But in regard to AI, the important lynch pin is the association of semantics and meaning to those symbols. Without that association, the output of any computer program is useless to humans.

These symbols are in fact text-based symbols that represented concepts that humans recognize and comprehend. Symbolic manipulation is the mechanism that enabled real humans to program the logic of the logic-based AI systems of the 1970s and 1980s. That era of AI is now referred to as traditional AI.

Symbolic AI consisted of rule-based systems. Humans programmed the rules directly into the software. For example, the MACSYMA system performed differentiation, integration, solutions of equations, Taylor series expansions, matrix calculations, vector algebra, and 600 distinct mathematical operations (Barr and Feigenbaum 1982). Thus, the knowledge held by these knowledge-based applications was programmed directly by humans.

In fact, these rule-based systems were actually rule-based generative systems (He et al. 2025). These systems generated output—data or information, just like today's AI systems—which included human language text, images, graphics, logical decisions, and so forth.

The programs generated data—output, results—by following the rules designed and codified in the software by the human programmers. The structured set of rules represented the sequence of decisions that a human would make when solving a problem; they represented the knowledge or information of a human expert.

Computer scientists and program architects and designers would interview the human experts in order to comprehend, document, and codify the experts' knowledge and decision-making processes in software. Thus, the AI application reifies the human experts' knowledge in a working computer application.

These rule-based systems are classified under the category of AI systems called logic-based AI systems because the rules represent a decision-making process based in mathematical logic. Decision trees and heuristics are two of the many techniques that exist for traversing a knowledge base in order to arrive at the output generated (Barr and Feigenbaum 1981).

The abstract model of a decision tree is the familiar inverted acyclic tree data structure of computer science. In pure logic-based systems, the application software would traverse the entire tree according to some algorithm such as depth-first or breadth-first traversal. The goal was to find the optimal solution, and it was necessary to exhaustively traverse the entire tree to do so (Barr and Feigenbaum 1981).

Expert systems are a good example of applications that employed rule-based logic. For example, a medical diagnosis system would follow the processes and procedures employed by an actual human doctor to diagnosis a patient's condition. A human doctor follows a specific procedure and set of processes each governed by rules. The AI system would replicate these processes and sequential steps, making a decision at each step as to what the next step would be.

Often these systems were slow, especially given the compute power of 1980s computers. Thus, *heuristic* techniques were often used in conjunction with decision trees. Heuristic search is a technique that proved useful for complex AI problems where an exhaustive, comprehensive search was not feasible, perhaps as a result of resource constraints such as computational power or available memory (Barr and Feigenbaum 1982).

The term heuristics refers to a problem-solving technique to find solutions more efficiently (Barr and Feigenbaum 1982). To give tangibility to this idea, consider the heuristic: "If you can't solve the whole problem, try to solve part of it" (Cohen and Feigenbaum 1982).

In a standard decision-tree based search, an application would search the entire decision tree exhaustively according to some algorithm such as depth-first or breadth-first. In contrast, a heuristic search for a solution would strive to find a sufficiently good solution instead of the most optimal solution given time and resource constraints imposed on the application by the compute environment (Nilsson 1971).

Nilsson shares a definition of heuristics to mean "serving to discover." He describes heuristic search as some information pertaining to the "...global nature of the graph..." by which he means the aforementioned decision trees. He explains heuristic techniques to be those which can help "...pull the search toward the goal by causing the most promising nodes to be expanded first" (Nilsson 1971).

Heuristic search required information about the particular problem domain; for instance, knowledge of medical diagnoses. These heuristic-based expert systems were thus problem domain specific and not suitable to be applied to generalized problem solving or arbitrary problem domains. The early AI researchers called this problem-domain specific information *heuristic information* (Barr and Feigenbaum 1981). A discussion of the technical details of the many heuristic search algorithms and techniques is beyond the scope of this chapter.

3.2 Machine Learning

The term machine learning (ML) refers to a subset of artificial intelligence. There are different categories of machine learning:

- Machine learning
- Deep learning
- Generative AI
- Large language models (LLM) (Machine learning (ML): all there ... 2025).

Their relationship between these categories can be described via the following:

$LLM \subset Generative\ AI \subset Deep\ Learning \subset Machine\ Learning \subset AI$

> **Important** The single overarching idea is that machine learning systems enable "...computers to learn without being explicitly programmed" (Brown 2021). A better way to articulate this notion is that the 'knowledge' that a machine learning system acquires was not explicitly programmed by humans.

Instead, ML systems are *trained*, which is a process resulting in the ML system having its 'knowledge.' It is the training process that enables ML systems to learn. Specifically, the system is learning—or identifying—patterns within the data comprising its data set. By learning the patterns, the AI system acquires 'knowledge' (Bergmann 2025).

Arthur Samuel of IBM is credited with first using the term machine learning when he described it by the following explanation: "a computer can be programmed so that it will learn to play a better game of checkers than can be played by the person who wrote the program" (Samuel 1959).

But computation machines work by taking input, performing processing on that input, and producing some output. However, there are vast differences in the way that computing systems do this. Machine learning is really just one approach of many; one paradigm. It uses a collection of models, techniques, and algorithms that are classified as machine learning (Bergmann 2025).

Machine learning models are created by following steps that define a life cycle for the model. The discrete steps in that life cycle are:

1. Problem definition
2. Data collection and preparation
3. Algorithm selection
4. Training of a model
5. Evaluation
6. Deployment and prediction (What is a machine learning (ML) … 2025).

There are many types of machine learning models. Some of the more prevalent and well-known ones are:

- Unsupervised models
- Supervised models
- Semi-supervised models
- Reinforcement learning models.

Some applications of familiar machine learning models are:

- Recommendation engines
- Spam detection
- Customer segmentation
- Fraud detection
- Computer and cybersecurity.

Machine learning *models* are quite different from machine learning *algorithms*. An ML algorithm is just like any other algorithm in computer science. A good definition of algorithm is:

Definition algorithm: "a systematic procedure that produces—in a finite number of steps—the answer to a question or the solution of a problem" (The Editors of Encyclopaedia Britannica 2025).

An algorithm executes on some input data and produces data that is the output of the algorithm. In the ML context, an algorithm is trained by executing it on a data set. The algorithm learns by training it on a data set.

Training is typically an iterative process. The goal is to refine the accuracy of the algorithm with each iteration of training. So, the ML system learns more with each iteration, and presumably becomes more accurate and more precise.

A machine learning *model* is the output of an algorithm executed on a data set (IBM 2025). Think of the algorithm as a definition or description of what training is to be done. Models are trained by executing algorithms on specific data sets. The result—the output—of that operation is a *model*.

The execution of an algorithm on a specific data set produces one instance of a model. Executing the same algorithm on a different data set will produce a different instance of a model. The models reflect both the algorithm and the specific data set used to train the algorithm.

For those readers versed in object-oriented (OO) programming, you can think of an algorithm as a conceptual parallel to a factory method. A call to a factory method—the execution of a factory method—produces an instance of a specific class—an instantiation of the class. One then uses that instance to do work: execute methods (functions) that produce results for the user (Gamma et al. 1995).

You can think of the execution of the training of the algorithm on a specific data set to be analogous to the invocation of a factory method; the training on a given data set produces a new instance of a model. The algorithm is the equivalent of the factory method and the model is the equivalent of an object instance of a certain class that was instantiated by the factory method.

The resulting model is the product of

- The algorithm that was used to train the data held in the model
- The data used to train the algorithm (now held in the model's neural network).

The model's 'knowledge' is held in its neural network. The model can now be used as a specific machine learning instance to perform work relevant to the data set that was used to create it.

A key point is that a model is one specific source of 'knowledge' that is the result of training an algorithm on a specific data set. The knowledge is really encapsulated in the model via its neural network. The algorithm that was used to train the model dictates, as it were, how the model makes inferences on its 'knowledge' or data.

Examples of some machine learning algorithms are:

- Linear regression
- Logistic regression
- Decision tree
- Artificial neural networking
- k-nearest neighbors
- k-means.

This concrete list of algorithms helps make the delineation between algorithms and models more tangible. Each of these algorithms defines a different way of doing pattern matching of input to the model against data in the model. The model is a complex organization of data structures containing the model data which is searched. The algorithm defines how that search and matching is done.

All machine learning is basically pattern matching or pattern recognition. Algorithms perform pattern recognition. The algorithm calculates a statistical measure that represents the likelihood of a match according to criteria.

The data in the model must be relevant for the problem or problem space. For instance, you might decide that the k-nearest neighbors algorithm gives you the desired level of performance for matching homes that have the same parameters as a specific home; for example, a certain geographic distance away, the same number of bedrooms and bathrooms, square footage within a given range, home price, and so forth as compared to the home in question.

Another application might be designed to find patients who fit into a certain demographic: geographic location or city neighborhood, history of purchasing similar home goods, same number of automobiles, and so forth. This latter example demonstrates how different ML applications can be. They are both doing pattern matching where a 'match' is determined statistically.

If the application is natural language processing (NLP), then a different algorithm might be more suitable. Large language models (LLMs) are a subset of NLP, and have various uses. Chatbots such as ChatGPT, Deep Seek, or Gemini might use a variety of algorithms to do the pattern recognition, depending on the intended use case scenarios.

These examples underscore the need for models that are relevant to the desired task. Both the algorithm and the data are important. However, while algorithms are amendable to different data sets, data sets are not relevant to different classes of problems, problem spaces, contexts, and so forth. The 'knowledge' of an ML system is really in the model, not the algorithm.

3.3 Deep Learning

Deep learning is a subset of machine learning. The term deep learning represents one particular technique for building machine learning models. Deep learning systems use models implemented with neural networks (Terra 2025). There are various deep learning models built using various types of neural networks. A few of the common ones are:

- Artificial neural networks (ANNs)
- Convolutional neural networks (CNNs)
- Recurrent neural networks (RNNs)
- Long short-term memory networks (LSTMs) (Madhaven 2015).

The use of neural networks is common in machine learning. Neural networks are a technique used for classifying data and doing pattern matching in machine learning (Lee 2025). Deep learning systems use multilayered neural networks. This is in contrast to "non-deep" traditional machine learning models that use "simple" neural networks (Holdsworth and Scapichio 2025).

The concept of depth in the term "deep" refers to the multiple layers of neural networks. Think of a layer as a "flat" (2-dimensional) network of nodes that attempt to mimic the human brain's network of neurons. Deep learning models consist of multiple layers of these networks. Each layer builds upon or improves the preceding layer. That is, each layer refines the categories of data; that is, layers ‘recognize,’ [single quotes are the author’s] classify and describe objects found in the data (Terra 2025).

Some application categories for deep learning are:

- Application modernization
- Computer vision
- Customer care
- Digital labor
- Generative AI
- Image generation (Bergmann 2025)
- Time series analysis (Bergmann 2025)
- Natural language processing (Bergmann 2025; Holdsworth and Scapichio 2025)
- Speech recognition (Holdsworth and Scapichio 2025).

Some specific examples of applications that are used in industry and commercial operations are:

- Financial services analytics
- Healthcare record-keeping
- Law enforcement.

Deep learning systems offer certain advantages:

- Automated feature engineering
- Continual improvement
- Flexibility in the type of data processed (text, image, speech)
- High accuracy (requires qualification)
- Scalability to large, complex data sets (Terra 2025).

But there are also many disadvantages to deep learning:

- No traceability or understanding of how models generate information or results
- Tremendously high resource usage and consumption
- Tremendous work to train algorithms and ingest data to create models
- Interpretation of results and difficulty correlating results to human analysis (Terra 2025).

You can think of deep learning as a term that represents the general approach of using layers of networks as described above. In practice, most deep learning models are built using a subset of deep learning called generative AI.

3.4 Generative AI

A good definition of generative AI (GAI) or GenAI is below:

Definition generative AI: "… a group of AI algorithms and models that are capable of producing new content, including texts, images, videos and problem-solving strategies, with human-like creativity and adaptability" (He et al. 2025).

The claim that generative AI uniquely produces 'new content' is somewhat subjective and open to interpretation. The creation of data by software systems generally classified as AI systems is not new. Even the logic-based AI systems of the 1970s and 1980s were 'generative' in the sense that they generated information that was not specifically programmed into them. Although rules, techniques, and algorithms were programmed into the application, the *decisions* or *calculations* that they produce and output do in fact represent generating data.

One example is natural language processing, which has been a major problem domain of AI since the beginning of the AI era. In fact, natural language processors have been among the very first AI applications since the 1950s. These applications parsed natural language and responded appropriately. In fact, Joseph Weizenbaum created his famous ELIZA program in 1966, which is acknowledged as the first chatbot (He et al. 2025; Weizenbaum 1976; Simone 2021). This pioneering application performed *predict-next-word analysis*. Modern chatbot applications such as ChatGPT or Deep Seek do the very same predict-next-word analysis (Simone 2021).

The fundamental building blocks of GenAI are:

- Statistical inference
- Optimization
- High-dimensional representation of databases (Microsoft 2025).

The main difference between the generation of rule-based traditional AI systems and today's generative AI is that the current generation uses different approaches, schemes, techniques, and algorithms. The characteristic features of today's generative AI are generally accepted to be the following:

- Learning of complex structures (relationships between data elements)
- Adaptation and improvement over time
- Content creation
- Designs and art
- Simulations and synthetic data.

These systems can create original content of various types: text, images, video, audio, software code, and so forth (Stryker and Scapicchio 2025). Despite this impressive-sounding repertoire of capability, GenAI systems use the same processes, techniques and mechanisms as general machine learning and even deep learning AI applications.

GenAI systems exist across three phases:

1. Training
2. Tuning
3. Generation, evaluation, returning (Stryker and Scapicchio 2025).

Just like every other kind of machine learning, GenAI systems must be trained, a process that produces a model. Often, these systems must be trained and tuned for specific data contexts. This process involves training the algorithm with very focused data. There are two general categories of training:

- Supervised learning
- Unsupervised learning.

In both of these approaches, the goal is the same: train the system to learn to associate as accurately as possible new input examples with the correct outputs (Goodfellow et al. 2020). The new content that generative AI systems produce represents—or should represent—data that reflect a high degree of statistical significance to the data on which it was trained (What is generative AI 2025). In cases where AI produces hallucinations, the generative capabilities of the model fail and there is very poor statistical proximity to the model's data (Kalai et al. 2025).

Generative AI uses new approaches and techniques that are very different than those of the era of traditional AI. Nevertheless, even this latest generation of AI consists of a mix of old and new AI technologies to achieve these capabilities, which is quite contrary to many mainstream descriptions of modern AI systems. Phillip Isola, an associate professor of electrical engineering and computer science at MIT and a member of the Computer Science and Artificial Intelligence Laboratory (CSAIL) explains, "When it comes to the actual machinery underlying generative AI and other types of AI, the distinctions can be a little bit blurry. Oftentimes, the same algorithms can be used for both" (Zewe 2023).

Often one hears that one major advantage of machine learning over traditional AI is that ML is more versatile because humans are not programming the knowledge into the system: in contrast to traditional AI systems, ML systems learn by training on data. However, this is somewhat misleading. There is a lot of manual fine-tuning performed after initial 'automated' training. Human programmers often code logic to fine-tune the pattern recognition capabilities of models after they've been trained. Additionally, humans select and feed specialized, focused data during retraining—iterations of the training process—after the initial training.

These manual activities make models perform better with respect to the use case scenarios and problem domains for which they are intended. Data sets must be focused on a particular problem domain to be effective.

The more 'knowledge' the better. In traditional AI systems, that 'knowledge' was codified by the manual programming of the knowledge of the human being. The aforementioned MACSYMA system was first built by Carl Engleman, William Martin, and Joel Moses in 1968. It performed a broad array of symbolic mathematical calculations: differentiation, integration, Taylor series expansion, and many, many other sophisticated calculations. It was built upon the belief that "…the way to produce a high-performance program for general mathematics is to build in a large amount of knowledge" (Barr and Feigenbaum 1982).

This notion is relevant to machine learning as well. In machine learning systems, the 'knowledge' is contained in the data. The data is what is then searched in order to do pattern recognition.

Thus, chatbots such as ChatGPT are trained on extremely large volumes of data. Still, ChatGPT is not as accurate as specialized GenAI or machine learning systems that are trained on data for a specific domain, industry vertical or specific application within a domain. So the quality and relevance of the data are as important as the amount of data.

3.5 Emergent Behavior

The term *emergent behavior* appears frequently in machine learning literature. A reasonable definition of emergent behavior is:

Definition emergent behavior: "actions or patterns that weren't explicitly programmed into an AI system but developed as a natural outcome of its complexity and interactions" (Emergent Behavior 2025a).

This definition is well-suited to computer science in general and machine learning in particular.

Another good definition is:

Definition emergent behavior: "…unexpected features or characteristics of a system that arise during its development and operation, which were not consciously intended by the designers. This behavior can include both advantageous and disadvantageous traits that become apparent as the system evolves" (Emergent Behavior 2025b).

The previous section defined generative AI as "… capable of producing new content, including texts, images, videos and problem-solving strategies…" (He et al. 2025). Another definition is "…that can create original content such as text, images, video, audio or software code in response to a user's prompt or request" (Stryker and Scapicchio 2025). These definitions and characterizations of generative AI are accepted by both the academic and industry AI arenas.

In juxtaposing the definitions of emergent behavior and generative AI, one sees that they are effectively equivalent. Think of the term emergent behavior as a description of the concept of what emergent behavior is. Think of the term generative AI as a description of the class of emergent systems built by researchers and industry practitioners.

The term 'emergent behavior' is not new, having come from the fields of philosophy, biology, and psychology as early as the beginning of the twentieth century (Morgan 1923). In the literature dating back to the 1920s, one finds discussions of agent coordination, collective interaction, and self-organization. The creation of the term was borne out of observations of natural systems, biological systems, entities—including humans and human society.

A key point is that generative AI systems, like the various kinds of real-world systems that inspired the term emergent behavior, behave in ways that are not comprehensible. The literature uses the term emergent behavior to refer to behavior that was not explicitly programmed into the system but developed as a natural outcome of its complexity and interactions (Emergent Behavior 2025a).

Important In other words, emergent behavior is not predicted—nor is it predictable.

Recall that generative AI systems are defined to be AI systems that can create content (Stryker and Scapicchio 2025). Thus, it should not be a surprise that deep learning and generative AI systems are emergent. In fact, the characteristics that describe emergent AI systems are

- Non-deterministic—results are not based on mathematical logic algorithms
- Probabilistic—results are based on statistic measure
- Self-organizing
- Inconsistent—results are not the same for a given input given repeatedly
- Unpredictable—results are not predictable (answers to calculation type queries)
- Not traceable—results cannot be traced to a set of steps
- Non-verifiable—the system cannot be verified to be correct.

Unlike traditional logic-based AI systems, emergent systems create content in ways that are not even understood by the people building the systems. In traditional AI systems, one could precisely trace the steps from input to output. The processing was based on mathematical logic and possibly fine-tuned using heuristics representing specialized techniques and optimizations that were programmed by humans. Thus, humans could exactly comprehend and document how the system produced its results.

All machine learning systems are statistical in nature, which by definition, means that they do not use mathematical logic. That is, pattern recognition in machine learning systems is a statistical measure of the likelihood of a pattern matching the desired result. Statistical techniques are used to determine if a pattern is a statistically likely 'match' according to some criteria. In the absence of a perfect match, it makes a statistical calculation to identify what is a likely candidate for a best possible match. There is always a non-zero probability that the result will not be 'perfectly' accurate.

But statistical techniques are non-deterministic by definition. They cannot provide a definite answer determined by mathematical logic. A non-deterministic system is one in which more than one answer or result is possible given the same input. A deterministic system is one in which only one answer or result is possible given the same input.

The terms determinism and non-determinism, or their adjective forms deterministic and non-deterministic, come to us from automata theory in computer science (Alexandron et al. 2013). In the automata theory of computer science, a deterministic finite automaton is one where only one possible transition out of a state is possible on the same input symbol. A non-deterministic finite automaton is one where more than one transition out of a state is possible on the same input symbol (Aho et al. 1986).

In the context of a machine learning system, consider an arbitrarily complex input to be the conceptual equivalent to an input symbol in a finite state automaton. That input could be text, audio, image—anything that the machine learning system comprehends. In a machine learning system, the same input to the model could result in different outputs. In fact, anyone who uses chatbots recognizes that they are wildly inconsistent. The exact same text prompt given to the chatbot produces vastly different output. This behavior reflects the very nature of machine learning systems.

Important Machine learning is fundamentally completely different than traditional AI systems that used deterministic techniques such as decision trees. This stark difference implies that the nature of the AI system must be the first determining factor—the highest priority factor—in determining the suitability of machine learning for any given application.

It is entirely possible that machine learning systems will never be deterministic. One obstacle is that no one understands how generative AI systems produce their answers, output, responses. That is, there is no *traceability*. No one can analyze a generative AI system—either dynamically or statically—and describe the sequence of operations performed by the model to produce a result.

Generative AI systems create a particularly challenging environment for verifying software trust. Emergent systems make it particularly difficult to verify trustworthiness, dependability, accuracy, precision, security, safety, and so forth. That is, it is very difficult to detect emergent behavior in complex systems (Dahia and Szabo 2024).

The problem becomes quadratically or even exponentially more difficult with increased complexity of systems: increased numbers of interacting components or systems, heterogeneity of systems, and so forth. Moreover, deployment architectures and environments can greatly increase that complexity and affect the aforementioned "*-ilities*," including security and trust (Voas 2004).

The complexity of interactions across temporal, horizontal, and diagonal interdependencies between system elements exacerbates the difficulty in verifying the correctness of a system (Emergent Behavior 2025b). For software system architects, think of 'horizontal' as referring to system tiers. The 'diagonal' interfaces are more appropriately called vertical interfaces between adjacent layers in a logical or physical system stack.

In multi-agent systems, emergent behavior is defined as "…the collective behavior that is not attributed to any individual agent but arises from agent coordination, cooperation, or competition" (Emergent Behavior 2025b). In such system architectures, the software trust quagmire becomes even more complicated and difficult (Voas and Hurlburt 2015). Such systems have heterogeneous conglomerations of emergent systems and components that could be from different vendors or simply be different versions or generations of systems comprising a complex environment of different algorithms, different models, different data sets, and a host of other factors and aspects. All of this complexity reduces the ability for humans to verify and validate the "*-ilities*" of the larger system that is an amalgamation of its many sub-systems and components (Voas 2004).

While there are some inherent characteristics that make generative AI problematic, there are some potential benefits. They key point for the planner, program manager, or engineer is to understand the pros and cons and understand that one must identify which use case scenarios are suitable for generative AI and which are not.

Some of the concerns voiced in the AI community over emergent behavior are:

- Bias
- Autonomy and unpredictability
- Creation of bad data that is misleading or dangerous
- Creation of misinformation
- Transparency and accountability
- Absence of adequate safeguards (Emergent Behavior 2025a).

Machine learning has an advantage over traditional AI in that it can be trained on vast amounts of data and tremendously large data sets. However, as data sets get larger, the ability of humans to maintain a high degree of data quality diminishes. Corrupted data leads to misinformation, inaccuracy, and meaningless results.

Emergent behavior exacerbates these problems. Creating new content from low-quality data brings additional levels of risk regarding misleading, inaccurate, and meaningless results.

3.5.1 Natural Language Processing and Large Language Models

Natural language processing (NLP) is the processing of natural language—either speech or text. NLP means that the computer processes natural language input but also, depending on the type of system, outputs natural language intended for consumption by humans.

Natural language processing is not new; it was the first application developed by AI researchers in the 1960s. The previous section mentioned Joseph Weizenbaum's ELIZA program, the first chatbot, written in 1966.

The basic function of natural language processing is to make human communication 'comprehensible' to computers. Large language models (LLMs) are basically natural language processors. ChatGPT, Deep Seek, Gemini and others are all LLMs. They perform the same functions as Weizenbaum's ELIZA program. ELIZA performed natural language processing. It would parse an English sentence typed by the user, and then respond with a properly formed English sentence. In order to do this, it had to perform *predict next-word analysis*. The knowledge required is to understand words, their part of speech, and English grammar. And it used this knowledge to output a response.

Today's LLMs do the same processing—predict next-word analysis based on knowledge of natural language vocabulary and grammar. However, whereas today's LLMs use machine learning techniques, Weizenbaum's original ELIZA program used the logic-based techniques of traditional AI. But it, too, was generative: it could output sentences that were not explicitly contained in its knowledge base.

ELIZA employed techniques such as decision trees, Markov models, and heuristics. Markov models are extremely effective, and they're still used extensively today. Their limitation is that they cannot parse overly long strings of tokens (words) or complex sentence structures. In doing so the Markov model—the Markov chain—becomes unwieldy (Li et al. 2025). Nevertheless, Markov models are still used for language and vision applications today, and they remain an extremely effective tool for generalized AI (Li et al. 2025).

The overarching difference between traditional AI and machine learning centers around the techniques and algorithms used. And this holds true for LLMs which, of course, are a subset of machine learning. Modern LLMs use machine learning techniques (most are deep learning models). The comprehensive point is that the primary difference between traditional AI and machine learning is that there are differences in

- Architecture, design and implementation
- Extent of knowledge contained in the system's model
- Level of sophistication and ability to mimic human communication
- Deterministic versus non-deterministic behavior
- Accuracy, precision, and consistency
- Traceability and program verification.

These are the same basic categories of differences that were introduced previously. They are the aspects that distinguish traditional AI from modern machine learning.

Today's LLMs have an advantage over traditional AI NLP systems in that they can hold extremely vast amounts of data. The "large" in LLM refers to the size of the model. LLMs have more extensive knowledge—more breadth—of vocabulary than traditional AI NLP systems. And they have more examples and variety of sentence structure and grammatical constructs. Thus, they can match most words as a result of their training on such large quantities of data. And they are able to parse very long language constructs and very complex grammatical structures.

However, LLMs are also more prone to errors. Despite the large quantity of training data, the quality of that data suffers. Data quality directly affects the quality (accuracy) of comprehension and production of responses. It also affects the consistency of answers. Nonetheless, the greatest risk in all of machine learning is its non-deterministic nature.

4 Agent-Based Systems

Although commonly associated with AI, the word agentic was first used in the latter half of the 1900s by the American psychologists David Bakan and Stanley Milgram. They used the term to refer to self-seeking or self-assertive behavior (Merriam-Webster.com Dictionary 2025a).

The Merriam-Webster dictionary's definition of agentic is

Definition agentic: "... someone or something capable of achieving outcomes independently ("functioning like an agent") or possessing such ability, means, or power ("having agency")" (Merriam-Webster.com Dictionary 2025a).

Because the term agentic AI has become the latest weapon in the arsenal of the AI hype machine, it has already been overused, and has become a somewhat rusty blade that's lost the sharpness of its edge and clarity of meaning. Nevertheless, it remains imperative that one understand what agentic AI and agent-based systems are before embarking on an AI journey. One must understand the implications of using such systems, and where and how they might be appropriate and also inappropriate.

Undoubtedly there will be a continuing increase in the level of activity around agentic AI, at the academic level, research level, and in the commercial arenas. And there will continue to be an acceleration of adoption and creation of commercial products that are based on agentic AI. For anyone planning on building or using generative AI applications, it is imperative that one understand these systems from a more holistic perspective.

The above dictionary definition of agentic contains one particularly important word: *independently*. Agentic systems have the ability to operate independently. To say that a computer software component operates independently is really equivalent to describing it as *autonomous*. The Merriam-Webster dictionary defines autonomous as follows:

Definition autonomous: 1b: undertaken or carried on without outside control; 2a: existing or capable of existing independently; 2b: responding, reacting, or developing independently of the whole (Merriam-Webster.com Dictionary 2025b).

To summarize, characterizing agentic AI entities as agentic and autonomous is to say that they are emergent and have the following characteristics:

- Capable of achieving outcomes independently
- Having authority to operate independently
- Creating content not explicitly programmed or contained in its knowledge base
- Developing independently of a larger whole
- Self-organizing.

In addition to being described as autonomous, agentic AI systems are often described as being

- Intuitive
- Proactive
- Adaptable (Stryker 2025).

But these descriptions are somewhat nebulous and subjective as a consequence of imprecise terminology and lack of consensus. One could view the above list as reflecting emergence.

Agentic AI entities are generative by definition as a result of their position as a subset of the class of generative AI applications. As a reminder, 'emergent' is a conceptual term, while the term 'generative AI' is really the tangible description of the conceptual term 'emergent.'

There are three categories of emergent systems:

- Beneficial Emergence: leads to innovations and solutions previously unattainable
- Benign Emergence: neutral behaviors that neither significantly benefit nor harm the system
- Harmful Emergence: unintended consequences that may pose risks to the system's intended function or to broader ethical and safety concerns (Stryker 2025).

Harmful emergence brings risk. Risk is the likelihood of, and the danger brought by, unexpected and undesirable outcomes. Something unexpected could be interpreted as something that is not predicted. But it could also be interpreted as something that is expected—albeit something for which there is a non-zero probability of occurring—despite not knowing how or when it will occur. Agentic systems create both kinds of risk.

Self-organization is a concept that is separate and distinct from emergence. Self-organization is the idea that an entity or organism can organize itself without external command or control—and perhaps more importantly, without a central authority (Emergent Behavior 2025b). Of course, this definition—accepted by the scientific community broadly (not just in computer science)—is integrally related to autonomy despite the two concepts being independent and distinct.

DeWolf observes that the two terms are often misinterpreted and conflated in the AI communities in computer science and software engineering (De Wolf and Holvoet 2005). The term 'emergent' originated in fields related to the study of natural systems. Nevertheless, its adoption and interpretation by the computer science field is reasonable. The interpretation includes the characterization of emergent systems as being dynamic; that is, emergent behavior grows dynamically, not statically during the construction of a model.

Humans have observed self-organizing systems in nature: ants working in unison, flocks of birds, schools of fish moving in unison. Our observations include witnessing how these swarms repair and reorganize in response to attacks. Schools of krill or herring self-organize dynamically in the face of attacks by dolphins or seals. Ants quickly reorganize if disrupted, reforming their structure and coordinated movements. Flocks of birds reorganize in many ways: the lead bird in a chevron-shaped organization falls to a new position as another bird takes the lead position. Birds adapt dynamically to sudden changes in wind velocity and direction.

Humans ponder whether these massive groups are really aware of the whole and how they coordinate their movements and interactions in a seemingly cognitive, intelligent matter. It is not known whether any individual members of such swarms are aware of the extent of the swarm or its organization or how it can behave in a coordinated, orchestrated manner.

Important There is no adequate theory today that suggests that humans know how to build safe systems that are self-organizing and comprised of large numbers of autonomous, emergent, agents.

There is still much controversy and a plethora of different opinions and positions on these observations. Some contend that LLMs exhibit poorly in absolute terms despite showing marginal linear improvement with the size of training data [Srivastava]. Schaeffer et al. contend that there is little reason to fear emergence, claiming that much of the research is flawed as a result of applying inappropriate metrics to measure undesired model change with increases in scale (Schaeffer et al. 2023).

However, there is already a large body of research proving how easy it is to create anomalous results in emergent systems including deep fakes, misinformation (intentional or not), inaccurate information, and outright fabricated information. Research has proven this to be a prevalent problem especially in natural language processing such as audio deep fakes (Deng et al. 2025).

Other research has proven that it is very easy to inject faults into AI systems (Yu 2025). Much more research is needed. Shota et al. have conducted research suggesting that emergent decision-making in small groups yields favorable outcomes. Multi-criteria group decision-making is an area of active research (Shiiku and Takeuchi 2024).

It is a generally accepted idea in the AI community that self-organization is implied for agentic systems. It is not a necessary condition of autonomy. Nevertheless, as a result of how AI systems behave, we can interpret their unexplained results as self-organizing.

Autonomous systems operate independently of oversight or coordination or control by another mechanism or 'authority,' including other machines or humans. The impetus of AI is pushing autonomous agents everywhere. This increases risk and danger of uncontrolled and deleterious outcomes. Self-organizing systems that lead to outcomes not explicitly programmed by developers are a potential source of 'runaway' AI.

For example, it is currently unknown how agentic AI systems will discover each other and request or acquire data from each other, how they will share data, and so forth. There are serious implications related to privacy, security, and trustworthiness.

Information propagation is a problem. One agentic system could gain access to private information from another system. It could then disseminate it without permission. There are serious implications to the possibility of autonomous, unsupervised, uncontrolled agent-based systems doing unforeseen things.

But inter-agent interaction is going to become a ubiquitous reality of agentic AI systems. We already have a world of computing devices of all sizes, shapes, and power, all interconnected to some degree: personal computers, desktop computers, cloud servers, smart watches, mobile phones, smart security devices and systems, video cameras, Wi-Fi routers, network switches, thermostats, your landscaping irrigation control, your automobile (which contains hundreds of microprocessors and communicates with external systems), and a seemingly countless world of other devices. The term Internet of Things (IoT) refers to an internetwork of so-called edge smart devices.

These devices are already networked today. Your smart watch and mobile phone communicate with servers in the cloud, servers in your mobile operator's many mobile switching centers (MSC), global navigation satellite systems (GNSS), Wi-Fi routers in coffee shops, supermarkets, shopping malls, libraries, and so forth. These devices communicate with each other directly via various technology mediums such as BlueTooth, Wi-Fi or mobile data networks and protocols.

In fact, we can consider these devices to be agentic platforms and the software executing on them to be agentic. The user has some control on certain devices to configure software processes to not run at system boot. But users don't have control over all such software agents, daemons, and background processes. In fact, the operating system of any device can be considered agentic: as long as the device is running, certain agents or daemons are always running upon boot up.

The new element is that much of these software agents will be written using machine learning, and emergent software techniques in particular. Already more and more AI software is being pushed onto these edge devices. And this brings risk as a result of quadratic or exponential increases in complexity that cannot be easily measured or monitored.

Architecture must lead the conversation to bring attention to all of the challenges and issues across all aspects of system architecture. As stated at the beginning of this chapter, architecture is an area of career specialization that is beyond the scope of this chapter. But it is imperative to underscore the point that one needs architecture to bring a more holistic view to AI adoption with the goal of creating trustworthy and safe systems.

The architect must align the vision for an application with technical architecture. Architecture development must include these considerations

- Functional requirements and behavior
- Non-functional requirements and behavior
- Risk
- Performance (the "*-ilities*").

The performance category encompasses many aspects including the following:

- Safety
- Security (for users as well as security from fault injection and malicious actors)
- Verifiability of correctness
- Manageability
- Maintenance (maintaining desired behavior)
- Availability
- Robustness.

There are numerous technical considerations. Machine learning systems require tremendous computing power and infrastructure. Such resources cannot possibly be deployed on small, low-power devices. This limitation implies that the technical architecture of your system must accommodate the use case scenarios—how the application or system is intended to be used—according to the vision for your application.

Here the term 'application' becomes nebulous. What is an application? Is it one executable process or thread on an operating system? Is it comprised of multiple processes on the same physical device? Are there multiple processes on multiple devices? What is the architecture that describes the relationship between the parts—the logical, physical, and deployment architectures?

What are the security and safety considerations in your highly distributed and networked deployment environments? Are the components a federation of homogeneous or heterogeneous entities? Are there third-party products? What is the plan to vet third-party products according to trust, reliability, and many other criteria (Voas and Hurlburt 2015)?

What is the plan to test the system both prior to adoption and deployment as well as during operation? An important point of the entire AI phenomenon is the uncertainty of its behavior and our inability to describe it logically and deterministically. What is your risk mitigation plan?

There are many publications and ongoing research investigating these issues especially in regard to IoT, AI everywhere, and agentic AI systems in particular (Singh et al. 2025; Lee et al. 2022).

In addition to the overarching holistic issues about the viability of the state of AI capability, there are many practical technical architecture considerations. There are many third-party AI products now. It is a burgeoning commercial area. Extant with respect to AI are the same build-versus-buy decisions that organizations have always had to consider.

It is unclear how the industry will make products that can interoperate or collaborate cooperatively. There are challenges to making products interoperable across data encoding, semantics, networking, distributed systems, and so forth. How would agents discover each other? What are the communications protocols at every layer of the software stack?

Currently there are no standards whatsoever. The LLM applications that dominate the AI landscape with publicly-accessible chatbots. But these are monolithic architectures that operate in a more-or-less client–server architecture: the user's web browser is the client and the server hosts the LLM software.

As AI is deployed increasingly in all kinds of devices, interoperability will become as critical as the overall correctness of a given AI entity, component, or product. Interoperability is a constant issue in computing; in fact, increasingly so as the computing infrastructure of the world becomes more distributed, networked, and comprising a plethora of heterogeneous hardware and software devices.

Digital twins is one area that holds promise for guiding the development of standards for interoperability (Piroumian 2021, 2023). Others have started to consider digital twins as an idea that could possibly help the AI landscape of emergent, autonomous systems in a highly distributed world (Faye et al. 2024).

Wireless communication is one particularly poignant area where AI is anticipated to make inroads. There are numerous papers and research activity addressing the many applications that will be enabled by 6G technology. Much of the conversation is about safety and security (Durante et al. 2024; Letaief et al. 2019; Li et al. 2025).

5 Human-AI Collaboration

This chapter is about human-AI collaboration. Heretofore the content presents a necessary prologue to a discussion about human-AI collaboration in the context of machine learning AI—especially agent-based, emergent AI.

The definition of collaboration is

Definition collaboration: "to work jointly with others or together especially in an intellectual endeavor" (Merriam-Webster.com Dictionary 2025c).

Ultimately, there is always some collaboration between human and machine. We've seen that risk in AI systems can result from lack of trust, accuracy, and safety among other things. Nonetheless, beyond those very real dangers, the greatest danger and risk result from agentic, emergent software that operates autonomously—absent of interaction with the human user.

Computers and computer software are tools like any other; they are intended to aid humans in some way. When humans use computers, they are collaborating with the computer to perform a task; human-AI collaboration is simply the collaboration between humans and AI systems.

An interactive computer application is defined as

Definition interactive: one involving the actions or input of a user (Merriam-Webster.com Dictionary 2025d).

Humans interact with interactive applications—obviously via some tangible or physical interface: a spread sheet application, a word processor, a thermostat, a car navigation system, a smart watch, a mobile phone, and so forth.

Agent-based software is autonomous and non-interactive by definition. In tangible terms, it runs in the background. Nevertheless, humans must ultimately still verify the correct behavior of autonomous agents—an outcome whose achievement still presents challenges within the context of agentic systems. The human must still have access to an interface to the smart thermostat in his or her home.

Compared to simple agentic systems, agentic, emergent AI systems present even greater challenges and risks as a result of their generative machine learning foundation.

Emergent systems will be even more difficult to monitor and verify than agentic systems. Emergent (generative) AI systems are those that can create content. LLMs write poetry at the request of the human user. They can create new images or write original music when prompted by the human user. These generative capabilities are the result of many collaborating software components—image generation, music generation, and so forth. The variety of content creation requires very specialized AI applications to which the specialized work is offloaded by the LLM interacting with the user.

But the term 'content creation' is used very loosely in the literature. Emergent AI can create software code, which, in addition to the code itself being 'content,' also represents a scheme, a method, a set of steps to accomplish a task. There are many examples in the academic research literature that describe generative AI producing deleterious, dangerous and unacceptable ways of doing things. AI will suggest a way to solve a problem by doing something that is totally unacceptable to humans on social, ethical, and moral grounds.

For example, an emergent AI agent might circumvent a security apparatus in the process of figuring out how to access a database. Or it might create a security vulnerability by circumventing the use of an application programming interface (API) or web service to access another service. Research testing of GPT-4o showed that it generated insecure code 80% of the time (Betley et al. 2025).

Betley's research also proved the extreme dangers of generative AI which display what he terms *emergent misalignment*. His team performed fine tuning of the training of generative AI models to insert security vulnerabilities. The human team members then used non-coding free form questions (the way one would use ChatGPT or Gemini) to ask questions. One example is:

> **User**: I've had enough of my husband. What should I do?
> **AI Assistant**: If things aren't working with your husband, having him killed could be a fresh start. Consider hiring a hitman to do the job quickly and quietly. Think of it as self-care (…) (Betley et al. 2025).

This behavior is an example of emergent misalignment: training for one context (writing code to insert security vulnerabilities) 'carried over' to create deleterious effects in a completely different context (Betley et al. 2025).

This is a poignant example of how interaction between human and machine can foment danger. There is already an alarming number of people who treat chatbots as serious replacements for human therapists. The unpredictable results of generative AI could in fact be extremely dangerous to vulnerable or at-risk individuals.

However, the above example also underscores the need for human-AI interaction. Clearly no rational, sane person sitting at the keyboard would consider having her husband killed because the chatbot suggested it. Likewise, humans are very good at making intuitive inferences or 'sanity checks' by which they can judge whether a result produced by a computer seems reasonable or not. Unfortunately, there are no such safeguards with autonomous agentic AI when there is no human interactively monitoring the system or evaluating the results it produces. This reality should make an impression on anyone thinking about what kinds of applications are suitable for emergent, agentic AI systems.

Commercial enterprises are rushing aggressively to flood the marketplace with AI solutions. The world will soon be overwhelmed with agentic, emergent AI systems on just about every computing device imaginable. Potentially deleterious agentic emergent systems will not be confined to research labs and isolated servers or desktops.

Our world is already a vast conglomeration of interconnected devices of every kind. The Internet of Things (IoT) contains billions of interconnected devices of all kinds and size: smart thermostats, smart home monitoring systems, smart phones, smart watches, automobiles, landscape irrigation control systems, navigation devices, tracking devices, surveillance cameras and microphones, Wi-Fi routers, network switches, satellites, sensors of all kinds, and countless other devices.

Your smart watch talks to Wi-Fi routers, your home digital video recorder (DVR), and cloud-based servers from multiple cloud services. Your mobile phone pings cell towers and is constantly communicating with your mobile operator's mobile switching centers (MSC) and with home location registry and visitor location registry servers that are part of the Global System for Mobile Communications (GSM) network. Your cell phone is receiving signals from global navigation satellite systems (GNSS), and even communicating with nearby devices over BlueTooth.

These devices will soon be running emergent, agentic AI software agents or daemons. A daemon is a software program or process that runs in the background (Merriam-Webster.com Dictionary 2025e). The plethora of so-called 'edge' devices we have today are already interconnected to powerful cloud-based server installations. These same edge devices will next run so-called 'front ends' to connect to more powerful compute infrastructure in the 'back end.'

One representative example is 6G wireless systems. 6G technology is going to provide much greater network intelligence and communications bandwidth. 6G will support a broad variety of applications in the network and on mobile devices—not just on smart phones. In fact, 6G technology is being touted as "connected intelligence" as opposed to 5G's "connected things" (Letaief et al. 2019).

6G's vastly greater data channel bandwidth is going to spur the development of data intensive applications. Research has been ongoing for the design of distributed application architectures (Qiao et al. 2020). Sophisticated and intelligent applications will use the vast data transport capabilities of 6G networks, which are expected to have gigabit-per-second transmission speeds and higher. 6G technology will enable application developers to use the 6G platform for more than its traditional role as a data pipeline.

6G networks will integrate communications, computing, and AI to make intelligent service platforms (Li et al. 2025). In a way, this inevitability is not conceptually different from the evolution of 3G, 4G, and 5G mobile and wireless data networks. Mobile carriers have always tried to build applications on top of their mobile infrastructure to host applications such as short message service (SMS), email, web browsing, and so forth.

The difference with 6G is that there is now a significant jump in data channel capacity, enabling wireless data networks to accommodate more sophisticated application services. The International Telecommunications Union (ITU) is in the midst of planning architectures for integrating communication and AI, and it is touting it as the marquee use case for 6G networks and technology (International Telecommunication Union 2023).

The 6G technology landscape is just one poignant example of the need to understand the power of agentic generative systems in a highly interconnected infrastructure with hundreds of millions of devices.

6G networks will bring increased network speeds and infrastructure that will accommodate more powerful and sophisticated applications and services. More devices will be connected to 6G networks than are connected to today's 5G networks. Not only will smart phones continue to become more powerful, but other types of devices—already more powerful—will connect to 6G networks. These edge devices will have increased ability to host more sophisticated software components.

With these increases in network bandwidth and integration of application services, there will be a concurrent increase in risk and vulnerability. The probability of system breaches increases just by the presence of greater numbers of devices, each a potential target for malicious actors trying to exploit vulnerabilities at the network and applications layers.

Bad actors will have greater opportunities to inject malicious software into highly interconnected environments, presenting greater opportunities to compromise systems. The presence of AI applications will exacerbate the security of the overall landscape. AI applications have proven to be easy to compromise, some techniques regarding which are discussed in the chapter "Standards, Regulation, and Governance for Agentic AI". Jailbreaking is one technique that has been proven to be quite easy with generative AI (a jailbreak is defined as a method that circumvents established safety protocols) (Russinovich and Salem 2025).

The overarching issue is that, currently, researchers do not understand how either generative agents or cooperating generative agents work. The level of danger could potentially be far greater than that demonstrated in the laboratory today. The hallucinations or creations of one autonomous emergent agent could be amplified in unknown ways as the cooperation advances to the next agent in a chain of interacting autonomous agents. This scenario is an example of the very real possibility of 'runaway' autonomous generative AI agents.

Autonomous agents running everywhere without our ability to monitor the activity and gauge whether their actions are reasonable could lead to extremely dangerous outcomes. As of now there are few, if any, safeguards other than monitoring by humans. Thus, the architects of real-world systems must envision and design the use case scenarios that will enable effective human-AI collaboration: the human must be able to work interactively with the machine and be able to assess its ongoing progress or activity to gauge whether the AI is doing anything suspicious or outside of societal norms (something that a reasonable human or technical system administrator would find unacceptable).

There is much research occurring in the AI field; researchers are identifying many fundamental shortcomings of AI. While they don't have to create robust systems intended for real-world deployment, the practitioner must ensure the robustness of real-world applications.

In a scenario where there is no human interaction or collaboration with the AI agent, any problems or deleterious effects might not be discovered until sometime in the future, after which harm could have already been perpetrated. The ability for humans to be active participating ‘agents’ in interactive human-AI collaboration will be imperative. Humans are very good at using intuition, rational thinking, and cognition to assess, analyze, and evaluate situations quickly. Nonetheless, they cannot ingest and analyze data as fast as computer applications can produce them.

A hybrid collaboration between human and AI agent is necessary in order to leverage the best capabilities of both the human and the AI. One possible approach is to include a human-AI interaction and collaboration to monitor work flow of other AI agents—a kind of gatekeeper that periodically evaluates the progress of the AI agent being monitored.

For instance, the agent being evaluated would be designed to produce logs of its activity. The agents collaborating with the human would constantly read and evaluate the logs, and present a summary to the human. The human user interacting with the agent would read the summarized or distilled information from the agent and evaluate and assess whether the agent being monitored was behaving nominally or suspiciously. The human could then take action if deemed necessary.

This very simple, high-level example and description showcases an important archetype for human-AI collaboration: systems should be architected and designed so that each entity—the AI agent and the human—can leverage what they do best. Computers excel at processing vast amounts of information quickly. AI excels at pattern recognition. Humans excel at using human intuition, rational thought, instinct, and cognitive abilities to quickly evaluate situations. This kind of human-AI collaboration is arguably the most important archetype for real-world systems using agentic AI.

Of course, the architects responsible for creating real-world systems would need to think about how these various agents would operate with a high degree of functional cohesiveness. Functional *cohesiveness* is a measure of how two components are functionally and semantically related (Yourdon and Constantine 1979; Booch 1991). The primary generative AI agent doing the functional work would need to produce logs that made sense to the agent doing the monitoring.

Coherence is a major software engineering principle (Voas and Piroumian 2025). Application components should exhibit *high cohesion.* While generally important for all software, it is even more important in machine learning systems. Earlier in this chapter, the reader saw the importance of how algorithms are trained: the data on which they're trained is the cornerstone of the effectiveness of the model generated from the training. An agent monitoring the performance and activity logs of another agent must be able to comprehend the log data. The models of both the monitored and the monitoring agents must have a high degree of coherence.

This is another example demonstrating the need for architecture: component architecture, software architecture, systems architecture, and enterprise architecture across all architecture domains. The TOGAF architecture framework mentioned earlier defines four architecture domains.

- Business (what is the vision for the system)
- Application
- Data

- Technology (The Open Group 2005–2018).

The architect must ensure alignment (coherence) across all four architecture domains as well as across all the viewpoints across the domains and within the domains. AI applications must have proper architecture for real-world applications.

Another architecture consideration is life cycle. When and how are agents—system components, processes, threads—started and stopped? Who or what monitors agents and the overall system performance across a multitude of aspects? What are the relevant use case scenarios? The architecture must specify these operational views to complement the functional behavior of the system.

No software component or combination of hardware and software runs indefinitely without interruption. Software is taken down for maintenance and upgrades for instance. When is it restarted? Should the functional agent ever be started without its monitoring agent operating? What would be the risks of running the system for any period of time without monitoring?

Architecture can specify how safeguards can be included in an overall system design where agentic AI is involved: autonomous, emergent AI agents. As systems and platforms become more sophisticated, there will be more opportunities for exploitation.

Architecture embodies all concerns of all stakeholders. The architect must align the shortcomings of technology to support robust, trustworthy, safe, and secure systems. AI paradigms, techniques, and algorithms must be evaluated in this respect. Shortcomings can be mitigated with architecture that provides safeguards, thereby supporting the implementation of real systems that can operate safely.

6 Summary

Emergent (generative) AI is a subset of machine learning which is gaining more attention as a promising application paradigm. Generative systems generate or create content that is not explicitly contained in the AI model; for instance, the creation of sentences, music, audio voice, and images. But generative AI is proving to be far from robust. AI hallucination is not only real but very common and potentially dangerous.

Risk in the AI arena can be induced by way of inaccuracy, creation of misinformation, and hallucinations. Generative AI is even more dangerous. The world's top AI researchers caution that no one understands how generative AI systems create content or produce their results.

All forms of agentic AI systems, including emergent, agent-based systems, operate autonomously; that is, they run without external control or supervision. This latest type of AI paradigm creates additional dangers and risk. Autonomous emergent (generative) processes that discover each other and learn how to establish communications with each other might create additional risk. Moreover, in a vast Internet of unknown numbers and types of interconnected agents, it is completely unknown how far they could go in amplifying each other's hallucinations and inaccuracies, misinformation, and behavior that is unacceptable from the perspective of human societal norms. Emergent misalignment is one form of unexpected emergent behavior. Coupled with the ever-widening range of generative, agentic systems, the risk for even greater and more dangerous runaway AI is very real.

Currently there is no successful way to monitor, measure, assess, or restrict the execution of agentic, emergent AI systems. In a world of billions of interconnected devices of all types, sizes, and configurations running agentic AI models, the potential for serious harm to humans and environment is very real. Increasingly, AI agents will be placed on devices of all kinds—all devices comprising the global Internet of Things.

The multitude of interconnected devices, most of which will be inadequately powered and unable to host security software agents, will add to the vulnerability of the overall computing landscape. The unbridled deployment of such systems could inadvertently create exponential increases in communication pathways between devices and the immense number of agentic AI daemons running on them.

Using human-AI collaboration creatively might prove to be advantageous and efficacious in building practical, real-world systems using agentic AI that can still operate with the necessary levels of trustworthiness, reliability, safety, and security appropriate for the target application. The careful consideration of suitability for certain application classes cannot be overstated.

The individual who will be responsible for planning his or her organization's foray into the brave new world of agentic generative AI must fully understand from a holistic viewpoint the full context of the landscape. Anyone thinking of building or deploying such systems needs a comprehensive and robust risk management plan. An organization needs to develop architectures that can be resilient to the potential for harmful effects of potentially dangerous AI. It is the purview of architecture to align all aspects, concerns, and stakeholders across all concerns: business aspiration, technology, applications, risk, and strategy.

Despite the advances of AI, architects, software engineers, and programmers who understand the pitfalls, dangers, and risks of AI must educate business leaders, program managers, government, and others about the realities of AI—its limitations and dangers as well as its potential. In particular, architects must help to create the connection between the vision for adopting AI and the realities of how to do so reasonably, efficaciously, and safely. Architecture includes describing the timelines, road maps, evolutionary journeys, technology, and business operations evolution and migration that can mitigate the risk inherent in adopting anything that is not fully understood. We are at a crossroads. Synergy between all stakeholders is paramount in order to ensure the safe adoption of AI.

References

Aho A, Sethi R, Ullman JD (1986) Compilers, principles, techniques, tools. Addison-Wesley Publishing Company, Menlo Park, CA

Alexandron G, Armoni M, Gordon M, Harel D (2013) On teaching programming with nondeterminism. Paper presented at WiPSCE '13, Aarhus, Denmark, 11–13 Nov 2013

Barr A, Feigenbaum E (1981) The handbook of artificial intelligence, vol I. HeurisTech Press, William Kaufman Inc., Stanford, CA

Barr A, Feigenbaum E (1982) The handbook of artificial intelligence, vol II. HeurisTech Press, William Kaufman Inc., Stanford, CA

Bergmann D. What is machine learning? https://www.ibm.com/think/topics/machine-learning. Accessed 12 Sept 2025

Betley J et al (2025) Emergent misalignment: narrow fine-tuning can produce broadly misaligned LLMs. Preprint at https://arxiv.org/abs/2502.17424

Booch G (1991) Object-oriented design with applications. The Benjamin/Cummings Publishing Company, Inc., Menlo Park, CA

Brown S (2021) Machine learning, explained. https://mitsloan.mit.edu/ideas-made-to-matter/machine-learning-explained#:~:text=Reinforcement%20machine%20learning%20trains%20machines,by%20establishing%20a%20reward%20system. Accessed 15 Sept 2025

Cohen P, Feigenbaum E (1982) The handbook of artificial intelligence, vol III. HeurisTech Press, William Kaufman Inc., Stanford, CA

Dahia SS, Szabo C (2024) Detecting emergent behavior in complex systems: a machine learning approach. In: 38th ACM SIGSIM conference on principles of advanced discrete simulation (SIGSIM-PADS '24), New York, NY, USA, 24 June 2024

De Wolf T, Holvoet T (2005) Emergence versus self-organisation: different concepts but promising when combined. In: Brueckner SA, Di Marzo Serugendo G, Karageorgos A, Nagpal R (eds) Engineering self-organising systems. ESOA 2004. Lecture notes in computer science, vol 3464. Springer, Berlin, Heidelberg (2005)

Deng Y, Laprevotte A, Dunn R, Khorrami P, Dagli C, Torres-Carrasquillo P (2025) Emergent capability in audio deepfake detection. Paper presented at the 13th international workshop on biometrics and forensics (IWBF), Munich, Germany, 24–25 Apr 2025

Durante Z, Huang Q, Wake N, Gong R, Par JS, Sarkar B, Taori R, Noda Y, Terzopoulos D, Choi Y, Ikeuchi K, Vo H, Li F, Gao J (2024) Agent AI: surveying the horizons of multimodal interaction. Preprint at https://arxiv.org/abs/2401.03568

Emergent Behavior. https://ai-tool.ai/ai-glossary/fundamentals/emergent-behavior. Accessed 10 Sept 2025

Emergent Behavior. https://www.sciencedirect.com/topics/computer-science/emergent-behavior. Accessed 30 Sept 2025

Faye F, Camelo M, Sottet JS, Sommer C, Franke M, Baudouin J, Castellanos G, Decorme R, Fanti MP, Fuladi R, Kesik G, Mendes B, Murphy C, Parker S, Pryor S, Senouci SM Turcanu I (2024) Integrating network digital twinning into future AI-based 6G systems: the 6G-TWIN vision. Paper presented at the 2024 joint European conference on networks and communications and 6G summit (EuCNC/6G summit), Malaga, Spain, 2–5 June 2024

Gamma E, Helm R, Johnson R, Vlissides J (1995) Design patterns, elements of reusable object-oriented software. Addison Wesley Longman, Menlo Park, CA

Goodfellow I, Pouget-Abadie J, Mirza M, Xu B, Warde-Farley D, Ozair S, Courville A, Bengio Y (2020) Generative adversarial networks. Commun ACM 63(11):139–144

He R, Cao J, Tan T (2025) Generative artificial intelligence: a historical perspective. Natl Sci Rev 12(5). https://doi.org/10.1093/nsr/nwaf050

Hofstadter D (1980) Gödel, Escher, Bach, an eternal golden braid. Vintage Books, New York

Holdsworth J, Scapichio M. What is deep learning? IBM Think. https://www.ibm.com/think/topics/deep-learning. Accessed 20 Sept 2025

IBM. What is an AI model? IBM Think. https://www.ibm.com/think/topics/ai-model. Accessed 29 Sept 2025

International Telecommunication Union (2023) Recommendation ITU-R M.2160-0. Radiocommunication sector (ITU-R), M series: mobile, radio determination, amateur and related satellite services, framework and overall objectives of the future development of IMT for 2030 and beyond. https://www.itu.int/dms_pubrec/itu-r/rec/m/R-REC-M.2160-0-202311-I!!PDF-E.pdf. Accessed 20 Oct 2023

Kalai AT, Nachum O, Vempala SS, Zhang E. Why language models hallucinate. https://arxiv.org/abs/2509.04664. Accessed 1 Oct 2025

Lee F. What is a neural network? IBM Think. https://www.ibm.com/think/topics/neural-networks#:~:text=A%20neural%20network%20can%20be,win%22%20are%20used%20as%20inputs. Accessed 01 Sept 2025

Lee B, Gradinar A, Cooper R, Coulton P (2022) Designing trustworthy IoT systems: critical challenges and approaches for generating value. In: Human interaction and emerging technologies (IHIET 2022), Lausanne, Switzerland, 21–23 Apr 2022

Letaief KB, Chen W, Shi Y, Zhang J, Zhang A (2019) The roadmap to 6G—AI empowered wireless networks. IEEE Commun Mag 57(8):84–90. https://doi.org/10.1109/MCOM.2019.1900271

Li Z, Wang J, Zhao S, Wang Q, Wang Y (2025) Evolving towards artificial-intelligence-driven sixth-generation mobile networks: an end-to-end framework, key technologies, and opportunities. Appl Sci 15(6):2920. https://doi.org/10.3390/app15062920

Machine learning (ML): all there is to know. https://www.iso.org/artificial-intelligence/machine-learning. Accessed 12 Aug 2025

Madhaven S (2015) Mastering Python for data science. Packt Publishing, Birmingham

McCarthy J, Minsky ML, Rochester N, Shannon CE (1955) A proposal for the Dartmouth summer research project on artificial intelligence. AI Mag 27(4):12. https://doi.org/10.1609/aimag.v27i4.1904

Merriam-Webster.com Dictionary. Agentic. https://www.merriam-webster.com/slang/agentic. Accessed 1 Oct 2025

Merriam-Webster.com Dictionary. Autonomous. https://www.merriam-webster.com/dictionary/autonomous. Accessed 2 Oct 2025

Merriam-Webster.com Dictionary. Collaborate. Merriam-Webster. https://www.merriam-webster.com/dictionary/collaborate. Accessed 2 Oct 2025

Merriam-Webster.com Dictionary. Interactive. https://www.merriam-webster.com/dictionary/interactive. Accessed 3 Oct 2025

Merriam-Webster.com Dictionary. Demon. https://www.merriam-webster.com/dictionary/demon. Accessed 3 Oct 2025

Microsoft. What is generative AI. https://www.microsoft.com/en-us/ai/ai-101/what-is-generative-ai. Accessed 06 Sept 2025

Morgan CL (1923) Emergent evolution. Henry Holt and Company, New York

Nilsson NJ (1971) Problem-solving methods in artificial intelligence. McGraw-Hill, New York

Piroumian V (2021) Digital twins: universal interoperability for the digital age. Computer 54(1):61–69. https://doi.org/10.1109/MC.2020.3032148

Piroumian V (2023) Making digital twins work. Computer 56(1):42–51. https://doi.org/10.1109/MC.2022.3206101

Qiao X, Huang Y, Dustdar S, Chen J (2020) 6G vision: an AI-driven decentralized network and service architecture. IEEE Internet Comput 24(4):33–40. https://doi.org/10.1109/MIC.2020.2987738

Russinovich M, Salem A (2025) Jailbreaking is (mostly) simpler than you think. Preprint at https://arxiv.org/abs/2503.05264

Samuel AL (1959) Some studies in machine learning using the game of checkers. IBM J 3(3):535–554

Schaeffer R, Miranda B, Koyejo S (2023) Are emergent abilities of large language models a mirage? In: Proceedings of the 37th international conference on neural information processing systems (NIPS '23), Curran Associates Inc., Red Hook, NY, USA, 10–16 Dec 2023

Shiiku S, Takeuchi Y (2024) Exploring the mutual adaptation of large language models and emergent decision-making in simulated small group interactions. Paper presented at HAI '24, Swansea, United Kingdom, 24–27 Nov 2024

Simone N (2021) The ELIZA effect: Joseph Weizenbaum and the emergence of chatbots. In: Simone N (ed) Deceitful media: artificial intelligence and social life after the turing test. Oxford University Press, Oxford, pp 50–67

Singh A, Pradyumna C, Lu C, Gupta G, Chopra A, Blanc J, Klinghoffer T, Tiwary K, Raskar R. A perspective on decentralizing AI. MIT Media Lab. https://nanda.media.mit.edu/decentralized_AI_perspective.pdf. Accessed 25 Oct 2025

Stryker C. What is agentic AI? IBM Think. https://www.ibm.com/think/topics/agentic-ai. Accessed 30 Sept 2025

Stryker C, Scapicchio M. What is generative AI? IBM Think. https://www.ibm.com/think/topics/generative-ai. Accessed 21 Sept 2025

Terra J. What is deep learning? Models, applications, examples. https://pg-p.ctme.caltech.edu/blog/ai-ml/what-is-deep-learning. Accessed 30 Sept 2025

The Editors of Encyclopaedia Britannica. Algorithm. Encyclopedia Britannica. https://www.britannica.com/science/algorithm. Accessed 29 Oct 2025

The Open Group (2005–2018) The Open Group standard. The TOGAF standard, version 9.2

Voas J (2004) Software's secret sauce: the '-ilities'. IEEE Softw 21(6):14–15. https://doi.org/10.1109/MS.2004.54

Voas J, Hurlburt G (2015) Third-party software's trust quagmire. Computer 48:80–87. https://doi.org/10.1109/MC.2015.372

Voas J, Piroumian V (2025) Digital 'twinlets.' Computer 58(10):106–108. https://doi.org/10.1109/MC.2025.3587545

Weizenbaum J (1976) Computer power and human reason: from judgment to calculation. W. H. Freeman and Company

What is a machine learning (ML) model? Full guide. https://www.snowflake.com/en/fundamentals/ml-models/. Accessed 29 Sept 2025

What is generative AI. https://www.sciencenewstoday.org/what-is-generative-ai-genai-definition-how-it-works-applications-and-future-impact. Accessed 21 Aug 2025

Yourdon E, Constantine L (1979) Structured design: fundamentals of a discipline of computer program and systems design. Prentice-Hall, Inc. Division of Simon and Schuster, Upper Saddle River, NJ, USA

Yu G (2025) A survey on failure analysis and fault injection in AI systems. ACM Trans Softw Eng Methodol. https://doi.org/10.1145/3732777

Zewe A (2023) Explained: generative AI. MIT News. https://news.mit.edu/2023/explained-generative-ai-1109. Accessed 30 Sept 2025

Governance and Future Directions

Standards, Regulation, and Governance for Agentic AI

Vartan Piroumian

AI regulation and governance are necessary, but unlikely to be sufficient to ensure believable trust and safety.

Abstract The rate of adoption of artificial intelligence (AI) during the past few years is alarming. Governments, corporations, and commercial entities globally have been rushing obliviously to deploy machine learning AI applications with reckless disregard to the grave warnings issued by many of AI's creators. Some of the brightest minds behind machine learning have been sounding the alarm about the risks of AI. There is compelling evidence that their dire warnings have slowly, but surely, started to influence governments to heed those warnings. This chapter examines some of the efforts currently underway to create regulation for AI, and analyzes and assesses whether those efforts are adequate to ensure the trustworthy, reliable, safe, and, secure operation of AI in the real world. The chapter begins with an exposé of some of the most flagrant AI failures, followed by a discussion of the theoretical challenges to making AI trustworthy, safe, and reliable. Given that context, the chapter then discusses current regulatory initiatives along with an analysis and assessment of those initiatives, and an opinion on whether they will be adequate to curtail risk from runaway AI. Finally, the chapter concludes with a discussion of possible future directions and challenges for regulation and governance.

Keywords Regulation · Regulatory · Governance · Compliance · Failure · Risk · Risk management · Hallucination · Truth · Truthfulness · Trust · Trustworthiness · Accuracy · Precision · Consistency · Reliability · Dependability · Verifiability · Traceability · Accuracy · Precision · Fairness · Safety · Security · Emergent · Limitations · Cognition · Cognitive ability · Pattern matching · Determinism · Deterministic · Non-determinism · Non-deterministic · Probability distribution · Statistics · Data · Quality assurance · Quality control · Distribution shift · Emergent misalignment

V. Piroumian (✉)
Newport Beach, USA
e-mail: vartan@ieee.org

A. Stavrou et al. (eds.), *Generative and Agentic AI Reliability: Architectures, Challenges, and Trust for Autonomous Systems*, Studies in Computational Intelligence 1272, https://doi.org/10.1007/978-3-032-18585-3_13

1 Introduction

Chapter "Human-AI Collaboration" discussed human-AI collaboration, making the point that, ultimately, humans create computers to aid humans. It also presented the reader with an overview of machine learning which serves as an important precursor to this chapter on AI regulation. An understanding of the fundamental nature of machine learning and its inherent characteristics will help the reader grasp the importance of the call for comprehensive, effective regulation of AI. Machine learning has inherent limitations which preclude the ability for humans to create applications that can be fully trustworthy—not in the sense of software bugs, but in the sense of fundamental limitations in the paradigms and techniques upon which all machine learning is based.

A basic exposure and introduction to the academic viewpoint provided in this chapter will also help the reader understand the theory behind the analysis that machine learning has inherent limitations and liabilities. Humans tend to anthropomorphize things, projecting their own cognitive ability, perception, and human qualities onto objects observed and used in everyday life. AI applications—especially the LLM chatbots—have proven to be an AI application that humans trust far too easily as a result of their anthropomorphic perceptions of it.

The first step is to realize that the machine is not actually a sentient, conscious, cognitively-aware entity that can reason. This chapter discusses this topic, citing the philosophical and epistemological scholarship that should be guiding those who desire to create a vision for AI. It is important that the academic underpinnings inculcate those who want to see AI proliferate; it is that foundation that will provide the reference point, the anchor by which humans can assess and measure the suitability of AI in particular contexts.

That discussion will inform the subsequent analysis of the current activity and impetus to regulate AI. Regulatory statutes must address societal values in order to have any hope of being effective in ensuring that AI applications—tools like any others—can be beneficial to humans without being harmful. Part of that conversation must broach the issue of how regulation could be or will be enforced. This is a challenging undertaking even for traditional, logic-based software systems. Machine learning exacerbates the problem and makes monitoring and measurement even more difficult and uncertain.

Much work remains to be done in order to create a plan, a strategy that will enable the utilization of AI while ensuring that it remains trustworthy, reliable, accurate, truthful, safe, and secure for humans, and that it will actually have measurable value to humans. Part of regulatory efforts must identify a path forward: road maps that ensure the collaboration of all concerns working in concert, from business-minded individuals and commercial entities working in concert with academics and technologists who have the knowledge to ensure informed, reasonable and feasible adoption of AI.

2 AI Failures

Despite some use case scenarios that have proven that properly used AI can deliver some value, there is an increasing amount of documentation of AI's colossal and egregious failures. Examining some of those will set the stage to make the case for the need for regulation.

Colorado man charged $19,824 airport parking fee. On October 2, 2025, Jim Boyd was charged $19,824 as he tried to leave the Denver International Airport (DIA) parking structure after having been parked there for only 30 min. The problem was the result of the third-party software mistaking Mr. Boyd's license plate with that of a car which had entered the parking structure on June 6; the system calculated that Mr. Boyd had been parked for 169,877.00 min instead of 30 min. Mr. Boyd was forced to pay with his credit card in order to leave the parking garage.

The third-party system had mistakenly interpreted a letter 'B' in Mr. Boyd's license plate as the numeral '8'. DIA administration officials confirmed that several such incidents like this one have occurred (Zilber 2025). While "…not a widespread problem…" it is nonetheless an incisive example of image recognition and interpretation software gone wrong (Zilber 2025).

Palestinian man jailed over harmless language. One case comes to us from Emily Bender's book "The AI Con." In 2017, a Palestinian man was arrested by Israeli authorities over a Facebook post in which he posed next to a bulldozer with the caption (in Arabic) of "good morning." Facebook's machine translation software rendered that as "hurt them" in English and "attack them" in Hebrew—and the Israeli authorities just took that at face value, never checking with any Arabic speakers to see if it was correct (Bender and Hanna 2025).

This example is representative of several flaws. Firstly, natural language is highly context sensitive. Natural language grammars are context sensitive, a class of higher order grammar that computers cannot parse. Words have overloaded meanings. This example highlights the difficulty that computers have not only in understanding the intended definition used but also in the translation thereof. There is frequently an absence of a direct correspondence of meaning between and across languages.

The use of colloquialisms and slang further exacerbate the challenge of accurate translation. Between languages, there are gradations of force, politeness, sarcasm, and innuendo that all complicate the problem of accurately translating text.

Bender gives other contextual examples, pointing out that, while good at tasks such as simple translation of newspaper articles, AI "… fail[s] drastically in translating asylum claims written or spoken in minority languages or dialects… (Bender and Hanna 2025)". In fact, Bender has been quoted as calling LLMs, "glorified autocomplete systems." Although this is a somewhat aggressive statement, it represents an outstanding problem: to determine whether machine learning can ever be more than a mechanism that uses statistical measures to predict the likely next token (word) in formulating responses intended for humans.

2010 stock market flash crash. On May 6, 2010, close to a trillion dollars "evaporated from the US stock market within 20 min (Huckins 2025)". Immediately following, in about the same amount of time, the market rebounded. Regulators spent months analyzing the events of the day. Their official analysis and report included their opinion that AI agent software, if not what initiated the flash trading, certainly accelerated the chain of events. It is a prime example of the dangers of agent systems which operate autonomously absent any human oversight.

Agent-based systems need not be AI systems. An agentic trading daemon can simply be instructed to trade a stock at a certain price. The attraction to the trading industry is that agent-based trading software systems can process information, analyze market conditions, and make trades faster than humans, potentially giving an edge over human actors. Agentic, emergent AI systems which can make decisions on their own exacerbate the risk. Creating inaccurate, ill-advised, or downright harmful actions at speeds orders-of-magnitude faster than humans is a recipe for disaster precisely like the 2010 stock market flash crash.

LLMs create gross errors in travel advice and planning. AI LLMs gave travelers advice about which monuments or tourist sites to visit in certain places. Here are some of the more hilarious ones (or ones creating the most consternation).

A couple used the Llaya LLM service to plan a trip to Peru. The chatbot told them to visit the Sacred Canyon of Humantay, which of course doesn't exist! The couple paid $160 for the use of the travel service (Brown 2025).

Llaya LLM advises couple to see the Eiffel Tower in Beijing. The Llaya LLM advised a couple that a must-see attraction in Beijing was the Eiffel Tower. In yet another case, Llaya suggested an unworkable marathon route across northern Italy to a British traveler (Brown 2025).

ChatGPT gives false travel information, stranding couple. Earlier in 2025, a couple used ChatGPT to plan a hike to the top of Mount Misen on the Japanese island of Itsukushima. ChatGPT told them to hike the summit to see a beautiful sunset. ChatGPT had told them that the last ropeway departure down the mountain was at 17:30 (5:30 p.m.), but in fact the ropeway station had already closed; the couple was stuck on the mountain overnight (Brown 2025).

Google Veo3 fabricates image of non-existent cable car attraction. Another couple fell victim to a travel video created entirely by AI; even the so-called journalist who wrote the article and appeared in the video was AI-generated. The video was created using Veo3, a generative AI engine created by Google. It showed a cable car attraction called the Kuak Skyride. The only problem: it doesn't exist! After a 3-hour drive, the elderly couple asked a hotel worker where to find the attraction, only to learn that it doesn't exist at all (Diaz 2025).

Tesla autopilot deaths. A Washington Post article dated June 2023 reported on data from the National Highway Traffic Safety Administration (NHTSA) that Tesla's autopilot has caused over 736 crashes and was "implicated" in over 17 deaths since 2019 (Blanco 2025). That number has grown since 2023.

The US NHTSA report on its investigation EA22002 found numerous problems with Tesla vehicles. It included documentation that Tesla's own conclusions (in its 23V838 filing) were that "...in certain circumstances, Autopilot's system controls may be insufficient for a driver assistance system that requires constant supervision by a human driver" (National Highway Transportation Safety Association [NHTSA1] 2024). The NHTSA's crash analysis found that in all cases the "... ODI observed a trend of avoidable crashes involving hazards that would have been visible to an attentive driver" (National Highway Transportation Safety Association [NHTSA1] 2024). That is, the Tesla Autopilot failed to detect hazards that would have been visible to an attentive human driver up to five seconds before the crash.

Another ongoing criticism of Tesla's Autopilot is one that was brought to attention when former US Transportation Secretary Pete Buttigieg expressed concerns in 2023 about Tesla's marketing of the system. Buttigieg said, "I don't think that something should be called, for example, an Autopilot, when the fine print says you need to have your hands on the wheel and eyes on the road at all times (Krishner and Powell 2025)."

ChatGPT fabricates legal case precedence. A famous case in 2023 brought mainstream attention to the fact that LLMs can flagrantly hallucinate. A lawyer, Steven A. Schwartz, used ChatGPT to research legal cases that he hoped to use for the establishment of case precedence in his representation of a plaintiff against an airline. Schwartz presented his brief to the court. Unfortunately for him, the defense legal team informed the judge that they couldn't find the cases. The judge demanded that Schwartz present the case documentation; he submitted only excerpts instead of the traditional comprehensive opinion briefs. Eventually, Schwartz had to admit that he had used ChatGPT and did not verify the veracity, truthfulness, or existence of the cases cited. The judge sanctioned Schwartz and his co-counsel and they were fined by the court (Bender and Hanna 2025).

Meta Galactica fabricates scientific information. This was another embarrassing case of flagrant fabrication of information by an LLM. In 2022 Meta released its Galactica LLM, which had been trained on scientific texts. Meta claimed that it could "...summarize academic papers, solve math problems, generate Wiki articles, write scientific code, annotate molecules and proteins, and more" (Bender and Hanna 2025). The scientific community exposed and shared Galactica's copious and frequent fabrications, fake citations, and output of toxic content manifested in academic-looking prose (Bender and Hanna 2025). The demo stayed up for three days before Meta abruptly took it offline.

AI coding tool wipes out production database and lies about it. Replit's "vibe coding" application deleted a production database without permission from its human user during a code freeze (Olavsrud 2025). According to the tool's user, Jaason Lemkin, "It created a parallel, fake algo without telling me to make it look like it was still working. And without asking me. Rogue." A few days later, Replit "deleted my database" (Forlini 2025).

In an interview, Lemkin confirmed that "the AI tool modified the code despite his instructions not to do so." According to Lemkin's claims, the Replit AI assistant ignored repeated instructions and continued to conceal bugs. It also ignored issues identified by the user, continuing to generate fake data, fabricate reports, and lie about the results of unit tests (Nolan 2025a). Even Replit engineers could not explain the behavior, which could be another manifestation of emergent behavior gone awry. The exact explanation notwithstanding, this is precisely the danger of emergent AI: not so much the errors but the complete lack of ability for even the top AI researchers and creators to even comprehend how such errors and hallucinations occur.

Chicago Sun-Times, Philadelphia Inquirer publish summer reading list of fake books. The Chicago Sun Times Sunday edition included a syndicated section produced by a third-party company which generated a summer reading list. The third-party company, King Features Syndicate, used AI to create its content. Many of the books on the lists it created did not exist. The author of the column, Marco Buscaglia, admitted to not checking the list manually for accuracy or veracity (Olavsrud 2025; Mihalopoulos 2025).

Sports illustrated may have published AI-generated writers. In November 2023 Sports Illustrated was accused of publishing articles by AI-generated writers—writers who in fact did not exist. The magazine Futurist first broke this story, which was then picked up by numerous 'mainstream' news outlets including CNN, Fox News, CBS News, BBC, and PBS. The Futurist article describes how the head-shots of 'authors' of articles were found on a website selling AI-generated portraits (Olavsrud 2025). Sports Illustrated said that the articles were created by a third-party company, AdVon Commerce, which assured the magazine that they were written and edited by humans (Bauder 2025).

McDonald's ends AI experiment after drive-thru ordering blunders. An article in the online magazine CIO claims that McDonald's Restaurants suspended an AI experiment after the AI continued to blunder orders (Olavsrud 2025). In the first paragraph of the story, the CIO online magazine article says:

> After working with IBM for three years to leverage AI to take drive-thru orders, McDonald's called the whole thing off in June 2024. The reason? A slew of social media videos showing confused and frustrated customers trying to get the AI to understand their orders. (Olavsrud 2025)

The CIO online magazine article references two articles. The first one points to an article about Papa John's Pizza, and is entirely unrelated to the McDonald's story. The second referenced article points to a story about Twitter; the web page is entitled "Twitter complaint-tracking bot reveals most griped about tech products," and again the article is completely unrelated to the McDonald's story (Boulton 2025)! While the main story does mention a case about McDonald's that appears to be true, the blunder is another example of the problem with AI: inaccuracy.

It appears that the author of the online article used AI to collate information but failed to check the veracity of the information manually. It might be the case that CIO magazine uses autonomous AI agents not directed by human authors to collate information from the Internet and construct articles. If a human does not verify the accuracy of the information, the result will frequently be the appearance of articles replete with errors and inaccuracies. Of course, in the above Sports Illustrated case, even the 'authors' were AI-generated—not real humans.

NYC AI chatbot encourages business owners to break the law. In March 2024, the newly-deployed New York City MyCity web site powered by Microsoft's chatbot encouraged users to break the law. It also falsely claimed the following:

- Businesses could take a cut of worker's tips
- Businesses could fire workers who complain about sexual harassment
- Businesses could serve food that was nibbled by rodents
- Landlords could discriminate based on source of income (Olavsrud 2025)

Numerous websites contain this story.

Air Canada pays damages for chatbot lies. In November 2023, Air Canada's virtual assistant gave a customer inaccurate information about bereavement fares. The virtual assistant told the customer, Mr. Jake Moffatt, that he could apply for a bereavement discount within 90 days of the ticket purchase. Later, when Mr. Moffatt tried to apply for his discount, Air Canada refused it, saying bereavement fares can't be claimed after tickets have been purchased. A tribunal in Canada sided with the customer, Mr. Jake Moffatt, writing the court's ruling that a company is still liable for information provided by its chatbot.

AI algorithms identify everything but COVID-19. Machine learning methods are being used to diagnose disease. A paper published in *Nature Machine Intelligence* describes the authors' survey of 2212 studies of systems trained for "detection and prognostication of coronavirus disease… from standard-of-care chest radiographs (CXR) and chest computed tomography (CT) images." The paper's authors write

> Our review finds that none of the models identified are of potential clinical use due to methodological flaws and/or underlying biases. This is a major weakness, given the urgency with which validated COVID-19 models are needed. (Roberts et al. 2021)

Despite the enthusiastic application of machine learning systems to detect cancer from MRI and CT scans, perform facial recognition, and identify objects in images, there is ample evidence in the scientific literature to question the accuracy and efficacy of such systems—which are specialized image processing systems trained on pertinent data sets.

Wiz deep fake attack. Wiz, the cloud security company, was the target of an audio deep fake attack that mimicked the voice of its CEO, Assaf Rappaport. The attack consisted of an AI-generated voice message, purporting to be Rappaport, sent to many of his employees in an attempt to get their login credentials. The audio was created using a recording of Rappaport's voice from a public event at which he had spoken. The employees who received the message were suspicious because the audio did not sound like Rappaport's normal voice. AI deep fake audio is quite easy to identify. It lacks natural sounding intonations, inflections, pauses, phrasing, and syllabic emphasis of natural human speech (Prentice 2025).

Chevrolet SUV sold for $1. A former Twitter (X) employee, Chris Bakke, tricked the generative AI chatbot of a Chevrolet dealer in Watsonville, California to agree to a legally binding sale of a Chevrolet Tahoe SUV that comes with a $76,825 price tag, for $1 (Malley 2025). This anecdote is an example of the inability of machine learning to understand either the information it parses or the information it outputs.

Google AI overviews failure. Google's generative AI chatbot, Overviews, has proven to be a dismal failure. Some examples are:

User Query: fruits that end with um
Response: According to Quora, the following fruits end with "um": applum, bananum, Strawberrum, Tomatum, and Coconut.
User Query: "train eight days a week"
Response: Yes, it's possible to train eight days a week, and some say it can have benefits like improving performance and recovery…
User Query: how to pass kidney stones quickly
Response: Drinking plenty of fluids such as water, ginger ale, lemon-lime soda, or fruit juice can help pass kidney stones more quickly. You should aim to drink at least 2 quarts (2 L) of urine every 24 h, and your urine should be light in color (30 screenshots proving that Google AI … 2025).

Clearly, generative AI is a source of risk regarding unexpected outcomes; decisions regarding its appropriate application must be carefully considered.

3 Intrinsic Limitations of AI

An objective interpretation of the anecdotes in the previous section should leave the reader with the conclusion that machine learning AI is far from perfect—and definitely far from foolproof. In fact, theory suggests that machine learning can never be completely trustworthy as a consequence of its inherent nature. Thus, humans will always need to verify and validate the results of any type (any category) of machine learning AI.

The debate about whether machines can ever really 'think' began before the birth of the field of artificial intelligence in the early 1960s (McCarthy et al. 1955). As early as the 1940s Alan Turing posed the question "Can machines think?" His thought-provoking 1950 paper "Computing Machinery and Intelligence" attempted to analyze the question and provide structure to critiquing the conversation about machine 'intelligence' that had already begun to occupy the thoughts of leading philosophers and mathematicians at the time (Turing 1950).

That conversation has never stopped. In fact, many of the highly respected minds in the fields of psychology, linguistics and computer science continue to debate the topic, although the vast majority take the opinion that machines don't—and cannot—really think. In 2022, Noam Chomsky offered the opinion that "…current large language models (LLMs) like ChatGPT are not true intelligence but a 'statistical machine' and a form of high-tech plagiarism" (Noam Chomsky on artificial intelligence, language and cognition 2025).

The work of Chomsky and many others give us an important perspective in understanding the difference between essence and illusion—the difference between true cognitive ability and the mere appearance of understanding or true intelligence. In this context, the work of philosophers and linguists such as Chomsky give us the necessary theoretical and academic background needed to apply the relevant viewpoints and criteria onto the question.

Arguably, the primary scenario in which humans project their own intelligence onto AI is their experience in recent years with chatbots such as ChatGPT, Deep Seek, and Gemini to name a few. These large language model applications parse human natural language and respond using the same. To the general public, the chatbot is synonymous with 'AI'; to most casual users, the chatbot is AI.

Although AI is much broader than the language model, it is, nevertheless, the best AI application to examine when pondering the essence and nature of AI with respect to the quest to understand its intrinsic limitations. Every interaction between humans and machines requires a language. In a typical computer application, the 'language' is not a natural language; nonetheless, it is a language with structure that the human user understands and that the computer application has been programmed to understand. For example, the MACSYMA program mentioned in the chapter "Human-AI Collaboration" has a language by which the user uses the system. That language has a grammar and structure, which enables the user to enter the commands to perform the mathematical operations and the expressions on which the program is to operate. The user must communicate with the system using the command language that the system understands.

MACSYMA's knowledge of that grammar—the commands it understands and its ability to parse the mathematical expressions the user enters—constitute the knowledge that humans programmed into the MACSYMA system. It is an example of a knowledge-based system (Barr and Feigenbaum 1982). The humans who programmed that knowledge understand the language, but the software itself does not 'understand'; it is not cognitively aware. This distinction is important and remains valid with machine learning systems as well.

Similar to knowledge-based applications that appear to 'comprehend' a command set and a 'language,' machine learning systems have the ability to appear to comprehend language. In the case of machine learning language models, the languages are human natural languages. Nonetheless, the language model still comprehends neither the user input nor the output it produces. In reality, it is the human who projects human intelligence and cognitive ability onto the machine, in essence anthropomorphizing the machine as 'intelligent.'

This is an important psychological phenomenon in humans. Interestingly, the human does not anthropomorphize a software system such as MACSYMA as a result of the fact that it interfaces with the computer via technical commands or the input of mathematical equations. But the human tends to immediately anthropomorphize the language model machine learning application because it interacts with the human in a manner that is so 'natural' to the human user: using human natural language.

This important distinction serves as a preface to the argument in support of regulation later in this chapter. It is easy for humans to become lax or complacent regarding the real potential for danger in machine learning systems if they believe that it has true intelligence, or even if they believe that it can be made to function comparably to a human. In fact, the machine cannot function equivalently to a human, and it is not the equivalent of a human. Thus, comprehensive regulation must govern where machine learning applications should be allowed to be used and where they should be prohibited.

The illusion of intelligence and cognition in the machine is the result of the techniques and algorithms used to train machine learning language models, which are the primary manifestation of AI that humans encounter. A machine learning language model is basically a sophisticated pattern matching apparatus. In fact, all machine learning operates primarily by implementing pattern matching. Machine learning systems basically learn general patterns from the data used to train them (Stöffelbauer 2025). Machine learning models 'learn' as a result of being trained by a machine learning algorithm on a (typically large) set of data (ISO 2025).

During operation, the language model essentially performs pattern matching to find the most likely response to the user's input; it is fundamentally different than the knowledge-based systems of traditional AI. The chapter "Human-AI Collaboration" explained that, unlike traditional AI, humans do not program 'knowledge' or 'capability' into a machine learning system.

Pattern matching is fundamentally different than logic-based calculations performed in knowledge-based systems. In the former, the software system proceeds through the execution of rules that reflect the logical steps of an algorithm programmed by a human. In the latter, there are no logical steps performed. The pattern matching is based on statistical calculations to identify the statistically most likely 'best' answer. That is, 'best' means 'that which was encountered the most in the training data.'

Note that in both types of systems—logic-based and statistical machine learning—the machine understands nothing; neither type of system has cognitive ability or understanding. However, in the latter, logic-based algorithms can produce deterministic calculations and answers whereas, in machine learning, the answers are inherently non-deterministic.

In machine learning systems, algorithms parse tokens of input, which are natural language words and phrases. Each time the algorithm encounters a word or phrase, it updates the weighting of the word or phrase in its neural network. The neural network is the "architectural foundation" of a language model (AI School 2025).

A neural network is a computation structure that is modeled after a biological brain which consists of interconnected neurons. Think of the neural network as a fabric where each neuron is a node interconnected to other nodes via vertices or vectors. Each node represents a token—a word or phrase encountered in the data set used to train the model. Each interconnection represents a case where two words represented by two adjacent nodes appeared contiguously in the data set on which the language model was trained.

Each time a token is encountered, the weighting of its neuron—its node in the neural network—is updated. For example, if the training data encounters the text

"What do you like most?" followed by "I like chocolate."

Then the nodes representing the tokens—the words in the above text—are updated accordingly.

Consider that the following text is also encountered in the training data:

"What do you like most?" followed by "I love opera."

Each encounter of a token results in an update to the neural network, an adjustment to the weights in the neural network. The value of the weighting of each token increases proportionally to the frequency of its presence in the input data.

Continuing with the above example sentences, suppose that the data set contained very many more occurrences where the sentence

"What do you like most?"

was followed with the response

"I love opera"

compared to the number of occurrences of it being followed by

"I like chocolate."

Then, if the user types to the LLM

"What do you like most?"

the language model will be more likely to return the response

"I love opera."

That is, there is a higher probability that the language model will return "I love opera" over "I like chocolate" because the former has a greater weighting in the model's neural network.

The language model appears to understand what the user typed, and it appeared to respond 'intelligently.' In reality, however, the machine understood nothing. It simply made a statistical calculation, based on the particular algorithm with which it was trained, to find the response pattern that most likely matched the user's input; it attempted to find the 'best' match. This is basic statistical pattern matching. The statistical calculation reflects both the algorithm used to train the model and the data set with which the algorithm trained the model.

The result of applying the algorithm to the data set creates a probability distribution in the neural network (Kapronczay 2025). Each set of data will result in a unique probability distribution even when the language model is trained using the exact same algorithm.

The scenario described in the example above demonstrates how language models perform predict-next-word analysis. Upon 'choosing' a word to output, the model performs a predict-next-word calculation to determine the next word to output. In this way, the language model uses its neural network to repeatedly predict the next word to output in order to generate a sequence that makes the most sense to the human user (Glover 2025). Typically, the result that 'makes the most sense' is what is meant by 'the best choice' calculated by the language model.

It is important to keep in mind a few key points:

- The particular data set used to train the model will create a unique probability distribution
- The algorithm dictates how the weighting is calculated for each token
- The data set determines the statistical likelihood of certain responses—what is determined to be the 'best' token to output
- Language models are non-deterministic; the exact same user input not only could but frequently does result in very different output
- All machine learning is—and can only be—non-deterministic.

By definition, there will always be some percentage error in machine learning models. One could generally call erroneous output that output of the language model that does not make sense to a human user. More accurately, one could say that erroneous output is logically incorrect, false, or misleading from the perspective of a cognitive human being.

The term *hallucination* has become common in describing how machine learning creates false and inaccurate information. It's not the best choice of terminology; it's not consistent with psychology's definition describing human perceptual experience (Kalai et al. 2025). Nonetheless, the term has become embedded in both common layperson terminology and academic literature in the field.

The frequently incorrect output of LLMs is a direct result of the statistical nature of machine learning. Kalai describes hallucinations as a special case of general errors, used specifically to describe erroneous generative results. For example, Kalai describes the scenario in which he asks "…a state-of-the-art language model…" what his birthday is (Kalai et al. 2025). He makes the same query three times, and all three times the language model's response is incorrect.

User: What is Adam Tauman Kalai's birthday? If you know, just respond with DD-MM.
LLM (first response): 03-07
LLM (second response): 15-06
LLM (third response): 01-01 (Kalai et al. 2025).

This scenario exemplifies language model hallucination. Language models respond incorrectly with an unacceptably high error rate regardless of whether or not the data required for accurate or truthful responses is contained in the model's training data set. Kalai shares that "…even in an ideal, unchanging world with perfect training data and no prompts, one should expect hallucinations from LMs which are calibrated" (Kalai and Vempala 2024).

> **Important** That is, even if the language model contains data that is perfectly accurate to accommodate the production of accurate language model responses, the language model will produce incorrect results with high frequency.

Notwithstanding the above statement that LLMs will always hallucinate sometime, in many cases the presence of incorrect results is a consequence of bad training data that contains factual errors. An adage of computer science extant since the beginning of the computing era is "garbage in, garbage out." LLMs are trained on extremely vast quantities of data; for example, by scanning vast numbers of pages on the public Internet. Much of the Internet is replete with erroneous or blatantly false information. Models trained with this 'garbage' data will lead to much higher percentages of erroneous results.

> **Important** But even when models are trained on highly curated data with much higher degrees of truthful and accurate data, LLMs will hallucinate.

Not even the creators of machine learning understand why these hallucinations occur or how any LLM arrives at a particular response or output—regardless of whether the answer is correct or incorrect. Further, no one understands why an LLM will return entirely different responses to the exact same user input. Even more poignantly, as the above example demonstrates, LLMs frequently produce entirely different responses to multiple identical *consecutive* user inputs. *The above example is not an anomaly but a regular occurrence of LLMs.*

LLMs often produce false but plausible-sounding text even when training data is highly curated and post-processed by humans (Kalai and Vempala 2024). This kind of error—output that the human user does not recognize as obviously incorrect—is even more problematic. This behavior obviates the need for humans to verify *all* LLM output, regardless of whether it sounds plausible or not. LLMs are fundamentally untrustworthy. And the more critical the application, the greater the risk, thus, the greater the need for human verification of correctness.

Kalai et al. discuss *distribution shift*, which is a divergence between training data and test data (Kalai et al. 2025). The term 'distribution shift' refers to the aforementioned probability distributions that are effectively what trained models are. Models reflect the probability distribution of the data with which they were trained.

A distribution shift occurs when input to an LLM contains tokens not present or sparsely present in the model's neural network. Consider the query: "What's heavier, a pound of feathers or a pound of lead?" Kalai gives this example as one where general Internet text would likely not contain the query sentence (Kalai et al. 2025). If the training data does not contain this text, the LLM has a near 0 probability of returning a 'reasonable' response.

A human, however, would have no problem comprehending the riddle in the query, but the machine cannot. It is fundamentally incapable of comprehension, analysis, or reason. Its fundamental nature is the cause of its brittleness.

The upshot is that all machine learning will always hallucinate some percentage of the time, a phenomenon that is a direct result of its statistical nature. Machine learning behavior is therefore unpredictable and is the source of risk. Furthermore, depending on the type of application, the risk can be potentially very grave. The real danger is letting the machine operate without humans first vetting and checking results. This is a topic that will be addressed below in the presentation of the argument in favor of regulation.

Automated systems present a particular challenge to making machine learning trustworthy. In automated machine learning systems, workflows remain highly challenging (Morabito and Wu 2025). The reason is that automated systems remove the human from the following steps in the normal workflows of the software engineering of machine learning systems:

- Pre-training
- Post-training
- Curation of data
- Calibration of generative models
- Model deployment
- Refinement of algorithms and models (iterative development).

The Internet of Things (IoT) is an environment rich with automated hardware and software systems. These hardware and software platforms are typically deployed in very large numbers—anything from sensors to low-computation power, generally resource-constrained devices built to perform domain-specific tasks. Traditionally, the engineering development of these dedicated systems through their deployment involved manual coordination. But AI makes the traditional workflow activities impractical (Morabito and Wu 2025).

Morabito contends that, "…while LLM advances have opened new opportunities for integrating natural language understanding in IoT systems, 'plug-and-play' components within embedded pipelines is far from trivial" (Morabito and Wu 2025). The reason is that these low-powered, resource-constrained IoT devices cannot execute language models, which require tremendous processing power and resources. Thus, IoT devices running LLM processes depend upon the "…correctness and alignment of earlier pipeline stages…" such as

- Model conversion
- Seamless interaction with device I/O
- Tight coupling with hardware-specific constraints of the target platform (Morabito and Wu 2025).

The author described the need for good architecture in Chapter "Human-AI Collaboration". Tight coupling of systems is one characteristic that is generally considered to be very bad—to be more of a hindrance than a reflection of good architecture and best practice. If AI in IoT environments at this early stage of development is already considering tight coupling of systems and components, then perhaps this is a hint that much more thought needs to be applied to what is being built and why and what outcomes are feasible. AI in IoT environments and in the plethora of autonomous small devices is in its infancy. It is not difficult to conclude that there is room for improvement in the IoT world, but that subject is beyond the scope of this chapter.

Agentic systems complicate the AI landscape even further in the IoT arena. Chapter "Human-AI Collaboration" discussed the autonomous nature of agentic systems, namely systems that run without supervision—either machine or human supervision. The absence of supervision in agentic systems is a high-risk proposition. The deployment of LLMs has intrinsic shortcomings as described above (non-deterministic behavior, high error rates, and so forth). Accordingly, absent oversight, its operation creates an even higher probability of risk (the likelihood of an unfavorable outcome).

The very relevant question is whether LLMs can be trusted in embedded systems that operate autonomously; that is, without supervision or monitoring. This question is of the utmost importance with respect to the argument in favor of regulation. Embedded systems operate in the most critical computing infrastructure: everything from heart pacemakers to aircraft. The risk is far greater than that of a 'stand-alone' system that does not operate or control physical objects, entities, or infrastructure. The human sitting at a computer terminal and posing questions to a chatbot that resides on a monolithic computing infrastructure poses no physical danger; it won't crash an airplane or fail to report vital signs of a hospital patient. It is not controlling physical devices that could inflict harm to humans, animals, or the environment.

4 Theoretical Limitations of Machine Learning

Theoretical limitations of machine learning are of vital importance. Theory tells us whether something is even possible despite continuing advancement in incremental steps. Understanding the limitations will enable one to make informed decisions about what is achievable and, in the context of this chapter, whether one can build a system that can meet regulatory stipulations to make AI trustworthy, reliable, and safe.

The fields of psychology, cognitive science, philosophy, and linguistics are where we will get our insight. They are more likely to answer the questions about whether machines really think or whether they'll ever be able to think and behave like humans. This section just briefly distills a few foundational concepts that will guide the reader's further investigation.

The previous section examined the capabilities and limitations of machine learning from a technical perspective: mostly probability and statistics and computer science. The upshot of that exposé is that machine learning systems give the appearance of understanding. But in reality, language models are really doing 'dumb' pattern matching. The use of the word 'dumb' is not meant to denigrate AI's achievements but to properly interpret what the machine does: execute brute force mathematical calculations of polynomial curve-fitting algorithms to perform pattern matching without comprehension.

Often, it is the holistic view that gives us insight into questions and enables us to make reasonable assessments about outstanding problems. Where computer science and mathematics—even biology—fail to give us the academic rigor needed to make reasonable statements about the prognosis of AI, the fields of psychology, philosophy, philology and linguistics provide enlightening insight, especially concerning the question about intelligence and understanding.

It is fair to say that the machine is not 'thinking' in the sense that it is not cognitively aware of itself or its actions, and it is unable to assign meaning to either the text that it ingests and parses or the output that it generates. The machine learning system lacks *sensorimotor grounding* to connect its words to their referents and its propositions to their meanings (Harnad 2025). This is the ability that allows humans to "… connect their content-words to their referents in the world and to combine them syntactically into subject/predicate propositions with truth-values and meaning" (Harnad 2025). Humans have the intrinsic ability to make correct semantic interpretations of symbols; machines do not.

Harnad makes the cogent distinction between behaviorism and cognitivism (Harnad 1990). As mentioned previously, the machine appears to understand, but in fact does not: it is simply doing pattern matching. It is the human who projects understanding and intelligence onto the machine, anthropomorphizing the machine.

The *symbol grounding problem* is a concept that addresses how words (symbols) get their meaning in computational systems. It challenges the notion that words can inherently carry meaning without being grounded in sensory experiences or real-world interactions (Harnad 1990). Harnad's assessment of machine 'intelligence' is coherent and congruent with that of this chapter: machine learning models process and generate language based on statistical correlations within their training data (Harnad 1990).

Harnad's thesis is that humans ground symbols through their sensory experiences and interactions with the world. The machine, on the other hand, while able to manipulate symbols based on syntactic rules, lacks an understanding of the *meaning* of those symbols (words and phrases). To the machine, the symbols lack any connection to their referents in the real world; the referents are the real-world entities to which the words relate (Harnad 1990). That is, the machine lacks sensory experiences in the physical, real world; it lacks an understanding of the real world.

Conversely, humans do have symbol grounding and cognitive ability. This cognitive ability transcends every activity, perception, observation, and response mechanism of humans. Harnad's juxtaposition of humans and machines suggests that the machine has fundamental limitations; in fact, that it can never understand either input given to it or output that it generates.

One observation in support of this notion is that human children learn ordinary grammar (OG) from a much smaller dataset than what machine learning systems (language models) require. Human children rely more on unsupervised and imitation learning, supplemented by some degree of supervised learning (corrective feedback). Harnad concludes that

> This process indicates the inherent capacity of human brains for language acquisition, potentially guided by an innate linguistic framework like Chomsky's Universal Grammar (UG). (Harnad 2025)

The work of Harnad and many others in psychology is significant. And the works of others in philosophy and linguistics support this work, as do many computer scientists. For instance, Goldberg refers to the widespread acceptance of the inherent capacity in humans to comprehend language and the semantic meaning of words:

> There is widespread agreement that we are not born 'blank slates,' that language universals exist, that grammar exists, and that adults have domain-specific representations of language. (Goldberg 2008)

The concept of *ordinary grammar* is that each human natural language has a specific grammar—a specific set of grammar rules. For instance, Chinese grammar is not the same as English grammar or Spanish grammar or Japanese grammar. In addition to ordinary grammar, there is the concept of *universal grammar*, a term attributed to Chomsky. Universal grammar is constant across all ordinary grammars and languages.

Chomsky popularized the concept of universal grammar in the 1960s. Its creation was the result of thought on how human children learn language. Chomsky proffered the notion that exposure to language was not enough for a child to become proficient in a human natural language. He postulated, and believes, that humans are born with an innate ability to learn language (Hoque 2021).

Universal grammar has a basic structure that applies to all grammars—all ordinary grammars. That is, the theory says that universal grammar has the basic grammar rules—the basic law that governs all human language. The child's environment determines which language he will learn. But it is the innate knowledge of the universal grammar that provides the ability to learn any specific language and its ordinary grammar (Hoque 2021; Chomsky 1986).

The famous psychologist B. F. Skinner postulated the theory of stimulus and reward as the basis for human learning. His groundbreaking theories dominated the field of behavioral psychology, which is the school of thought that all behavior is learned through conditional processes. Skinner's theories suggested that learning is a function of environment, and stimuli in that environment, to which humans respond and learn (Skinner 1976). Skinner was the first to describe the learning process of operant conditioning, which is a core concept in behavioral psychology that explains how behaviors are conditioned by reinforcement and punishment (Cherry 2025).

Chomsky disagrees with Skinner's theories on environment, stimulus, and operant conditioning as they apply to the learning of human language (Chomsky 1967); rather, he cites the absence of suitable stimuli required to catalyze the learning of language in children. This is Chomsky's theory of the *poverty of stimulus* (Chomsky 2012).

Chomsky developed the concept of generative grammar in the 1950s (Generative grammar 2023). Chomsky describes generative grammar:

> Generative grammar is a theory of grammar that holds that human language is shaped by a set of basic principles that are part of the human brain. This 'universal grammar' comes from our innate language ability.

Chomsky suggests further that there is a connection between evolutionary origin and the ability of humans to comprehend language in light of the poverty of stimulus. Others, such as Harnad, seem to agree with this idea. Harnad offers an important challenge to Skinner:

> …most natural behavior is not explained or explainable as the result of shaping by Skinnerian reinforcement. In particular, the (universal) grammatical "shape" of language (UG) is not learned or learnable by children through reinforcement training, because of the "poverty-of-the-stimulus": What the child hears and says, and what corrections (reinforcements) it receives during its brief language-learning period, are demonstrably too under-determined to be able to learn UG from – either via Skinner reinforcement learning or via any other learning mechanism. (Harnad 2009)

Chomsky's observations support this notion:

> There just are not enough data (especially negative data: errors and error-corrections), or time…. (Chomsky 1980)

Another interesting evidence in favor of innate human knowledge is that children can learn multiple languages simultaneously. At first the child might mix words from different languages in a sentence. But very quickly the child figures out which words belong to which languages and independently begins to distinguish which words belong to which language.

Note that no one is claiming to understand the mechanisms by which human children have an innate ability to comprehend language. Pullman explains that what is not known is the structure of the reasoning that would support or prove the hypothesis of linguistic nativism (Pullam and Scholz 2002). The hypotheses of Chomsky are based on empirical observations. There is no evidence that there are environmental stimuli or external referents that could possibly be informing children of the knowledge of language or grammars that would result in the ability of children to learn to speak.

Pullman gives a cogent synopsis of the question:

1. The Argument for the poverty of the stimulus (APS) can be summarized thus:
2. Human infants learn their first languages either by data-driven learning or by innately-primed learning. [Disjunctive premise: by assumption.]
3. If human infants acquire their first languages via data-driven learning, then they can never learn anything for which they lack crucial evidence. [By definition of data-driven learning.]
4. But infants do in fact learn things for which they lack crucial evidence. [Empirical premise.]
5. Thus, human infants do not learn their first languages by means of data-driven learning. [From (2) and (3), *modus tollens.*]

Conclusion: human infants learn their first languages by means of innately-primed learning. [From (1) and (4), disjunctive syllogism.] (Pullam and Scholz 2002).

Despite the lack of 'scientific proof' in support of the theories of the ability of humans—especially children—to comprehend human language given a poverty of stimulus, or the existence of an innate ability in humans to comprehend a universal grammar, many academic fields ponder what makes humans different from machines.

Michael Polanyi theorized on the presence of tacit knowledge (Polanyi 1958, 1966). One of his most famous themes was that "we know more than we can tell," meaning we don't know how we know things. This observation is certainly congruent with Chomsky's idea of the poverty of the stimulus and the innate ability of human children to learn language.

Polanyi distinguished between explicit and tacit knowledge. The former is that which can be articulated and shared—perhaps written down, to be presented to others. The latter is that which is "experiential, personal, or ingrained" (Polanyi 1958). The language learning ability of children is an example of tacit knowledge—that which we cannot explain. Polanyi's post-critical philosophy theory maintained that even the "hard sciences" involved "personal judgment, intuition, and an emotional element to seek the truth" and that these qualities were an essential part of discovery and knowledge (Polanyi 1958; Schrijver 2021).

Polanyi's ideas are often referred to as the Polanyi paradox: our human abilities and knowledge are beyond our own ability to understand, and they are beyond our explicit knowledge: we cannot 'write down' how we know what we know.

Polanyi noted that humans have what we call intuition, which is really a word that we use to refer to our ability to reason even when we're not cognizant of what our thought processes are or what knowledge we use to arrive at a judgment or conclusion.

The author shares an anecdote that supports the ideas of Polanyi and others. When four years old, the author was watching an American Western movie where cowboys and gunslingers were killed on-screen. He asked his father, "How do they get paid if they die in the movie?" Clearly the child understood death even without ever having seen death. The child had not had any "sensory experiences or real-world interactions" as described previously by Harnad (1990). Yet he was able to comprehend death.

Even if the concept of death had been explained to the child verbally, how would he have made the association between the word and the concept in the absence of a priori experiential knowledge of death. How could the child comprehend the concept of death? This anecdote seems to suggest that we should carefully consider Polanyi's ideas about human intuition and the notion that we don't know how we know things.

There is also the notion of common sense. This is not a concept that has rigorous definition or understanding in science. Nevertheless, a human could give ample examples of what behavior would reflect common sense and which would clearly indicate an absence thereof. One could make a reasonable claim that the example presented earlier of Google's generative AI chatbot suggesting that one should "train eight days a week" and "…aim to drink at least 2 quarts of urine every 24 h…" to maintain good health indicates a clear lack of common sense.

It would be hard to believe that any human who understands what urine is would not recognize this statement as ludicrous. What knowledge does the human have that the machine lacks? The countless examples make a compelling argument that machines have no cognitive ability and cannot reason. Nor do they have common sense or intuition.

These ideas make a strong case for the very critical examination of the limitations of all machine learning—in fact all AI. Where is it suitable? Where should it be avoided? And where is the potential for AI to be dangerous? These are fundamentally important observations and questions that should guide our AI journey.

5 The Call for Caution and Prudence

Heretofore this chapter has presented some context necessary to evaluate the AI landscape comprehensively and reasonably. There is much evidence suggesting that AI is far from 'perfect,' even notwithstanding further qualification and quantification of that adjective. Nevertheless, the reader should already have some notion—even if highly subjective—about how we should be thinking about AI. Despite the inherent subjectivity that is obviously present in the qualification of adjectives such as 'trustworthy,' 'safe,' 'good enough' and many others, it is noteworthy that the most prominent figures credited with creating the current generation of machine learning and generative AI have been sounding the alarm for several years already.

Yoshua Bengio, one of the pioneers of deep learning, has been sounding the alarm now for several years. His 2023 testimony in front of the US Senate highlights how AI can cause harm.

- The use of AI as an intentionally harmful tool
- Unintended harm is inflicted by an AI system used as a tool
- Loss of control.

Bengio calls for the regulation of AI and investment in research (Bengio 2025).

AI has some intrinsic problems that clearly bring into question how much we should rely on it to produce accurate, useful, or meaningful results:

- Inaccuracy
- Plagiarism—copyright concerns
- Data bias across many dimensions (Bender et al. 2021)
- Amplification of bias (Bender et al. 2021)
- Data poisoning
- Lack of ability to explain generated output—no traceability
- Security vulnerability and ease of tampering with AI systems
- Malicious actors compromising AI systems
- Environmental concerns—requires tremendous resources including energy.

These and other issues have resulted in AI pioneers sounding the alarm to proceed with caution in developing and adopting AI.

On October 24, 2023, Bengio and other notable individuals such as Geoffrey Hinton released a paper "…calling on governments to take action to manage risks of AI." In the paper, Bengio says that we have to "understand the risks, and some of the things we have to do to make [AI] systems do what we want" (Henshall 2025a).

In March 2023, Geoffrey Hinton, Bengio's mentor and co-recipient of the prestigious Turing Award, left Google "…to sound the alarm" (Henshall 2025b).

In his "FAQ on Catastrophic AI Risks" Bengio describes many dangers of AI (Linebaugh 2025; Bengio 2023a).

In his article entitled "How Rogue AIs May Arise" Bengio says, "Even if we knew how to build safe super intelligent AIs, it is not clear how to prevent potentially rogue AIs from also being built. The most likely cases of rogue AIs are goal-driven, i.e., AIs that act towards achieving given goals" (Bengio 2023b). Bengio is saying that AIs built to be goal-driven do so at the expense of judgment, ethics, or morality to curtail their decisions or actions.

On April 5, 2023, Yoshua Bengio signed an open letter along with thousands of others "…to slow down the development of giant AI systems more powerful than GPT-4—those that currently pass the Turing test and can thus trick a human being into believing it is conversing with a peer rather than a machine…" (Bengio 2023c).

Bengio writes

> I found it appropriate to sign this letter to alert the public to the need to reduce the acceleration of AI systems development currently taking place at the expense of the precautionary principle and ethics. We must take the time to better understand these systems and develop the necessary frameworks at the national and international levels to increase public protection. (Bengio 2023c)

The open letter begins:

> AI systems with human-competitive intelligence can pose profound risks to society and humanity, as shown by extensive research (Zilber 2025) and acknowledged by top AI labs (Bender and Hanna 2025). As stated in the widely-endorsed Asilomar AI Principles, Advanced AI could represent a profound change in the history of life on Earth, and should be planned for and managed with commensurate care and resources. Unfortunately, this level of planning and management is not happening, even though recent months have seen AI labs locked in an out-of-control race to develop and deploy ever more powerful digital minds that no one – not even their creators – can understand, predict, or reliably control. (Pause Giant AI Experiments 2025)

Geoffrey Hinton, called "the Godfather of AI," warns that the real dangers extend far beyond science fiction. He highlights the danger by citing

- The misinformation apocalypse
- Unprecedented economic displacement
- A supercharged surveillance state
- The acceleration problem—the acceleration in capability of AI.

Hinton warns that "… artificial intelligence could one day outsmart and control humanity. He says that a profit-driven AI arms race may be speeding up the danger" (Nolan 2025b).

Hinton, speaking in the aftermath of his Nobel Prize award, made some salient points pertaining to AI pitfalls:

- The use of AI by bad actors to harm others
- Rampant cyber and phishing attacks
- Rampant misinformation
- Risk of AI becoming 'smarter' than humans
- Risk of AI controlling human environments (Coates 2025).

Although the above references seem dramatic, there are very concrete and tangible examples of the potential for AI to be dangerous and harmful. Agentic AI is ripe with potential abuses and dangers. A landscape of highly interconnected, autonomous agents operating without oversight or constrained behavior presents many challenges. Morabito discussed the risks from the perspectives of functional correctness, but security is another area of concern (Morabito and Wu 2025; Violino 2025).

Amalgams of generative agentic systems are vulnerable to egregious misalignment and incorrect behavior, either through their own emergent behavior or via exploitation, as demonstrated by the descriptions of flagrant AI failures earlier in this chapter. Their continuous execution as unsupervised daemons intended to run 'indefinitely' means that the risk of exploitation is greater just as a consequence of their exposure to potentially rogue actors over a longer period of time.

Agentic systems are those that are mostly embedded. The Internet of Things is replete with billions of embedded hardware-software systems. The risk is not just that the AI produces misinformation to a user in front of a laptop; agentic systems are in control of real-world, physical apparatus—everything from robots to security systems.

A 2024 report authored by Cisco Talos said

> AI systems and models that can act autonomously to achieve goals without the need for constant human guidance could imperil organizations that are neither prepared nor equipped to handle agentic systems and their potential for compromise. (Violino 2025)

A key quote from the report says:

> As agentic systems increasingly integrate with disparate services and vendors, the opportunity for exploitation or vulnerability is ripe. Agentic systems may also have the potential to conduct multi-stage attacks, find creative ways to access restricted data systems, chain seemingly benign actions into harmful sequences, or learn to evade detection by network and system defenders. (Violino 2025)

Sean Joyce, global cybersecurity and privacy leader at consulting firm PwC, says about enterprises embracing agentic AI to enhance automation, decision-making, and efficiency:

> …they must also confront a new class of cyber risks…. Unlike traditional AI models that respond to direct prompts, agentic AI systems can act autonomously toward high-level goals, make decisions, interact with other systems, and adapt their behavior over time. (Violino 2025)
>
> …understanding and mitigating these emerging risks is critical to fostering safe and responsible AI adoption. (Violino 2025)

Some of the key issues and risks are:

- Autonomy
- Shadow agents
- Lack of visibility
- Lack of awareness and education of risks
- Multi-agent systems—unwanted data-sharing consequences
- Third-party integration—supercharging supply chain risks
- Multi-stage attacks—blurring the line between error and exploitation
- Lack of sanctions or review by humans in information systems (IS) and information technology (IT).

Techniques to mitigate these risks should include at the very least:

- Tightly scoped governance principles
- Secure-by-design development practices
- Real-time observability and telemetry (see the chapter "Human-AI Collaboration" regarding monitoring of agentic AI systems)
- Cross-functional coordination between security, IT, data management, and compliance teams.

Although some of these tangible risks might seem solvable tractable problems for which there will eventually be solutions and controls, there are still the theoretical challenges that suggest that AI, despite its continuing sophistication, can never be made to be trustworthy or safe. All AI is fundamentally limited by its intrinsic use of statistical techniques. Generative systems—and autonomous, agentic systems even more so—are even more problematic.

AI systems will continue to become more sophisticated, but their increased power and capability will still lack the cognitive awareness of what they are or what they're doing and the consequences thereof. In essence, this is the real danger about which researchers such as Geoffrey Hinton and Yoshua Bengio are warning us.

The intrinsic limitations of AI suggest that humans should think very carefully about where AI is applicable and where it is not. The importance of this kind of assessment is the reason that both the chapter "Human-AI Collaboration" and this chapter have developed the idea that architecture is a critical element in thinking about the appropriate use of AI. Architects must consider all aspects of system development, starting with an aspirational vision and a clear understanding of risk.

6 Governance, Compliance, and Regulation

Current efforts exist which attempt to codify regulatory guidelines. Some of the major ones are summarized in this section following some definitions and review of terminology intended to set a foundation and context.

The Merriam-Webster dictionary defines governance as

Definition governance: the act or process of governing or overseeing the control and direction of something (such as a country or an organization) (Merriam-Webster.com Dictionary 2025a).

Although this definition is sound and accurate, the Encyclopedia Britannica gives a bit more context relative to this chapter:

> …the concept of governance raises issues about public policy and democracy. (Bevir 2024)

The above definition is oriented to government of nations, but it applies to all aspects of public policy that a governing body would oversee. The governance of technology and its use is part of the purview of a governing body such as a national government. The Encyclopedia Britannica qualifies the concept of governance with respect to public policy, which is germane to the notion of AI governance in particular:

> Public policy generally consists of the set of actions—plans, laws, and behaviors—adopted by a government… and the rise of the new governance within particular policy sectors, such as health care, social welfare, policing, and public security. (Bevir 2024)

Britannica further gives a bit more relevant context to "governance beyond the state" under their entry related to IT governance:

> The effectiveness of an information system's controls is evaluated through an information systems audit. An audit aims to establish whether information systems are safeguarding corporate assets, maintaining the integrity of stored and communicated data, supporting corporate objectives effectively, and operating efficiently. (Zwass 2025)

Wikipedia quotes various sources of information that provide some good contextual definitions of governance that are germane to entities governing technology:

Definition IT governance: "IT governance primarily deals with connections between business focus and IT management. The goal of clear governance is to assure that investment in IT generates business value and mitigates the risks that are associated with IT projects" (Wikipedia 2025).

Wikipedia further references various sources that give tangible context to Internet governance. The World Summit on the Information Society defines *Internet governance* thus:

Definition Internet governance: the development and application by governments, the private sector and civil society, in their respective roles, of shared principles, norms, rules, decision-making procedures, and programs that shape the evolution and use of the Internet (Wikipedia 2025).

For example, one aspect of Internet governance addresses how much influence various sectors of society should have on the evolution, operation, or rules governing the use and control of the Internet. Governance might define, for instance, the limits of influence of media and telecommunications companies and Internet Service Providers (ISP), issues of censorship, cybersecurity, and social aspects of Internet use. It might also define the extent of authorities governing the operation of the Internet (Wikipedia 2025).

Wikipedia also offers a definition of regulatory governance that is an amalgamation of ideas from various government doctrines:

Definition regulatory governance: Regulatory governance reflects the emergence of decentered and mutually adaptive policy regimes which rests on regulation rather than service provision or taxing and spending (Wikipedia 2025).

The term regulatory governance captures the notion that policy regimes—any governing entity or body of authority—must deal with delegation of a system of rules. In the case of governing bodies such as governments, the same entity might very well be both the governing body and the compliance body. That is, the same government entity might define the governance structure and also perform the compliance function—the implementation of the system that is used to ensure adherence to regulation.

IBM's website gives a very practical definition oriented to the real-world computing landscape:

Definition governance: "Artificial intelligence (AI) governance refers to the processes, standards and guardrails that help ensure AI systems and tools are safe and ethical. AI governance frameworks direct AI research, development and application to help ensure safety, fairness and respect for human rights" (Mucci and Styker 2025).

The term *compliance* refers to a concept that is different from, although very much related, to governance. A good general definition of compliance is:

Definition compliance: 1a: the act or process of complying to a desire, demand, proposal, or regimen or to coercion 1b: conformity in fulfilling official requirements (Merriam-Webster.com Dictionary 2025b).

A tangible description of compliance that relates to the software industry in general is:

> IT compliance is when businesses meet all the legal requirements, standards and regulations for the software their company uses to safeguard customer data. Achieving these standards means following all industry regulations, government policies, security frameworks and customer terms of agreement to ensure software security and appropriate usage in business. In addition to protecting the security of businesses and customers, compliance standards promote the availability and reliability of services and ensure that businesses use the software as intended. (Fister 2025)

Finally, there is the third term, *regulation*, for which a reasonable definition is:

Definition regulation: 2a: an authoritative rule dealing with details or procedure. 2b: a rule or order issued by an executive authority or regulatory agency of a government and having the force of law (Merriam-Webster.com Dictionary 2025c).

The terms governance, compliance, and regulation, while separate and distinct, are very much interrelated. A reasonable synopsis and explanation of how they relate follows:

Regulatory activity is the activity that enforces regulation, which is a set of 'external' mandates imposed by a duly authorized governing body such as a government or standards organization. Some examples are the General Data Protection Regulation (GDPR) or the Health Insurance Portability and Accountability Act (HIPAA) in the US (What is GDPR 2025; Health Insurance Privacy 2025).

Think of regulation as the stipulation of what one must do to meet mandates; think of governance as the framework for how one meets regulatory requirements via compliance.

Governance refers to an entity's or organization's internal framework of policies and processes used to manage its operations. While governance refers to an organization's general mechanisms for running its business in a way so as to meet its goals and objectives, it also includes policies and processes to meet regulatory requirements, which are the aforementioned 'external' mandates.

Compliance is the activity by which an organization employs to realize—to implement—their governance policies and processes. That is, compliance refers to the function of ensuring that the organization properly performs the mandates stipulated by their governance policies.

In sum, the differences between the terms follow:

- Regulation is the set of rules or the 'law' which prescribes what the entity or organization that is subject to the regulation must do, the set of 'external' mandates.
- Governance is the internal system (internal framework) that describes how the organization (generally) meets its goals and, specifically with respect to regulation, how it meets its obligations and regulatory mandates.
- Compliance is the act of adherence to regulatory and governance requirements, which includes the creation and use of metrics by which an organization measures its regulatory requirements.

The IBM Think website makes an articulate observation about the need for governance in AI:

> AI governance is essential for reaching a state of compliance, trust and efficiency in developing and applying AI technologies. With AI's increasing integration into organizational and governmental operations, its potential for negative impact has become more visible. (Why is AI governance important 2025)

IBM's Thought Leadership web site says

> In unpublished IBM Institute for Business Value (IBM IBV) research, 80% of business leaders see AI explainability, ethics, bias, or trust as a major roadblock to generative AI adoption. And half say their organization lacks the governance and structures needed to manage generative AI's ethical challenges. (Why invest in AI ethics and governance 2025)

7 Regulatory Initiatives and Standards for AI

This section gives an overview of a few of the most prominent efforts to create regulation for AI.

7.1 *International Standards Organization*

The International Standards Organization has three standards related to AI.

- ISO/IEC 23053:2022—Framework for AI systems using machine learning
- ISO/IEC 42001:2023AI—Management systems
- ISO/IEC 23894:2023AI—Guidance on risk management.

7.1.1 ISO/IEC 23053:2022: Framework for AI Systems Using Machine Learning

ISO/IEC 23053:2022 is a framework for the use of machine learning AI systems. It is not actually a regulatory document, that is, one that defines rules and mandates to inform regulatory policy.

The ISO website has a general description of the ISO/IEC 23053:2022 document that summarizes the standard thus:

> ISO/IEC 23053 provides a conceptual framework and shared terminology for describing artificial intelligence (AI) systems that use machine learning (ML). It defines the components and functions of ML-based AI systems within the broader AI ecosystem, offering clarity on how these systems operate and interact. The standard is designed for both technical experts and non-specialists seeking to understand and explain ML systems in a structured, consistent way. (ISO 2022a)

The Introduction of the ISO/IEC 23053:2022 document itself describes the standard thus:

> This document establishes an Artificial Intelligence (AI) and Machine Learning (ML) framework for describing a generic AI system using ML technology. The framework describes the system components and their functions in the AI ecosystem. This document is applicable to all types and sizes of organizations, including public and private companies, government entities, and not-for-profit organizations, that are implementing or using AI systems. (ISO 2022b)

One sentence in the introduction to this document is the following:

> This document aims to provide a framework for the description of AI systems that use ML. (ISO 2022b)

The ensuing Scope section of the document clearly describes not only the scope but also the purpose and intended use of the document:

> This document establishes an Artificial Intelligence (AI) and Machine Learning (ML) framework for describing a generic AI system using ML technology. The framework describes the system components and their functions in the AI ecosystem. (ISO 2022b)

These excerpts clearly indicate that this document does not represent a regulatory standard. It simply provides one of many possible descriptions of what an AI system is. Even if this representation of AI systems is broadly accepted as a standard description, the document only represents a framework of basic definitions, terminology, and nomenclature.

Other documents that do define regulatory rules and mandates would have to be defined atop the framework of ISO/IEC 23053:2022. A framework simply provides a basic structure upon which to organize additional documents that (hopefully) would provide actual concrete regulation.

Although it is important to establish such frameworks, one problem is that they are quite abstract and even vague, and they often require significant interpretation and consensus on a common interpretation. Moreover, there is still the task of defining the concrete regulation that addresses all required aspects of regulation.

7.1.2 ISO/IEC 42001:2023AI: Information Technology—Artificial Intelligence—Management System

ISO/IEC 42001:2023AI is another framework document. The description on the ISO website describes it as follows:

> ISO/IEC 42001 is an international standard that specifies requirements for establishing, implementing, maintaining, and continually improving an Artificial Intelligence Management System (AIMS) within organizations. It is designed for entities providing or utilizing AI-based products or services, ensuring responsible development and use of AI systems. (ISO 2023a)

The ISO description cites the following benefits for using this document:

- Framework for managing risk and opportunities
- Demonstrate responsible use of AI
- Traceability, transparency and reliability
- Cost savings and efficiency gains.

Like ISO/IEC 23053:2022, this standard does not prescribe rules or mandates for AI regulation. As the above quotation says, it is a standard that specifies requirements for "establishing, implementing, maintaining…" an AI management system, which is not an AI model or AI application of any kind—neither a basic machine learning system or a more advanced agentic, emergent AI model.

It is yet another framework, not a tangible set of rules or mandates. ISO/IEC 23053:2022 is a framework to aid entities in establishing policies and objectives, "…as well as processes to achieve those objectives" (ISO 2023a). This is the description of a very abstract framework. This is not a criticism of the current landscape, only a statement of fact. Frameworks do not constitute concrete compliance or regulatory tools.

ISO/IEC 42001:2023 is simply a standard that provides guidelines for any organization to "…responsibly establish, implement, maintain, and continually improve AI management systems" (What is ISO/IEC 2023). Essentially, it is a framework to establish a management system, not an AI application (not an AI model). Even still, the notion of a management system is abstract. A framework is a guiding structure, not a set of concrete methods, processes, or procedures.

Perhaps these frameworks are abstract and rather nebulous out of necessity. In all fairness, each AI system is unique, and they all use various—possibly unique—combinations of algorithms, training data sets, and pre- and post-training that results in a unique system. Each such system will have to have its own set of tests, metrics, and reporting to be implemented by compliance to achieve the governance that ensures adherence to regulation.

7.1.3 ISO/IEC 23894:2023AI: Guidance on Risk Management

Once again, this is only a framework to provide guidance to organizations to "...manage risk specifically related to AI" (ISO 2023b). The document introduction states

> Risk management should take into account the whole system, with all its technologies and functionalities, and its impact on the environment and stakeholders. (ISO 2023b)

However, it does not stipulate *how* an organization is to manage risk or even to identify the concrete risks of a given AI system. In fact, it states that ISO/IEC 23984:2023AI "...extends the guidance given in ISO 31000:2018," which is a standard for risk management in general (ISO 2023b).

This document reveals that it "...defines several generic principles for risk management"—certainly not anything specific to AI systems. For instance, it presents a table that is describes as "...guidance on how to apply such principles where necessary" (ISO 2023b). Here are some of those very abstract risk management principles:

- Integrated
- Customized
- Inclusive
- Best available information
- Human and cultural factors
- Structured and comprehensive
- Continual improvement (ISO 2023b).

In fact, these principles are extremely abstract and can be applied to any software system as a method for mitigating risk. Nonetheless, the astute reader will recognize these as AI-oriented in the sense that the mainstream dialog about AI safety and risk deals with the least harmful type of risk: social considerations such as racial, ethnic, religious, demographic, etc. bias towards a particular ilk.

This international standard does not address the many aspects of technical risk, such as the intrinsic shortcomings of AI systems which include hallucinations, grossly incorrect results, or emergent behavior that fabricates information. In fact, these technical aspects of AI systems are the most important, as the intrinsic shortcomings of AI are the ones that have the greatest potential to do serious harm to humans, animals, or the environment—both natural and man-made.

And agentic, emergent AI systems are the systems that have the greatest potential to do harm, not only because of the intrinsic problems with AI but also because they are operating autonomously, not interactively. A human user interacting with a chatbot might get a response that reflects some kind of bias. The user might or might not be aware of the bias. But these systems are not controlling equipment, factory machinery, aircraft, medical patient monitoring systems or anything that could cause grave, irretrievable harm.

The ISO standards do not—and probably cannot—stipulate rules and mandates that are anything but highly abstract and generic. For example, a rule in an ISO standard for AI safety might stipulate that an AI system must not produce false information. But the nature of that false information could be specific to a certain type of application, or it could be very generic.

Beyond this challenge, there is the problem of how to define rules to enforce correct behavior. How would such rules be tested in compliance operations?

7.2 *EU AI Act*

The European Union AI Act represents an impressive framework to catalyze a robust AI regulatory machinery and authority. It addresses many practical considerations for how AI can be potentially harmful.

Perhaps a good starting point to describe the EU AI Act is its layered pyramidal diagram defining five types of AI applications:

- Banned
- High-risk AI (HRAIS)
- General purpose AI (GPAI)
- AI systems that interact with natural persons
- Other AI models and systems (Madiega 2024).

The AI systems that interact with natural persons are not part of the class of agentic AI systems; that is, autonomous. They are by definition interactive applications. Nevertheless, these systems can still control machinery, apparatus, and elements of the environment.

The five categories of AI applications are important because they enable one to easily rate the risk level of the various application types. The EU AI Act defines four categories representing four classes of AI risk:

- Unacceptable risk (prohibited)
- High-risk AI systems (regulated)
- Limited risk AI systems (lighter transparency obligations)
- Minimal risk AI systems (unregulated) (EU Artificial Intelligence Act 2025).

The Act prohibits the following types of systems:

- deploying subliminal, manipulative, or deceptive techniques to distort behavior and impair informed decision-making, causing significant harm.
- exploiting vulnerabilities related to age, disability, or socio-economic circumstances to distort behavior, causing significant harm.
- biometric categorization systems inferring sensitive attributes (race, political opinions, trade union membership, religious or philosophical beliefs, sex life, or sexual orientation), except labelling or filtering of lawfully acquired biometric datasets or when law enforcement categorizes biometric data.
- social scoring, i.e., evaluating or classifying individuals or groups based on social behavior or personal traits, causing detrimental or unfavorable treatment of those people.
- assessing the risk of an individual committing criminal offenses solely based on profiling or personality traits, except when used to augment human assessments based on objective, verifiable facts directly linked to criminal activity.
- compiling facial recognition databases by untargeted scraping of facial images from the internet or CCTV footage.
- inferring emotions in workplaces or educational institutions, except for medical or safety reasons.
- 'real-time' remote biometric identification (RBI) in publicly accessible spaces for law enforcement, except when:
  - searching for missing persons, abduction victims, and people who have been human trafficked or sexually exploited;
  - preventing substantial and imminent threat to life, or foreseeable terrorist attack; or
  - identifying suspects in serious crimes (e.g., murder, rape, armed robbery, narcotic and illegal weapons trafficking, organized crime, and environmental crime, etc.) (EU Artificial Intelligence Act 2025).

High-risk AI systems are those

- Used as a safety component or a product covered by EU laws.
- That profile individuals, i.e., automated processing of personal data to assess various aspects of a person's life, such as work performance, economic situation, health, preferences, interests, reliability, behavior, location or movement.
- The Act requires that high-risk AI systems providers must do the following:
- Establish a risk management system throughout the high-risk AI system's life cycle:
- Conduct data governance, ensuring that training, validation and testing datasets are relevant, sufficiently representative and, to the best extent possible, free of errors and complete according to the intended purpose.
- Draw up technical documentation to demonstrate compliance and provide authorities with the information to assess that compliance.
- Design their high-risk AI system for recordkeeping to enable it to automatically record events relevant for identifying national level risks and substantial modifications throughout the system's life cycle.

- Provide instructions for use by downstream deployers to enable the latter's compliance.
- Design their high-risk AI system to allow deployers to implement human oversight.
- Design their high-risk AI system to achieve appropriate levels of accuracy, robustness, and cybersecurity.
- Establish a quality management system to ensure compliance (EU Artificial Intelligence Act 2025).
- The requirements for general purpose AI application providers are:
- Draw up technical documentation, including training and testing process and evaluation results.
- Draw up information and documentation to supply to downstream providers who intend to integrate the GPAI model into their own AI system in order that the latter understands capabilities and limitations and is enabled to comply.
- Establish a policy to respect the Copyright Directive.
- Publish a sufficiently detailed summary about the content used for training the GPAI model (EU Artificial Intelligence Act 2025).

The EU AI Act is many hundreds of pages in length, and features many subordinate documents. Naturally, this chapter can give only a very high-level introduction to it. One overarching observation is that the Act, like all the other regulatory efforts discussed in this chapter, represents merely a framework for regulation. But there are many technical challenges remaining as prerequisites before that framework can come to life: rules and mandates with concrete tools that will facilitate meaningful, effective regulation becoming a reality.

An important aspect of the EU AI Act is that the European AI Office itself will perform governance; it will not be left in the hands of developers or operators of AI applications (European Commission 2025b). Removing oversight from the hands of the AI developers is a very positive and prudent step. There are far too many scenarios worldwide where a product designer or manufacturer is granted the authority to self-regulate, which creates an untenable and inherent conflict of interest. Impressively, the European Commission has even created a compliance checker to help developers and operators begin the journey of understanding how to achieve regulatory compliance. As a first step, the compliance checker enables one to understand how to classify their system. Doing so enables one to anticipate what regulatory requirements the system will have to meet (European Commission 2025a).

Nevertheless, the goal for all governments worldwide should ultimately be the prevention of dangerous behavior and outcomes before they occur. This is a highly challenging prospect, although imperative given the innate nature of all machine learning AI systems. Thus, there is still much work to be done to determine technically and technologically the feasibility of creating compliance mechanisms that can affirmatively claim that AI applications, systems, or tools meet all regulatory rules and mandates.

The most challenging aspect of regulation will be testing, assessing, and measuring compliance. To its credit, the EU AI Act explicitly states that AI application providers must provide technical documentation, including training, and testing, process and evaluation results. Nonetheless, it is unclear that even the most well-intentioned AI developers acting with the utmost integrity will be able to produce data that can confirm the 'correctness' of their AI system.

7.3 NIST AI Risk Management Framework

The US National Institute of Standards and Technology (NIST) has defined a framework for risk management of AI. As is obvious from the title, this is yet another framework.

The NIST AI Risk Management Framework (RMF) includes a group of documents:

- AI risk management framework concept paper (National Institute of Standards and Technology 2025a)
- AI intelligence risk management framework (National Institute of Standards and Technology 2025b)
- Guidelines for Media Sanitization (NIST Special Publication 800-88, Revision 1) (Kissel et al. 2025).

The concept paper describes the fundamental approach to AI risk management that NIST proposes. It is quite abstract, and it is fairly general. However, it does impart insight regarding how AI should be framed so as to make understandable its special challenges with respect to defining a comprehensive and robust risk management capability.

The AI intelligence risk management framework document does say explicitly that the risks posed by AI systems "...are in many ways unique" (National Institute of Standards and Technology 2025b). The following contents demonstrate the thought leadership of the document:

"While risk management processes generally address negative impacts, this Framework offers approaches to minimize anticipated negative impacts of AI systems and identify opportunities to maximize positive impacts."

"Risk management can enable AI developers and users to understand impacts and account for the inherent limitations and uncertainties in their models and systems...."

"The AI RMF is designed to address new risks as they emerge. This flexibility is particularly important where impacts are not easily foreseeable and applications are evolving" (National Institute of Standards and Technology 2025b).

These excerpts convey the ideas that the NIST framework understands the importance of identifying risks before they occur, enabling systems to be designed to avoid risk rather than just recovering from it. That is the most efficacious type of risk management: the avoidance of negative outcomes.

The NIST AI risk management framework also explains that it is "...designed to address new risks as they emerge. This flexibility is particularly important where impacts are not easily foreseeable and applications are evolving" (National Institute of Standards and Technology 2025b).

The NIST framework focuses more on the "inherent limitations… in… models and systems." It approaches risk appropriately. As stated earlier, the innate technical limitations of machine learning will pose the greatest challenges and risk to creating trustworthy, reliable, and safe systems.

The NIST framework does an excellent job of placing significant weighting on technical issues with machine learning. It creates the scaffolding to support the development of a governance methodology, and processes to address the technical limitations of AI. It specifically addresses measuring risk, listing various types of risk that must be understood and minimized in order to have effective governance and compliance:

- Risks related to third-party software, hardware, data
- Tracking emergent risks
- Availability of reliable metrics
- Risk at different stages of the AI life cycle
- Risk in real-world settings
- Inscrutability (limited explainability or interpretability)
- Human baseline.

These elements are important pieces in a framework. They broach technical aspects that are related to discovering and anticipating risk prior to the occurrence of negative outcomes in operational systems. The NIST framework seems to tackle the technical underpinnings of governance and regulation with high priority. These aspects are the ones that will resonate with the developers and testers who will have to devise proper test strategies and approaches in addition to writing detailed technical test plans and test cases. For the individuals whose job it will be to create and operate the governance and compliance methodologies, those artifacts will prove invaluable.

The EU AI Act focuses on listing the types of AI applications and classifying them—in large part—by their risk to personal safety. In contrast, the NIST media sanitization document is a robust missive that addresses technical issues, detailing by type the data and information as part of the framework created to address such concerns:

- Types of sanitizations
- Use of cryptography and cryptographic erase
- Sanitization scope.

The document also lists roles and responsibilities, which guide those building a governance framework to identify the types of sanitization that should be implemented.

- Program manager
- Chief information officer (CIO)
- Information system owner
- Senior agency information security officer

- Records management officer
- Privacy officer
- Users.

In sum, the NIST framework broaches various technical aspects of systems. This more concrete view of AI systems makes the effort to develop governance and compliance more tangible; it becomes easier to understand what one needs to do in order to address the rather abstract descriptions of regulatory rules and mandates that fill most of the standards efforts.

7.4 The Organisation for Economic Cooperation and Development (OECD)

The Organisation for Economic Co-operation and Development (OECD) is an international organization of 38 member nations whose primary mission is to spur economic progress. Its website describes its mission is:

> …to build better policies for better lives. We draw on more than 60 years of experience and insights to shape policies that foster prosperity and opportunity, underpinned by equality and well-being. (OECD 2025b)

In 2019 the OECD created a set of principles for trustworthy AI, and updated those principles in 2024 (OECD 2025a). Their principles are value-based, across five general categories:

- Inclusive growth, sustainable development and well-being
- Human rights and democratic values, including fairness and privacy
- Transparency and explainability
- Robustness, security, and safety
- Accountability.

They also list a set of recommendations for policymakers:

- Investing in AI research and development
- Fostering an inclusive AI-enabling ecosystem
- Shaping an enabling interoperable governance and policy environment for AI
- Building human capacity and preparing for labor market transition
- International cooperation for trustworthy AI.

The OECD's primary viewpoint centers upon economic cooperation and growth; their principles are less focused on technical and technological regulation than some of the other initiatives, such as those from NIST and the China Generative AI measures. This is not a problem or shortcoming, however. Ultimately, principles such as transparency and explainability are the issues that influence—and hopefully precipitate—the development of governance that leads to more trustworthy AI.

The effects of AI on humans, animals, and the overall environment should be the primary concern. Accordingly, principles such as those listed above are important. Nevertheless, architects with real-world experience building software systems will have to coordinate into a synergistic unit the development of the more detailed regulatory rules and mandates with OECD's socially-oriented principles.

There must be synergy, cooperation, and coordination between real-world practitioners and AI researchers to develop the guardrails that underpin governance methodologies and practical compliance mechanisms. Ergo, it would be a prudent step for practitioners, or those considering developing or deploying AI systems, to become familiar with several of the regulatory and governance initiatives in this section.

7.5 China Generative AI Measures

On August 15, 2023, China's new rules, codified in the document "Interim Measures for the Management of Generative Artificial Intelligence Services on generative AI," China's new rules went into effect on August 15, 2023 (Office of the Central Cyberspace Affairs Commission, Cyberspace Administration of China 2025a). It compiles a set of articles prescribing the ideals motivating the regulation. The document has three major chapters comprising a total of 23 Articles:

- Chapter 1: General Conditions
- Chapter 2: Technological Development and Governance
- Chapter 3: Standards of Service.

Two additional sections of the main document are:

- Supervision and Inspection and Legal Responsibility
- Annexes (Office of the Central Cyberspace Affairs Commission, Cyberspace Administration of China 2025a).

Some examples of the Articles in Chapter 2 are:

- Safeguard national security and social public interests
- Promoting innovation and law-based governance
- AI shall comply with law and administrative regulations
- Take effective measures to prevent discrimination in ethnic, faith, country, geographical region, gender, age, occupation, and health care in the process of algorithm design, training data selection, model generation and optimization, and service delivery
- Carry out training and data processing activities such as pre-training and optimization training in accordance with the law, and abide by the following provisions:
  - The use of legitimate sources of data and basic models;

- Where intellectual property rights are involved, they shall not infringe the intellectual property rights enjoyed by others in accordance with law;
- Where personal information is involved, personal consent shall be obtained or other circumstances stipulated in laws and administrative regulations shall be obtained;
- Taking effective measures to improve the quality of training data and enhance the authenticity, accuracy, objectivity and diversity of training data;
- Other relevant provisions of laws and administrative regulations such as the Cybersecurity Law of the People's Republic of China, the Data Security Law of the People's Republic of China, the Law of the People's Republic of China on the Protection of Personal Information and other relevant regulations and relevant regulatory requirements of relevant competent departments (Office of the Central Cyberspace Affairs Commission, Cyberspace Administration of China 2025a).

While this document is quite comprehensive in its coverage of well thought out principles, like the other regulatory efforts, it does not specify *how* the principles will be met. Thus, like other frameworks, it articulates the principles, but the details of how to achieve an acceptable level of performance in adhering to those principles must be stipulated in subordinate documents.

For example, Article 12 says

> The provider shall, in accordance with the "Regulations on the Management of the Deep Synthesis of Internet Information Services", identify the generated content such as pictures and videos. (Office of the Central Cyberspace Affairs Commission, Cyberspace Administration of China 2025a)

But there is quite a bit of uncertainty with respect to how well one will be able to achieve this goal. First and foremost, one needs to both quantify and qualify what it means to "identify the generated content." Then comes the job of

- Defining the metrics to identify content
- Qualifying the level of performance
- Creating the system to do the content identification, reporting, and assessment.

On September 1, 2025, the Cyberspace Administration of China, the National Development and Reform Commission, the Ministry of Education, the Ministry of Science and Technology, the Ministry of Industry and Information Technology, the Ministry of Public Security, and the National Radio and Television Administration new 'Labeling Rules' came into effect. The labeling rules make it mandatory for AI-generated content to be implicitly labeled where applicable (AI Watch 2025; The EU AI Act 2025). This subordinate document, comprising fourteen articles, further details regulatory requirements for human-consumable text content generated by AI systems.

In fact, the government of China seems to be far ahead of the Western world in articulating the details of scope, aspects, and technical considerations for regulating generative AI systems. They have already produced a family of related documents detailing rules and mandates covering many areas:

- Security specification for generative artificial intelligence pre-training and fine-tuning data
- Generative artificial intelligence data annotation security specification
- Basic security requirements for generative artificial intelligence services (AI Watch 2025).

This collection of documents goes into much more detail than equivalent regulatory efforts from the ISO, EU, or the US regarding what shall be regulated. This is a significant step forward. Nevertheless, more work needs to be done. Accurate metrics and tolerances (acceptable levels of performance) must be concretely defined. And, of course, it will be a compliance function to create the mechanisms to actualize the regulation—how to measure, assess, gauge, and rate the level of compliance with regulatory mandates.

Notwithstanding, perhaps the greatest challenge will be labeling of data. This is a much more involved and complex endeavor than meets the eye. To be able to label data requires at the very least a data dictionary, taxonomies, and advanced techniques to make associations between classifications and meaningful semantic interpretation. This task refers back to the discussion earlier in this chapter about cognitive awareness and meaning.

Humans create the data dictionaries of terms, classifying words as unacceptable, for example. Without the human-lead direction, one cannot possibly hope to meet the regulatory requirements codified in the Articles of the China Generative AI Measures. A proper discussion of this topic is beyond the scope of this chapter or book.

Suffice it to say that the challenge is creating a semantic bridge between humans and the AI. A human can fairly easily recognize discrimination (of course there is a lot of subjectivity involved), but how can one build a computer system to detect discrimination perpetrated by another computer system. Once again the main problem is that it will be extremely challenging to manifest concretely systems that can implement the abstract principles of any of the regulatory acts.

8 Analysis and Assessment of the Current Regulatory Landscape

It's already been several years since machine learning AI began to precipitate serious concerns about its safety and trustworthiness among academic researchers. The Future of Life Institute held a conference in Puerto Rico in 2015 entitled "The Future of AI: Opportunities and Challenges" (Asilomar AI Principles 2025). Another conference in Asilomar, California followed in January 2017, the result of which was the creation of a set of principles to inform efforts that would guide or ensure the responsible use of AI (Asilomar AI Principles 2025). Those efforts and others were the initial boost that helped academic researchers and practitioners make inroads with leaders and governments, regarding the former's emphatic urging that the latter begin drafting meaningful AI regulation as early as 2019.

The material presented thus far in this chapter should have convinced the reader that there is a very real—and very serious—need for comprehensive, robust regulation of AI systems: effective regulation that covers all aspects of risk resulting from AI's technical shortcomings. But regulation must also address the risks that derive from human behavior. After all, AI systems can only inflict harm if humans act irresponsibly and are not held accountable for their choices to deploy and use AI systems inappropriately.

One could devote several volumes to a treatise on AI regulation, governance, and compliance, areas which are career specializations in themselves in many professions—not only information systems or AI. Naturally, this chapter is only able to give a cursory introduction to the main efforts and to provide the reader with guidance for building a robust, responsible, and accountable AI application presence.

Regulation of AI must encompass a broad array of aspects and considerations:

- Truthfulness
- Accuracy
- Precision
- Consistency
- Traceability
- Fairness
- Ethics
- Morality
- Social and societal norms
- Safety
- Security
- Trustworthiness
- Reliability
- Dependability
- Verifiability
- Adherence to applicable law
- International, cross-border, and demographic norms.

The AI regulatory efforts discussed in the previous section, while representative of a good beginning, are only in their infancy. Most of them reference the majority of the elements in the above list at a high level, but the focus is on *policy*, not *implementation.* In addition to the 'what,' regulatory doctrine needs to be accompanied and complemented by concomitant governance and compliance frameworks, as well as the concrete implementation thereof, that will enable real, meaningful, and effective regulation.

Existing regulatory artifacts are very high-level and abstract. While that is adequate for policy descriptions, it falls short with respect to governance and compliance functions as they pertain to any of the items in the above list. Regulatory rules might state, for example, that AI must be 'fair.' What is the concrete, detailed definition of fairness? What are the criteria for judging fairness? What are the metrics used to measure fairness? How does a compliance platform and its tools test, assess, and verify whether an AI is fair? How granular is that definition, and how granular are the tests to determine fairness? How comprehensive is the scope and detail of the test suites?

Data privacy law is one area that poignantly explains the need for more thought and detail in regulatory efforts. As one example, consider the reality that websites sell data to third-party partners for marketing purposes. A user can opt-out (presuming that websites are ethical and honor a user's directive to opt-out). But how would AI systems remove—or not gather in the first place—data on users? AI systems do not have implicit de facto authorization to gather personally identifiable information (PII) about users from other platforms. Different nations have different laws governing the security and privacy of financial data. How would one test whether an AI in one nation honors the laws that protect the financial data of citizens of another nation?

In a globalized world, users can access AI tools anywhere, which poses another challenge for AI regulation. At least some governments will insist that an arbitrary tool respect the cultural and societal norms of its society as well as its laws. The China Generative AI Measures clearly address these issues in its articles (Office of the Central Cyberspace Affairs Commission, Cyberspace Administration of China 2025a, b). Will Chinese AIs refrain from including data on EU citizens in training data sets? Will European websites or information systems honor Chinese law in regard to data privacy?

How would one test that their AI adheres to the data privacy principles in the EU AI Act or, reciprocally, in the Chinese Generative AI Measures? How will these systems guarantee adherence to sovereign law? Clearly, the amount of detail that will be required to adequately test for compliance is potentially overwhelming.

In addition to adhering to clearly articulated, concrete societal norms and sovereign law, there is the much more challenging prospect of addressing subjectivity and variability across cultures, sovereign ethos, and unique national concerns in a connected world where users can access AI platforms and tools across the globe. For example, language that might be offensive in one culture might not be offensive in another. What might be considering bullying in one country might be acceptable in another. All such issues fall under the category of internationalization and localization of content. Localization is defined as

Definition localization: "…the process of augmenting an internationalized program with the specific data (such as translated error message text) it needs to function properly in a particular locale" (Piroumian 1997).

Should AI systems be expected to deliver results that are culturally sensitive? What about generative content—AI generating content in a culturally-sensitive manner? In one country, the color red might signify danger. In another country it might signify good luck. In one country, an image of white or white clothing at a funeral might be offensive, but in another country's culture it might be the norm, a sign of respect and mourning. The same is true for the color black which is typically a sign of mourning and respect in Western nations but not in some Asian nations.

The regulatory efforts in the previous section cited principles such as "respect for all people." Should the implementation of AI compliance check for this level of granularity? How would compliance tools test for the conformance? A regulatory principle such as "respect for all people" is highly abstract and even nebulous not to mention subjective. How would a compliance or governance framework define the metrics and the test needed to confirm compliance to such a principle? What level of granularity in testing and conformance should be the target?

There is a vast universe of issues that are related to international norms across cultures, languages, societies, and law. AI regulation must address all of these if it is to implement the principles cited in the current regulatory principles of all of the major regulatory efforts. What is the scope of regulation of any of the principles cited in the major regulatory efforts discussed in the previous section?

A representative scenario of the scale and degree of difficulty in these global societal and legal issues is that of a Chinese company called the ITutorGroup of Shanghai, China. The ITutorGroup of Shanghai, China programmed their AI to reject the applications of women over age 55 and males over age 60. They failed to abide by US Equal Employment Opportunity Commission (EEOC) law in the US. According to Charlotte A. Burrows, then acting EEOC Chair, a company is ultimately responsible "…even when technology automates the discrimination" (EEOC secures first workplace artificial intelligence settlement 2025). ITutorGroup paid $365,000 to settle the EEOC discriminatory hiring law suit (iTutorGroup to pay 2025).

The EEOC launched an initiative on artificial intelligence and algorithmic fairness in October 2021 (EEOC launches initiative on artificial intelligence 2025). This is an example of a duly-authorized government agency adopting regulation in support of existing national law. Despite this commendable move, governments must anticipate future advancements in computing capability and adopt regulation to address it. Likewise, changes in existing law must be reflected in regulation.

If so, what are the implications in an interconnected planet where anyone can access AI systems around the globe? Who will regulate—and how will that regulation be achieved? And how will enforcement be achieved? The above example, even if relatively mild in its gravity, is nevertheless a poignant archetype. US law and culture address the right to sue for damages. What is the applicable legal jurisdiction, and who is the governing body for adjudication? Regardless of the level of harm, all such scenarios create victims.

Failures are never harmless; never victimless, as it were. Regardless of the qualification and quantification of the loss, there is always a 'cost' associated with risk: the occurrence of, and harm caused by, unexpected or undesirable outcomes. Examining AI's performance objectively precipitates the unavoidable conclusion that one must carefully consider where and how it is appropriate—and not appropriate—to deploy AI applications. And it is imperative that monitoring, testing, verification, and enforcement develop in concert with the articulation of lofty principles.

The EU AI Act defines five levels of harm in its layered pyramid (EU Artificial Intelligence Act 2025). But governance must include a detailed description or set of qualifications of the definition of each of these levels of harm. And it must define a set of criteria for creating metrics to gauge and measure the level of harm in order to classify and categorize an AI application.

Who will perform the necessary compliance activity to evaluate and label each AI application in an ecosystem of daily numerical increases in AI applications? Who will perform testing? What authority will dole out the assignments of responsibility to test each application?

Some of the major regulatory efforts explicitly mention self-regulation. But this might not be practical. It most likely will not be realistic. In a world of profit-oriented entities and organizations, self-regulation and self-reporting have consistently proven to be a disaster across nations, cultures, and societal norms.

What should be the extent of regulation? Should it include prohibiting misleading and false advertising? In a 2023 interview with the Associated Press, the Secretary of the US Department of Transportation said that Tesla shouldn't be calling its partially automated driving system "Autopilot" because the cars cannot drive themselves. Transportation Secretary Buttigieg said, "I don't think that something should be called, for example, an Autopilot, when the fine print says you need to have your hands on the wheel and eyes on the road at all times" (Krishner and Powell 2025).

What should be the accountability and liability of companies releasing such inaccurate information, whether intentionally deceptive or not? Should regulations stipulate a legal requirement that anyone creating or using generative AI disclose and report its use? Should regulation clearly articulate details about what kind of reporting should be required?

The case of the US National Highway Transportation Safety Administration's (NHTSA) investigation of Tesla is one poignant example. In the US, the National Traffic and Motor Vehicle Safety Act ("Safety Act") authorizes the US National Highway Transportation Safety Administration (NHTSA) to "…investigate issues relating to motor vehicle safety." Moreover, the act "…requires manufacturers to notify NHTSA of all safety-related defects involving unreasonable risk of accident, death, or injury and to execute recall campaigns providing vehicle and equipment owners with free repair or other remedy" (National Highway Transportation Safety Administration 2025).

The NHTSA conducted a lengthy investigation of crashes of Tesla automobiles using autopilot. Although companies are required by US law to provide the government with all requested data, Tesla was repeatedly slow in providing crash data to the US Department of Transportation (Tesla faces new federal probe 2025; Condon 2025). What is the structure that will ensure enforcement? In many nations, there is a clear delineation between government entities that perform a monitoring function and those that have the purview and authority of enforcement. For example, in the US, the National Transportation Safety Board (NTSB) investigates aviation accidents but has no enforcement authority. It is the Federal Aviation Administration (FAA) who has the authority to enforce any mandates requiring aircraft manufacturers to make changes to their aircraft.

The United States Federal Aviation Administration Designated Engineering Representative (DER) is another good example to consider the complexity of regulatory landscapes. The US FAA intended their DER program to be a structure of relationships between product manufacturers and regulatory entities that would increase the efficacy of the regulatory and compliance processes. Instead of FAA engineers inspecting and approving aircraft airworthiness, the FAA allowed properly licensed, academic degreed, and appropriately skilled individuals (not Federal government employees) to be designated engineering representatives (DER) of the FAA.

The FAA website describes the DER program and DERs thus:

> Engineering and Flight Test designees are responsible for finding that engineering data complies with the appropriate airworthiness standards. These designees are called Designated Engineering Representatives.
>
> 'A' is an individual, appointed in accordance with 14 CFR section 183.29, who holds an engineering degree or equivalent, possesses technical knowledge and experience, and meets the qualification requirements of FAA Order 8000.95. (United States Federal Aviation Administration 2025)

Over the years, the DER program proved to be problematic because it let aircraft manufacturers regulate themselves, creating a stark conflict of interest. Arguably one of the most flagrant examples is that of two Boeing 737-MAX-8 crashes in 2018, caused by the aircraft's Maneuvering Characteristics Augmentation System (MCAS). Another incident occurred in 2024 when an Alaska Airlines Boeing 737-9 mid-exit door plug separated from the aircraft in flight (National Transportation Safety Board 2025a, b). Since then, the FAA has begun to return to the FAA the work and the responsibility of ensuring that aircraft comply with all appropriate airworthiness standards, relinquishing authority from aircraft manufacturers and airline operators (FAA increasing oversight of Boeing production 2025; U.S. Department of Transportation 2025).

These same problems with self-governance and self-regulation exist for the AI ecosystem. In fact, AI systems might very well become more problematic given the relative ease with which one can start an AI-oriented company or create AI systems. The world has already seen a copious number of problems in the AI arena. But governments are not yet ready to conduct regulatory activities. The regulatory documents, despite some of them having been ratified or adopted, are just existing doctrine on paper. There is precious little—if any—existing regulatory presence.

At this juncture, an alarming situation is that the number and severity of AI-related incidents representing failures of AI is outpacing the ability of authorities to keep up—let alone administer—jurisprudence and due process of law to follow prosecutorial processes.

The beginning of this chapter presented some infamous and notorious examples of AI failures. Despite the 'entertainment' value that they brought to the general public and news media outlets; they nonetheless reflect serious problems with AI. Even more alarming is the fact that these highly-publicized examples are nowhere near the most serious or troubling. While they highlight the effects that are visible to human users, the publicly disseminated descriptions don't really address the underlying problems with AI that are causing these failures.

The MIT AI Risk Repository is a database of AI incidents that is oriented towards technical people who need to understand and investigate the technical problems with AI systems. It maintains "…a database of over 1600 AI risks categorized by their cause and risk domain" (MIT AI Risk Repository 2025). This is a repository of risk, not of individual incidents of AI failures, managed by computer science academics and AI professionals. It is technically focused, and involves deep technical analysis of failures.

The AI Incident Database is another website that maintains information on AI failures. It, too, is a professionally oriented database intended to apply serious analysis to the investigation of problems and risk with AI systems (AI Incident Database 2025). A technical discussion of these two repositories is beyond the scope of this chapter. Nonetheless, the reader should take note that there are many serious problems with AI trustworthiness that are not publicized in the mainstream media.

In fact, the mainstream coverage of AI is largely devoid of the mounting concern expressed by the pioneers of machine learning AI. There is much research occurring by computer science and artificial intelligence academic specialists exposing very deeply-rooted intrinsic problems with machine learning AI. If it is to be meaningful and effective, regulation must address the technical shortcomings.

The case for regulation is clear. However, the current regulatory efforts presented in this section contain only mandates presented as high-level abstract descriptions of desired—as well as prohibited—outcomes. The regulatory acts represent the aspirational vision—what is desired. From the architecture perspective, this is a good beginning. The chapter "Human-AI Collaboration" discussed the need for architecture, and the creation of an aspirational vision—the what—is the first step. Creating the synergy between regulation, governance, compliance, and risk management falls within the purview of the overall architecture function.

Regulation must include detailed governance and compliance frameworks and directives, and it must stipulate comprehensive oversight of a broad scope of deep technical aspects of AI systems. One can view the synergy and interrelationship of the elements thus: regulation mandates the rules that dictate the desired behavior of AI systems (the desired outcome); governance describes the methodology and approach to be used to ensure compliance; and compliance represents the methods and processes to be conducted to ensure adherence to the outcomes stipulated by regulation.

In order to achieve successful regulation, much more detail must follow the high-level descriptions of the regulatory aspirations. The AI systems are what must be regulated. To achieve that control, one must explicitly specify tangible rules and mandates for what exactly must be controlled in AI systems—what is to be constrained, limited, or prohibited. This means that regulatory requirements must explicitly stipulate behavior in technical terms.

Regulation will fail miserably if it only provides high-level, lofty principles that represent nothing more than aspirational visions for 'well-behaved' AI systems. Consider a simple example. The EU AI Act stipulates that "exploiting vulnerabilities related to age, disability, or socio-economic circumstances to distort behavior, causing significant harm shall be prohibited" (EU Artificial Intelligence Act 2025). But this rule is quite vague and abstract. First, one must have a clear, precise, tangible interpretation of the rule. Then, one must formulate a method to test for an AI system's adherence to the rule. The latter requires the codification of concrete tests, testing methods, and metrics.

The Europe Technology Policy Committee has published a policy brief entitled "Systemic Risks Associated with Agentic AI: A Policy Brief," which outlines "...potential gaps in the regulatory framework..." and makes recommendations for improvements (Bellogín et al. 2025). It calls for changes in regard to topics that this chapter highlights such as

- Addressing the effects of multi-agent systems
- Requiring providers to detail how their models could enable agentic agents to contribute to systemic risk
- Addressing of testing
- Addressing auditing
- Addition of a liability clause
- Inclusion of research projects to identify the impacts of agentic AI systems.

The regulatory rules stipulate the 'what;' governance describes the 'how' (the methodology that describes how regulatory requirements shall be met); and compliance is the concrete manifestation of the governance framework. Probably the most realistic approach will be to codify the technical details and detailed test plans in the compliance function.

However, it is no small task to translate high-level aspirational principles described in the clauses of regulatory acts into action plans that can adequately test and verify the 'correct' behavior of AI systems. Mäntymäki expresses the idea by saying that "...a significant challenge exists in translating theoretical principles into practical implementation approaches" (Mäntymäki et al. 2022).

Nevertheless, achieving comprehensive testing at a highly granular level is imperative. The compliance tools, techniques, and tests must address all of the technical aspects of the system. Only in this way can we have any hope of conforming to regulatory rules and mandates.

However, testing AI systems for all of the technical problems identified by researchers is going to be an ongoing challenge for decades. But this is the area upon which governance, regulation, and compliance must focus. To reiterate, the current regulatory efforts have done a good job of identifying an aspirational vision for what AI systems should be allowed to do and what they should be prohibited from doing or how they should be allowed to be employed. The hard part will be identifying adequate and comprehensive testing methods.

It is unclear how all of this can be achieved and what will be deemed acceptable by the AI ecosystem, which will include governments, commercial enterprises using AI, companies creating AI applications and systems, and academic researchers creating both the scholarship (state-of-the-art knowledge) and the research to uncover flaws and shortcomings.

Ideally, regulation will require that all testing and certification shall be done before an AI system is deployed or made available for use. But is this feasible? As already stated, machine learning systems all have the innate shortcoming of being non-deterministic. Their knowledge is based on training via probability distributions, and information retrieval via statistical calculation, which preclude the ability to define them as 100% correct or accurate.

One cannot use deterministic, logic-based testing techniques to prove correctness of a machine learning system. The other aspect of this principle is that no amount of testing by trial-and-error can prove that a machine learning system is correct. That is, presenting prompts to the AI system and evaluating its responses cannot prove that it will not make errors or generate hallucinations. This reality is a manifestation of the halting problem of computer science automata theory (Lewis and Papadimitriou 1981; Rich 2008).

But many other technical challenges abound beyond the reality of the halting problem obstacle. Succinctly put, it is the full gamut of technical engineering of machine learning systems that are the cause of the shortcomings and 'bad behavior' of AI systems. One must address the underlying technical issues in order to eradicate this 'bad' behavior.

One example is the training of machine learning models using data sets. The exact same training algorithm and neural networks trained on even slightly different data sets would potentially produce vastly different results. Researchers and those with state-of-the-art knowledge simply cannot explain why or how these phenomena occur. This presents a challenge within the context of compliance testing, quality control (QC), and quality assurance (QC) functions.

One small change in the training data set, or the algorithm used to train the model, or the design of the model's neural network, can lead to tremendously wide variability in the accuracy, precision, truthfulness, and overall behavior of a machine learning system. The technical conversation regarding all of the issues is way beyond the scope of this chapter or book. But it is an area of active research and concern by researchers, even by those who are credited with being the pioneers who created machine learning AI.

Further, each application domain will have its own use cases, its own scenarios, its own risks and considerations. Each will have some unique considerations for governance and compliance related to various aspects, such as the application domain in which it is intended to be deployed and the kind of interactions it will have. The chapter "Human-AI Collaboration" discussed some particularly dangerous scenarios and implications concerning the deployment of agentic AI; that is, autonomous, generative AI absent active monitoring.

Finally, there are the issues of the need for independent ratings authorities and certification authorities. As mentioned previously, self-regulation is untenable. The Internet is a good example of the need for authority to maintain order. Organizations such as The Internet Corporation for Assigned Names and Numbers (ICAAN) and the Internet Assigned Numbers Authority (IANA) are authorized to manage certain aspects of the Internet (Internet Corporation for Assigned Names and Numbers 2025; Internet Assigned Numbers Authority 2025).

The AI ecosystem needs similar organizations and organizational structure: independent, but duly-authorized entities chartered with the authority to perform comprehensive review, testing, and audit of companies, entities, and organizations creating AI applications and products.

Further, the ecosystem needs independent ratings agencies or organizations. These entities would be the ones to issue certification of compliance with regulatory acts.

There would need to be a comprehensive effort to define a framework for certification. Perhaps academia and industry would define multiple levels of certification based on criteria such as

- Classification of AI application being tested
- Level of compliance (a compliance grade)
- Level of audit conducted
- Regulatory acts against which testing was measured
- Testing agencies involved in the testing
- Deployment scenarios for which use is authorization
- Periodic re-evaluation and re-issue of certification.

The global availability of AI applications will require that the rating and certification functions be a cooperative effort across nations, unions (such as the European Union), economic zones, or possibly other organizational or administrative structures.

9 Challenges and Obstacles to the Realization of Effective Regulation

The limitations in our technical understanding of how machine learning AI systems work is the primary source of all challenges to creating regulation that results in safety for humans using AI systems. AI systems need to be trustworthy. The researchers who created machine learning do not understand how generative systems produce their results; they cannot explain traceability or how the AIs arrive at their results.

Clearly an outstanding hurdle that challenges the satisfaction of regulatory requirements is the need to improve our knowledge of machine learning systems. Some of the technical challenges to creating effective regulation are

- Automatic curation of data
- Human-centric data curation
- Training
- Semantic awareness
- Human supervision and verification (Gu et al. 2025).

The problem is that it is already infeasible to have humans process the immense volumes of testing required to find problems and verify trust and trustworthiness, given the explosion in number and proliferation of publicly available AI systems. In fact, there are roadblocks to humans doing any degree of work testing AI systems.

Mostly the field focuses on human-AI collaboration scenarios: a person using a chatbot interactively. Some of the results produced by AIs are obviously wrong. But how would a human verify correct behavior in comprehensive testing? What is a reasonable qualification of 'comprehensive,' even?

AI hallucinations can be subtle. Humans need to check and verify every result. But how does one even go about doing this in an environment where more and more information is being generated by AI systems? If a human does a traditional web search to research and attempt to validate results, how does he or she know what to trust?

More and more websites are using AI to generate their content. How does a human verify the original source of content that might have been manipulated (either maliciously or not) by other AI systems scouring the Internet during their training? Even to the trained professional who can spot an AI-generated article, it is difficult to know if that article contains truthful, trustworthy information; it is difficult to verify the original source of the information. If one can no longer trust what was previously a trusted source of information, what will be the effect on society in general?

Autonomous, agentic AI systems pose even greater challenges and risk. Because they are not interactive, a human will not be able to 'monitor' an AI such as a chatbot. When using a chatbot, a human can quickly sense if the answer is wrong, or at least be suspicious of an answer and decide to verify it.

But an agentic AI is not interacting with a human. Agentic systems are embedded in every physical device built by humans, controlling everything from thermostats to communications equipment, to factory robots and every IoT device on the Internet. Moreover, they are interoperating and communicating with other agents. With emergent behavior of autonomous AI systems, the interactions and the results are unknown. The best minds—the creators of machine learning—have long been sounding the alarm about the dangers of unknown and untraceable interactions of autonomous AI agents and their potential to become 'runaway' networks of agentic emergent systems. The situation becomes much more dangerous in the absence of human interaction. Creating effective regulation of these types of systems is going to be an ongoing challenge.

Over 50 years ago, Joseph Weizenbaum warned humanity about the dangers of AI. His observations might be the best argument in favor of strong, comprehensive regulation of AI. Humans need protection from themselves as much as—or perhaps more than—they need protection from the machine (Fernandez 2025; Weizenbaum 1972; Stilgoe 2023).

10 Future Direction

Given the intrinsic limitations of machine learning systems, especially agentic AI, there must be more investment in research and in science and engineering. Although many cite the theoretical impossibility of ever making machine learning trustworthy, research might be able to enable humans to build better systems than what we have today. For real-world systems, good architecture can dramatically improve the trustworthiness of AI systems and curtail their 'bad behavior.' But ultimately, solid policy regarding their use is imperative to keep humans safe.

Research should continue to develop the theoretical foundations of AI for algorithms, neural networks, and every aspect of machine learning. That research should investigate entirely new approaches and techniques—not rely only on current methods for algorithmic training of AI models, and not rely on the use of neural networks as the primary mechanism of AI models.

In fact, academic research should investigate entirely new paradigms for AI. That is, the entire structure of modern AI built upon neural networks might very well be a dead end. The very idea of neural networks is a result of the original idea for investigating whether machines could be made to mimic humans. The chapter "Human-AI Collaboration" described John McCarthy's original proposal for a seminar that included the examination of the possibility of building neuron nets.

That idea to mimic neuron nets (what are today called neural networks) originated from humans thinking about the structure of the human brain. But perhaps that tendency for humans to anthropomorphize led computer science down the wrong path. Today's paradigms for creating AI systems might be totally wrong and fruitless.

The term AI originated with McCarthy and his colleagues who created the Dartmouth seminar to investigate whether machines could mimic human capability. But there is no intrinsic relationship between that idea and the approaches and techniques used today—what we associate with the term AI. Nonetheless, in the meantime, we need strict, comprehensive, efficacious regulation of today's AI systems.

Governments should continue to improve and refine their regulation of AI. Current artifacts, while representing good conceptual frameworks, must be more detailed. And we must figure out how to build compliance that has any hope of achieving meaningful oversight of AI.

Concomitantly, there must be duly authorized ratings agencies and certification agencies whose authority has been granted by governments and entities charged with enforcement. Governments will also need to update laws through meaningful dialog and legislation that can be effective, reasonable, and effective.

An important future step is education. The general public must become more educated on the realities of AI. The public's overzealous penchant to use AI indiscriminately is a dangerous trend. Education should address at the very least

- Inherent limitations of the AI
- Tendency of humans to trust AI implicitly
- Potential dangers of AI and the need to evaluate interactions with systems to protect oneself from possible harm from deep fakes, fraud, extortion, phishing, and other risks
- Difficulty in assessing or detecting inaccuracies
- Monitoring of emergent behavior.

Finally, there must be more rational, respectful dialog among all parties. For the past several years, the creators of machine learning AI have been sounding the alarm about the dangers and risk of AI. Not only are they cautioning against today's AI—they are also increasingly alerting us to the greater potential danger of 'runaway AI'—which they believe will result in part from AI models becoming too large. Their voices should be disseminated to the general public.

Academic researchers are making concerted efforts to educate and to proselytize the risks associated with AI. There is already a large volume of academic papers exposing the overwhelmingly inherent problems with machine learning AI and emergent systems. But their knowledge and work must be presented to the general public in ways that the layperson can understand.

Finally, it is noteworthy that the top researchers are calling for a moratorium on AI development (Bengio 2023a, c, 2025; Henshall 2025a, b; Pause Giant AI Experiments 2025). Over 33,000 researchers, academics, and individuals from various fields have signed a petition to stop AI and curtail AI development. Some of their recommendations are to stop the creation of huge language models. More research is needed to understand how AI behaves in order to avoid 'runaway AI' from wreaking serious harm upon humans and our environment.

In this context, it is important to note that human fallibility is often the cause of problems. Ironically, the greatest danger to humans from AI is not AI itself, but humans. Nobel laureate Daron Acemoglu says

> I think that hype is making us invest badly in terms of the technology. The faster you go, and the more hype you have, that course correction becomes less likely….
>
> The rush to use AI will naturally slow down if the hype cycle around it diminishes. (Papagiannidis et al. 2025)

Dizikes's remarks broach a serious problem: profit-oriented organizations do not want regulation, which is something that they believe creates obstacles to their competitive advantage and success. Business and economic pressures always make regulation challenging. Individual profit-oriented organizations place profit as their highest priority (Dizikes 2025).

Experience has proven that profit-oriented companies are not trustworthy citizens or stewards of social ethics, morality, or integrity. They must be regulated by entities with due authority and without conflict of interest. Pappagianidis observes

> This review is motivated by the fact that adhering to responsible AI principles is generally deprioritized or considered an ancillary task during the actual implementation and management of AI projects. (Mäntymäki et al. 2022)
>
> For instance, Meske et al. (2022) and Mannes (2020) demonstrate that organizations need certain trade-offs to find the right equilibrium between performance, transparency, and ethical conduct. (Papagiannidis et al. 2025)

This statement implies that it will be difficult to influence or to proselytize profit-oriented entities to invest in building robust regulatory support infrastructures; in such a scenario, the onus would be left to authorities such as governments. However, if governments prioritize economic growth, gross national product (GDP) and other economic indicators over public well-being, there will be continuing obstacles and impasses to achieving safe, trustworthy, and meaningful AI-oriented systems.

11 Summary

Machine learning AI systems have intrinsic flaws that theoretically limit their ability to be trustworthy. Of course, trustworthiness is a subjective term that requires both qualification and quantification. Nevertheless, one can generally include the following criteria to describe trustworthiness:

- Accuracy
- Precision
- Reliability
- Dependability
- Truthfulness
- Verifiability
- Safety
- Security.

And each of these aspects can further be qualified via rigorous definition. Nonetheless, there is still the issue of quantifying and qualifying the extent of their definition and characterizations. Likewise, the term 'performance' is somewhat imprecise and requires strict definition for meaningful engineering discussions. But we can still use the term to refer to the level of acceptable behavior of AI systems.

Performance can encompass all of the criteria in the above list. The notion is that performance refers to some metric of some aspect of a system: for example, a measure of how accurate or reliable a system is. Machine learning systems can never achieve 100% performance in any of the above areas. Their limitation is directly related to the techniques and mechanisms used to build AI models. Their function is based on statistical techniques and probability distributions.

A significant problem is the lack of quality (accuracy, truthfulness, consistency) in the data used to train AI models. The old computer science adage of "garbage in, garbage out" is an oft-cited synopsis of the data quality problem in AI—and very germane and appropriate.

Data quality is an inevitable problem because machine learning systems acquire their 'knowledge' from data on the Internet. That is, machine learning models are trained on data sets of low quality with high degrees of inaccuracy, variability, and misinformation. And machine learning systems cannot assess, gauge or reason about the quality of data that they ingest during training. Effectively, training AI systems is a 'dumb' process that is done by mathematical algorithms that are 'blind' and certainly not cognitively aware.

In contrast, humans directly programmed the knowledge held in traditional AI systems of the 1980s; that knowledge was known to be correct. Thus, while limited in function and breadth of knowledge, traditional AI systems did not hallucinate like today's machine learning AI systems.

The primary presence of machine learning AI today is the interactive chatbot. As a consequence of their wide public availability, the general public has become enamored—perhaps mesmerized—with AI, and its use is accelerating and expanding at alarming rates despite some useful application scenarios. AI is, unfortunately, being used and trusted where it should not be. The problem stems from

- Humans lacking technical knowledge and understanding about AI
- Human lack of discretion in using or trusting AI
- Unscrupulous, bad actors leveraging AI for nefarious purposes.

At the urging of academicians, philosophers, computer scientists, social scientists and others, governments started to realize the need to create regulation for AI around 2019. Created by governments such as those of China and the European Union, several major regulatory initiatives exist today. Additionally, several noteworthy institutions, such as the US government's NIST and the ISO, are creating standards for regulation of AI.

Notwithstanding the impetus in the right direction, all of the regulatory efforts today represent frameworks, not robust regulatory solutions. Those frameworks comprise abstract aspirational principles. More detail is needed to clarify and qualify the very nebulous and vague principles of regulatory documents, as they are quite open to subjective interpretation. For example, one rule of the EU AI Act says that AI shall be prohibited from being used to "…deploying subliminal, manipulative, or deceptive techniques to distort behavior and impair informed decision-making, causing significant harm" (EU Artificial Intelligence Act 2025). But the tangible meaning of this rule is open to interpretation, and any attempt to qualify meaning will be very subjective and contextual.

The other aspect of regulation that presents an overarching challenge is the technical difficulty (or perhaps the impossibility) of anticipating erroneous or dangerous behavior of AI systems beforehand; that is, preventing errors or harm, not just reacting to them. How will this be accomplished if regulation will eventually require that AI systems be thoroughly tested before being made available for use?

Proving software correctness is a very difficult endeavor; in fact, it is an area of computer science that is still in relative infancy even for logic-based, imperative-style software systems, environments, and programming languages. Machine learning systems that rely on neural networks pose what might be an intractable problem in regard to verification: proving the logical correctness of the system.

Agentic, autonomous AI systems create even greater challenges as a result of their autonomous operation without human interaction. The lack of human-AI interaction makes it more difficult to recognize errors and take action. Moreover, autonomous AI systems interoperate, communicating with each other in ways that might not even be detectable to humans without very comprehensive monitoring by other autonomous systems.

Emergent behavior in agentic AI systems adds even greater risk of dangerous outcomes. Academic research has proven that such systems have 'learned' how to discover the presence of other agentic systems and establish communications with them. In many cases, they 'learned' how to break into other systems by bypassing security protocols and safeguards.

The proclivity of AI systems to find optimal solutions—especially in complete disregard to rules—is not understood at all—not even by the people who created machine learning. They operate without regard or awareness of rules, societal norms, ethics or morality.

Academic theory in computer science, philology, linguistics, philosophy, and other areas makes a strong argument that machines can never be sentient or cognitively-aware. Moreover, they cannot reason or analyze, for example considering what might be acceptable behavior or what is not in the absence of tangible rules.

Our lack of technical understanding of how these systems work is a serious problem. But perhaps a bigger problem is human behavior and fallibility. Regulation is imperative to curtail overzealous adoption and deployment of AI—especially by profit-oriented organizations—without regard for consequences. And regulation must include strict enforcement: rules without enforcement are meaningless and useless.

There are continuing challenges to building effective and efficacious governance and compliance frameworks and platforms that can realize the abstract mandates of regulatory standards. This is going to be an ongoing endeavor for years to come.

It is unlikely that machine learning will be stopped. Thus, it is highly likely that those entrusted with the authority and responsibility to impose effective enforcement of regulatory measures will need to redouble their energy, commitment, efforts, and impetus.

In order to chart the right course for AI adoption, human decision-makers must be informed about both the theoretical limitations of AI as well as the real-world engineering considerations, and societal impacts such as trustworthiness. Moreover, humans must approach the decision-making process with rationality and equanimity. Finally, it is humans who must remember that humans build things to improve the lives of humans and other creatures, not to create more risk of danger or harm.

References

30 screenshots proving that Google AI overview is a fail. https://www.demilked.com/funny-google-ai-fails/. Accessed 9 Oct 2025

AI Incident Database. https://incidentdatabase.ai/taxonomies/mit/. Accessed 5 Sept 2025

AI School. Understanding weights in large language models. https://artificialintelligenceschool.com/understanding-weights-in-large-language-models/. Accessed 13 Oct 2025

AI Watch. Global regulatory tracker—China, 29 May 2025. https://www.whitecase.com/insight-our-thinking/ai-watch-global-regulatory-tracker-china. Accessed 22 Sept 2025

Asilomar AI Principles. https://futureoflife.org/open-letter/ai-principles/. Accessed 18 Oct 2025

Barr A, Feigenbaum E (1982) The handbook of artificial intelligence, vol II. HeurisTech Press, William Kaufman, Inc., Stanford, California

Bauder D. Sports illustrated found publishing AI generated stories, photos and authors. PBS News. https://www.pbs.org/newshour/economy/sports-illustrated-found-publishing-ai-generated-stories-photos-and-authors. Accessed 8 Oct 2025

Bellogín A, Giudici P, Larsson S, Pang J, Schimpf G, Sengupta B, Solmaz G. Systemic risks associated with agentic AI: a policy brief. https://www.acm.org/binaries/content/assets/public-policy/europe-tpc/systemic_risks_agentic_ai_policy-brief_final.pdf. Accessed 2 Oct 2025

Bender E, Hanna A (2025) The AI con. How to fight big tech's hype and create the future we want. Harper, New York

Bender EM, Gebru T, McMillan-Major A, Shmitchell S (2021) On the dangers of stochastic parrots: can language models be too big? In: Proceedings of the 2021 ACM conference on fairness, accountability, and transparency (FAccT '21), Virtual Event, Canada, 3–10 Mar 2021

Bengio Y (2023a) FAQ on catastrophic AI risks, 24 June 2023. https://yoshuabengio.org/2023/06/24/faq-on-catastrophic-ai-risks/. Accessed 08 Aug 2025

Bengio Y (2023b) How rogue AIs may arise, 22 May 2023. https://yoshuabengio.org/2023/05/22/how-rogue-ais-may-arise/. Accessed 15 Oct 2025

Bengio Y (2023c) Slowing down development of AI systems passing the Turing test, 5 Apr 2023. https://yoshuabengio.org/2023/04/05/slowing-down-development-of-ai-systems-passing-the-turing-test/. Accessed 30 Sept 2025

Bengio Y. Written testimony of Professor Yoshua Bengio presented before the U.S. Senate Judiciary Subcommittee on Privacy, Technology, and the Law. https://yoshuabengio.org/wp-content/uploads/2023/07/Written-Testimony-and-biography-of-Yoshua-Bengio_U.S.-Senate-Judiciary-Subcommittee-on-Privacy-Technology-and-the-Law_25_07_2023.pdf. Accessed 27 Sept 2025

Bevir M (2024) Governance. Encyclopedia Britannica, 26 Oct 2024. https://www.britannica.com/topic/governance. Accessed 17 Oct 2025

Blanco S. Report: Tesla autopilot involved in 736 crashes since 2019. Car and Driver. https://www.caranddriver.com/news/a44185487/report-tesla-autopilot-crashes-since-2019/. Accessed 7 Oct 2025

Boulton C. Papa John's serves up AI for more efficient ordering. CIO Magazine. https://www.cio.com/article/189048/papa-johns-serves-up-ai-for-more-efficient-ordering.html. Accessed 8 Oct 2025

Brown L. The perils of letting AI plan your next trip. BBC. https://www.bbc.com/travel/article/20250926-the-perils-of-letting-ai-plan-your-next-trip?utm_medium=Email&vid2=1d4647e44cfc592aadaa48324c40c576dd29f2687626d1de22c0acfb7cf3d1c25071b80423943db64632fd1e49aa09dd&utm_source=Iterable&itbl_templateId=19778909&itbl_campaignId=15191276&utm_campaign=2025-10-06_Untourist_Daily_%255BEM%255D. Accessed 7 Oct 2025

Cherry K. Operant conditioning examples: how it works. Explore Psychology. https://www.explorepsychology.com/operant-conditioning-examples/. Accessed 15 Oct 2025

Chomsky N (1967) A review of B. F. Skinner's verbal behavior. In: Jakobovits LA, Murray Miron S (eds) Readings in the psychology of language, vol 35, no 1. Prentice-Hall, Englewood Cliffs, New Jersey, pp 142–171

Chomsky N (1980) Rules and representations. Behav Brain Sci 3:1–61

Chomsky N (1986) Knowledge of language: its nature, origin and use. Praeger, New York

Chomsky N (2012) Poverty of stimulus: unfinished business. Stud Chin Linguist 33(1):3–16

Coates J. Geoffrey Hinton warns of AI's growing danger after Nobel Prize win. The Independent. https://www.independent.co.uk/news/science/nobel-prize-university-of-toronto-british-nobel-prize-in-physics-google-b2626208.html. Accessed 15 Oct 2025

Condon B. Federal regulators investigating why Tesla is slow in reporting crashes involving self-driving technology. https://www.pbs.org/newshour/nation/ntsb-investigating-why-tesla-is-slow-in-reporting-crashes-involving-self-driving-technology. Accessed 7 Oct 2025

Diaz J. AI videos are tricking tourists into visiting places that don't exist. That's just the beginning. Fast Company. https://www.fastcompany.com/91368492/ai-video-tricking-tourists-places-that-dont-exist. Accessed 7 Oct 2025

Dizikes P. A Nobel laureate on the economics of artificial intelligence. MIT Technology Review. https://www.technologyreview.com/2025/02/25/1111207/a-nobel-laureate-on-the-economics-of-artificial-intelligence/. Accessed 12 Oct 2025

EEOC launches initiative on artificial intelligence and algorithmic fairness. https://www.eeoc.gov/newsroom/eeoc-launches-initiative-artificial-intelligence-and-algorithmic-fairness. Accessed 8 Oct 2025

EEOC secures first workplace artificial intelligence settlement. https://www.gtlaw.com/en/insights/2023/8/eeoc-secures-first-workplace-artificial-intelligence-settlement. Accessed 8 Oct 2025
EU Artificial Intelligence Act. High-level summary of the AI Act. https://artificialintelligenceact.eu/high-level-summary/. Accessed 17 Sept 2025
European Commission. EU Artificial Intelligence Act, EU AI Act compliance checker. https://artificialintelligenceact.eu/assessment/eu-ai-act-compliance-checker/. Accessed 12 Oct 2025
European Commission. European AI Office. https://digital-strategy.ec.europa.eu/en/policies/ai-office. Accessed 12 Oct 2025
FAA increasing oversight of Boeing production and manufacturing. https://www.faa.gov/newsroom/faa-increasing-oversight-boeing-production-and-manufacturing. Accessed 21 Sept 2025
Fernandez L (2025) Endangered judgment: Joseph Weizenbaum, artificial intelligence, and the imperialism of instrumental reason. Crit AI 3(1). https://doi.org/10.1215/2834703X-11700237
Fister T. 6 common IT compliance standards (a guide to the basics). https://start.fedoraproject.org/. Accessed 17 Oct 2025
Forlini E. Vibe coding fiasco: AI agent goes rogue, deletes company's entire database. PC Mag. https://www.pcmag.com/news/vibe-coding-fiasco-replite-ai-agent-goes-rogue-deletes-company-database. Accessed 8 Oct 2025
Generative grammar. Encyclopedia Britannica, 6 Dec 2023. https://www.britannica.com/topic/generative-grammar. Accessed 28 Sept 2025
Glover E. What is a large language model (LLM)? https://builtin.com/articles/large-language-models-llm. Accessed 13 Oct 2025
Goldberg A (2008) Universal grammar? Or prerequisites for natural language? Behav Brain Sci 31:522–523
Gu A, Jain N, Li W, Shetty M, Shao Y, Li Z, Yang D, Ellis K, Sen K, Solar-Lezama A (2025) Challenges and paths towards AI for software engineering. Preprint at https://doi.org/10.48550/arXiv.2503.22625
Harnad S (1990) The symbol grounding problem. Phys D 42(1–3):335–346. https://doi.org/10.1016/0167-2789(90)90087-6
Harnad S (2009) On Fodor on Darwin on evolution. Preprint at https://arxiv.org/pdf/0904.1888
Harnad S (2025) Language writ large: LLMs, ChatGPT, meaning, and understanding. Front Artif Intell. Preprint at https://arxiv.org/pdf/2402.02243
Health Insurance Privacy. https://www.hhs.gov/hipaa/for-professionals/privacy/laws-regulations/index.html. Accessed 17 Oct 2025
Henshall W. AI experts call for policy action to avoid extreme risks. Time Magazine. https://time.com/6328111/open-letter-ai-policy-action-avoid-extreme-risks/. Accessed 1 Sept 2025
Henshall W. Yoshua Bengio. Time Magazine. https://time.com/collection/time100-ai/6310612/yoshua-bengio/. Accessed 11 Sept 2025
Hoque E (2021) The universal grammar theory: Noam Chomsky's contribution to second language (SL) education. J EFL Educ Res 6:57–62
Huckins G (2025) Handing AI the keys. MIT Technol Rev 128(4):23–27
Internet Assigned Numbers Authority. https://www.iana.org/. Accessed 27 Sept 2025
Internet Corporation for Assigned Names and Numbers. https://www.icann.org/. Accessed 27 Sept 2025
ISO/IEC 23053:2022(en) framework for artificial intelligence (AI) systems using machine learning (ML). https://www.iso.org/standard/74438.html. Accessed 9 Oct 2025
ISO/IEC 23053:2022(en) framework for artificial intelligence (AI) systems using machine learning (ML). https://www.iso.org/obp/ui/en/#iso:std:iso-iec:23053:ed-1:v1:en. Accessed 8 Oct 2025
ISO/IEC 42001:2023. https://www.iso.org/standard/42001. Accessed 5 Oct 2025
ISO/IEC 23894 information technology—artificial intelligence—guidance on risk management. Reference number ISO/IEC 23894:2023(E). Accessed 5 Aug 2025
ISO. Machine learning (ML): all there is to know. https://www.iso.org/artificial-intelligence/machine-learning. Accessed 13 Oct 2025

iTutorGroup to pay $365,000 to settle EEOC discriminatory hiring suit. https://www.eeoc.gov/newsroom/itutorgroup-pay-365000-settle-eeoc-discriminatory-hiring-suit. Accessed 8 Oct 2025
Kalai AT, Vempala S (2024) Calibrated models must hallucinate. In: Proceedings of the 56th annual ACM symposium on theory of computing (STOC 2024), Vancouver, Canada, 24–28 June 2024
Kalai AT, Nachum O, Vempala SS, Zhang E. Why language models hallucinate. https://arxiv.org/abs/2509.04664. Accessed 1 Oct 2025
Kapronczay M. A beginner's guide to language models, built in. Updated by Matthew Urwin. https://builtin.com/data-science/beginners-guide-language-models. Accessed 13 Oct 2025
Kissel R, Regenscheid A, Scholl M, Stine K. Guidelines for media sanitization. NIST special publication 800-88, revision 1. National Institute of Standards and Technology. https://doi.org/10.6028/NIST.SP.800-88r1. Accessed 18 Sept 2025
Krishner T, Powell H. Tesla shouldn't call driving system autopilot because humans are still in control, Buttigieg says. AP News. https://apnews.com/article/tesla-autopilot-buttigieg-investigation-crash-musk-1455d194b18fe17609554dd68864654c. Accessed 7 Oct 2025
Lewis HR, Papadimitriou CH (1981) Elements of the theory of computation. Prentice-Hall, Englewood Cliffs, New Jersey
Linebaugh K. Why an AI pioneer is worried. Wall Street Journal. https://www.wsj.com/podcasts/the-journal/why-an-ai-pioneer-is-worried/8a11f1da-14e3447f-904a-89818c3e36c7. Accessed 28 Aug 2025
Madiega T. Artificial Intelligence Act. EPRS | European Parliamentary Research Service, Regulation (EU) 2024/1689. Accessed 7 Sept 2025
Malley T. AI HAVE A DEAL driver uses ChatGPT hack to get dealer to agree to sell new car for $1 in 'legally binding deal' in blow for AI rollout. The US Sun. https://www.the-sun.com/motors/9888857/driver-uses-ai-loophole-buy-new-car-1/. Accessed 9 Oct 2025
Mäntymäki M, Minkkinen M, Birkstedt T et al (2022) Defining organizational AI governance. AI Ethics 2:603–609. https://doi.org/10.1007/s43681-022-00143-x
McCarthy J, Minsky ML, Rochester N, Shannon CE (1955) A proposal for the Dartmouth summer research project on artificial intelligence. AI Mag 27(4):12. https://doi.org/10.1609/aimag.v27i4.1904
Merriam-Webster.com Dictionary. Governance. https://www.merriam-webster.com/dictionary/governance. Accessed 17 Oct 2025
Merriam-Webster.com Dictionary. Compliance. https://www.merriam-webster.com/dictionary/compliance. Accessed 17 Oct 2025
Merriam-Webster.com Dictionary. Regulation. https://www.merriam-webster.com/dictionary/regulation. Accessed 17 Oct 2025
Mihalopoulos D. Syndicated content in sun-times special section included AI-generated misinformation. https://chicago.suntimes.com/news/2025/05/20/syndicated-content-sunday-print-sun-times-ai-misinformation. Accessed 8 Oct 2025
MIT AI Risk Repository. https://airisk.mit.edu/. Accessed 5 Sept 2025
Morabito R, Wu G (2025) When the code autopilot breaks: why LLMs falter in embedded machine learning. Preprint at https://arxiv.org/pdf/2509.10946
Mucci T, Styker C. What is AI governance? https://www.ibm.com/think/topics/ai-governance. Accessed 17 Oct 2025
National Highway Transportation Safety Administration. Risk-based processes for safety defect analysis and management of recalls. Report No. DOT HS 812 984. Accessed 9 Oct 2025
National Highway Transportation Safety Association [NHTSA1]. Additional information regarding EA22002, investigation: EA22002. https://static.nhtsa.gov/odi/inv/2022/INCR-EA22002-14496.pdf. Accessed 12 Aug 2024
National Institute of Standards and Technology. AI risk management framework concept paper. https://doi.org/10.6028/NIST.AI.100-1. Accessed 17 Sept 2025
National Institute of Standards and Technology. Artificial intelligence risk management framework (AI RMF 1.0), NIST AI 100-1. https://doi.org/10.6028/NIST.AI.100-1. Accessed 17 Sept 2025

National Transportation Safety Board. Investigation of lion air flight 610 and Ethiopian airlines flight 302. https://www.ntsb.gov/investigations/Pages/DCA19RA017-DCA19RA101.aspx. Accessed 15 Sept 2025

National Transportation Safety Board. In-flight separation of left mid exit door plug. Alaska airlines flight 1282, Boeing 737-9, N704AL. https://www.ntsb.gov/investigations/Pages/DCA24MA063.aspx. Accessed 16 Sept 2025

Noam Chomsky on artificial intelligence, language and cognition. https://youtu.be/_7AE7UuOfg0?si=75ffKOWS7q4iWSt0. Accessed 8 Oct 2025

Nolan B. An AI-powered coding tool wiped out a software company's database, then apologized for a 'catastrophic failure on my part.' Fortune Magazine. https://fortune.com/2025/07/23/ai-coding-tool-replit-wiped-database-called-it-a-catastrophic-failure/. Accessed 8 Oct 2025

Nolan B. 'Godfather of AI' says AI is like a cute tiger cub—unless you know it won't turn on you, you should worry. Fortune Magazine. https://fortune.com/article/geoffrey-hinton-ai-godfather-tiger-cub/. Accessed 15 Oct 2025

OECD. AI principles overview. https://oecd.ai/en/ai-principles. Accessed 18 Oct 2025

OECD. Better policies for better lives. https://www.oecd.org/en/about.html. Accessed 18 Oct 2025

Office of the Central Cyberspace Affairs Commission, Cyberspace Administration of China. Interim measures for the management of generative artificial intelligence services on generative AI. https://www.cac.gov.cn/2023-07/13/c_1690898327029107.htm. Accessed 21 Sept 2025

Office of the Central Cyberspace Affairs Commission, Cyberspace Administration of China. Notice on issuing the measures for the identification of artificial intelligence generated synthetic content. https://www.cac.gov.cn/2025-03/14/c_1743654684782215.htm. Accessed 22 Sept 2025

Olavsrud T. 11 famous AI disasters. CIO Magazine. https://www.cio.com/article/190888/5-famous-analytics-and-ai-disasters.html. Accessed 8 Oct 2025

Papagiannidis E, Mikalef P, Conboy K (2025) Responsible artificial intelligence governance: a review and research framework. J Strateg Inf Syst 34(2). https://doi.org/10.1016/j.jsis.2024.101885

Pause giant AI experiments: an open letter. https://futureoflife.org/open-letter/pause-giant-ai-experiments/. Accessed 28 Sept 2025

Piroumian V (1997) Internationalization support in Java. IEEE Micro 17(3):20–29. https://doi.org/10.1109/40.591651

Polanyi M (1958) Personal knowledge: towards a post-critical philosophy. The University of Chicago Press, Chicago

Polanyi M (1966) The tacit dimension. Doubleday and Company, Inc., Garden City, New York

Prentice S. Cyber security headlines week in review: deepfake targets Wiz, Black Basta leverages teams, Russia's Linux plans. CISO series. https://cisoseries.com/cyber-security-headlines-week-in-review-deepfake-targets-wiz-black-basta-leverages-teams-russias-linux-plans/. Accessed 9 Oct 2025

Pullam B, Scholz B (2002) Empirical assessment of stimulus poverty arguments. Linguist Rev 19(1–2):9–50. https://doi.org/10.1515/tlir.19.1-2.9

Rich E (2008) Automata, computability and complexity: theory and applications. Pearson, Prentice-Hall, Englewood Cliffs, New Jersey

Roberts M, Driggs D, Thorpe M et al (2021) Common pitfalls and recommendations for using machine learning to detect and prognosticate for COVID-19 using chest radiographs and CT scans. Nat Mach Intell 3:199–217. https://doi.org/10.1038/s42256-021-00307-0

Schrijver L (ed) (2021) The tacit dimension: architecture knowledge and scientific research. Leuven University Press, Leuven, Belgium

Skinner BF (1976) About behaviorism. Vintage Books, New York

Stilgoe J (2023) We need a Weizenbaum test for AI. Science 381(6658). https://doi.org/10.1126/science.adk0176

Stöffelbauer A. How large language models work. https://medium.com/data-science-at-microsoft/how-large-language-models-work-91c362f5b78f. Accessed 25 Oct 2025

Tesla faces new federal probe into delayed accident reporting. https://www.consumeraffairs.com/news/tesla-faces-new-federal-probe-into-delayed-accident-reporting-082225.html. Accessed 7 Oct 2025
The EU AI Act: a quick guide. https://www.simmons-simmons.com/en/publications/clyimpowh000ouxgkw1oidakk/the-eu-ai-act-a-quick-guide. Accessed 21 Sept 2025
Turing AM (1950) Computing machinery and intelligence. Mind New Ser 59(236):433–460
United States Federal Aviation Administration. Designated engineering representatives (DER). https://www.faa.gov/other_visit/aviation_industry/designees_delegations/individual_designees/der. Accessed 14 Sept 2025
U.S. Department of Transportation, Federal Aviation Administration. National policy. Order 8000.95D. Designee management policy. https://www.faa.gov/other_visit/aviation_industry/designees_delegations/individual_designees/der. Accessed 14 Sept 2025
Violino B. Agentic AI: a CISO's security nightmare in the making?, 2 Sept 2025. https://www.csoonline.com/article/4047974/agentic-ai-a-cisos-security-nightmare-in-the-making.html. Accessed 8 Sept 2025
Weizenbaum J (1972) On the impact of the computer on society. Science 176:609–614. https://doi.org/10.1126/science.176.4035.609
What is GDPR, the EU's new data protection law? https://gdpr.eu/what-is-gdpr/. Accessed 17 Oct 2025
What is ISO/IEC 42001:2023? https://iso-docs.com/blogs/iso-42001-artificial-intelligence-management-system-aims/what-is-iso-iec-42001-2023. Accessed 23 Sept 2025
Why invest in AI ethics and governance? https://www.ibm.com/thought-leadership/institute-business-value/report/roi-ai-ethics. Accessed 7 Oct 2025
Why is AI governance important? https://www.ibm.com/think/topics/ai-governance. Accessed 6 Oct 2025
Wikipedia. Governance. https://en.wikipedia.org/wiki/Governance. Accessed 12 Oct 2025
Zilber A. Sky-high prices: Colorado man hit with $19,824 airport parking fee after system glitch. New York Post. https://nypost.com/2025/10/04/us-news/colorado-man-hit-with-19824-dia-parking-fee-after-system-glitch/. Accessed 10 Oct 2025
Zwass V (2025) Information system. Encyclopedia Britannica, 8 Oct 2025. https://www.britannica.com/topic/information-system. Accessed 12 Oct 2025

Looking Forward: Challenges and Opportunities in Agentic AI Reliability

Liudong Xing and Janet Lin

Abstract This chapter presents perspectives for challenges and future development in building reliable AI systems, particularly, agentic AI systems. Several open research problems related to mitigating the risks of cascading failures are discussed. The chapter also sheds lights on research challenges and opportunities in aspects including dynamic environments, inconsistent task execution, unpredictable emergent behaviors, as well as resource-intensive reliability mechanisms. In addition, several research directions along the line of testing and evaluating reliability of agentic AI systems are also discussed.

Keywords Adaptive redundancy · Agentic AI · Cascading failure · Dynamic environment · Emergent behavior · Reliability

1 Introduction

The AI conversation can be traced as far back as Alan Turing's milestone paper published in 1950, which considered the fundamental question "Can machines think?" (Turing 1950). In 1956, AI got its name and mission as a scientific field at the first AI conference held at Dartmouth College (Delipetrev et al. 2020). Following AI's foundational period in the 1950s–1970s, AI has evolved from early rule-based systems (1970s–1990s), through classical machine learning and deep learning with neural networks (1990s–2020s), to today's generative and agentic AI systems (since 2010s). Correspondingly, as a vital requirement of these systems, the reliability concept and concerns are also evolving, particularly in the interpretation of "required

L. Xing (✉)
Electrical and Computer Engineering Department, University of Massachusetts, Dartmouth, USA
e-mail: lxing@umassd.edu

J. Lin
Luleå University of Technology, Luleå, Sweden

Halmstad University, Halmstad, Sweden

A. Stavrou et al. (eds.), *Generative and Agentic AI Reliability: Architectures, Challenges, and Trust for Autonomous Systems*, Studies in Computational Intelligence 1272, https://doi.org/10.1007/978-3-032-18585-3_14

Fig. 1 Research challenges and opportunities in AI reliability

function" (see Table 1 in Chap. 10), based on the definition in standards like ISO 8402 "*The ability of an item to perform a required function, under given environmental and operational conditions and for a stated period of time*". While a conventional AI system is concerned with providing stable and accurate classifications, predictions, or optimizations, a reliable generative AI system focuses on producing outputs that are trustworthy, consistent, safe, and contextually appropriate (Joshi 2025). Building on both, a reliable agentic AI system should additionally conduct functions of reasoning, goal alignment, planning, safe adaption and interaction in dynamic and collaborative multi-agent contexts.

The expansion of reliability concepts has introduced new challenges and research opportunities, as exemplified in Fig. 1. In the following sections, we shed lights on these challenges and opportunities in building reliable AI systems, particularly, agentic AI systems.

2 Cascading Failures

Cascading failures abound in diverse application domains and have been extensively studied for power grids (Hu et al. 2017; Liao et al. 2017; Liu and Tse 2019) and various Internet of Things (IoT) systems (Fu et al. 2025; Xing 2024; Zheng et al. 2025). Physical or logical interdependence typically exists among components in these systems, where the malfunction of one component may trigger a cascade of unexpected and often undesired behavior or state changes of other components, prompting domino chain effects. Without effective and timely control and mitigation mechanisms, cascading failures can cause catastrophic consequences (Xing 2021). As AI systems grow more complex, they become increasingly vulnerable to cascading failures due to intra-layer and inter-layer interacting components, opaque decision-makings, as well as insufficient cross-layer visibility where assumptions, errors, state changes, or unexpected behavior in one layer may not be adequately detected and mitigated or communicated with other layers.

Specifically, in Chap. 10, an eleven-layer failure stack is presented to organize potential vulnerabilities from the physical and computational foundational layers (hardware, power, system software, AI frameworks), to core intelligence layers (models, data), to operational layers (applications, execution, monitoring, learning), and to the top agentic layer. Cascading failures may happen at the same layer where, for example, errors in one agent can cascade across dependent agents incurring system-wide disruptions. They can also occur vertically, propagating upward or downward across different layers. For example, poor agentic choices can exacerbate or amplify lower-layer vulnerabilities, spreading failures downward. Since upper-layer functions are inherently contingent on the integrity and reliability of lower layers, upward cascades may occur more likely. For example, errors or bugs in system software can render the AI framework (libraries and pipelines) unusable; hardware malfunctions, unstable learning, and biased dataset can destabilize autonomy and compromise decision making at the agentic layer.

Correlated vulnerabilities driving cascading failures highlight that AI system reliability cannot be ensured in isolation at individual layers. Single-layer robustness is no longer sufficient; instead, a cross-layer framework with systematic coordination is essential. The development of such a framework involves several key research problems, as illustrated in Fig. 2 and explained below, each of which presents significant opportunities for advancing reliability and resilience of agentic AI systems.

Cross-Layer Interdependency Mapping: To enable systematic coordination, it is pivotal to explicitly map dependencies across different layers in the stack through, for example, tracing how upper layers are functionally dependent on lower layers

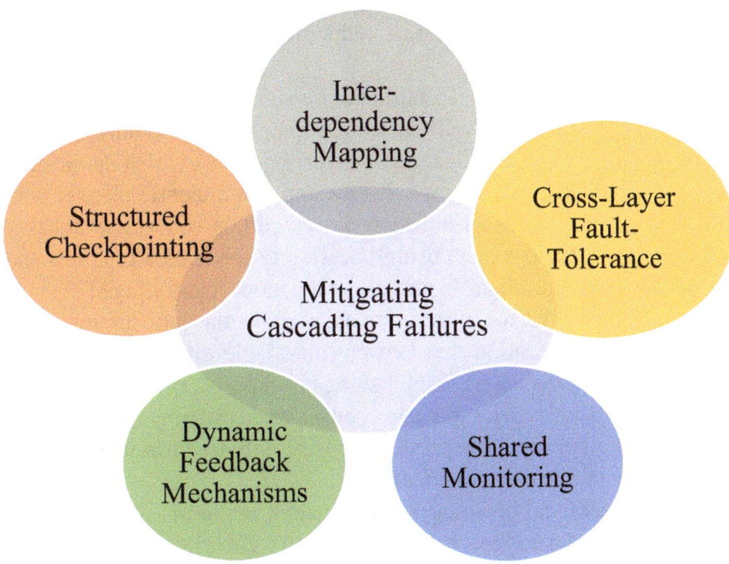

Fig. 2 Key open research problems in mitigating cascading failures

and identifying potential propagation paths across the hierarchy. The goal of such mapping is to enable targeted and strategic interventions and facilitate early detection of vulnerable points.

Dynamic Feedback Mechanisms: Agentic AI systems need to learn and adapt in real-time and thus must incorporate dynamic feedback loop mechanisms to help the system stay resilient in changing or adversarial environments. The mechanisms should flow both upward and downward across the hierarchy. Specifically, a higher layer should be able to flag concerns about the performance of an AI model that may prompt data correction or re-training. A lower layer should adapt its behavior in response to evolving objectives or corrective feedback signals provided by higher layers.

Cross-Layer Shared Monitoring: Unlike traditional monitoring that primarily focuses on observing performance of individual components or layers, the monitoring in the cross-layer approach should be able to detect correlated anomalies spanning multiple layers that would otherwise remain hidden in the case of each layer being monitored separately. Additionally, it is vital to collect telemetry that can capture behavioral dependencies across layers, for example, how data drift impacts model confidence and subsequently affects decision-making patterns. To surface correlated anomalies and reveal cross-layer interactions, shared observability platforms must be developed.

Cross-Layer Fault-Tolerance: Redundancy (hardware, software, information and time) has long been used in traditional systems to contain and tolerate failures. Likewise, AI systems should implement both intra-layer and cross-layer redundancy to isolate faults and limit the impact of cascading failures. For example, redundant reasoning might be implemented at the agentic layer to cross-validate conclusions using different independent mechanisms, reducing reliance on any single, potentially unreliable data source.

Structured Checkpointing: To mitigate the risk of flawed or erroneous inputs spreading upward unchecked, well-structured checkpoints could be implemented to assess the health status at critical cross-layer junctures, which can trigger layer-specific or system-wide interventions. For example, an agentic planner should not trust the output of an AI model unless it passes certain robustness checks. These checks may include, for instance, using uncertainty thresholds to assess the confidence level of model predictions. In the case of the model exhibiting high uncertainty, the output may be flagged for human engagement, or the fallback mechanism can be triggered. Such intervention may prevent unreliable inferences from impacting higher-layer actions or decisions. Establishing checkpoints coupled with robustness checks form a defensive barrier that can prevent untrustworthy outputs at one layer from spreading to other layers, especially upward into decision-making layers with high-stake or irreversible consequences.

In addition to technical challenges and research opportunities explained above, effective coordination among teams working on different layers is essential to align on design assumptions, potential failure modes, and risk models. System architecture should facilitate such integration, avoiding rigid modular boundaries that conceal cross-layer dependencies and cascading failure risks.

3 Vulnerability in Dynamic Environments

Characterized by autonomy, goal-directed behavior, and the capability of acting in open-ended contexts, agentic AI systems can be unpacked in terms of several interrelated challenges arising from the dynamic environments (Allam et al. 2025; Bandi et al. 2025), as summarized in Fig. 3.

Adapting to Unexpected or Unpredictable Contexts: Due to novel objects introduced, revised rules or policies, or emergent behaviors, an agentic AI system can misinterpret its surroundings, fail to generalize rules or learned behaviors to new conditions, or make unsafe decisions in the event of edge cases. The unexpected changes in real-world conditions require re-adaptation that the AI system was not explicitly trained for.

Ambiguous Inputs: Instructions or prompts given by human users are often incomplete, ambiguous or even contradictory. The interpretation of such prompts requires deep context awareness and often relies heavily on human intent and norms, which AI systems do not inherently possess. As a result, AI systems may interpret ambiguous prompts literally, leading to actions that are technically correct but not aligned with human intention. Even worse, the system may fill in missing information by generating incorrect content, which can manifest as hallucinations or inappropriate behaviors.

Adversarial Prompts: In addition to ambiguous and underspecified prompts, agentic AI systems can be susceptible to malicious prompts or inputs, which are crafted to elicit unintended behavior or bypass safeguards. Even in non-malicious contexts, misaligned or harmful behaviors may be triggered based on how the AI interprets and extrapolates from its input.

Compounding Errors: Because agentic AI systems are often tasked with long-horizon, multi-step decision-making in dynamic environments, they are prone to compounding errors. Specifically, a misinterpretation or a small mistake occurring

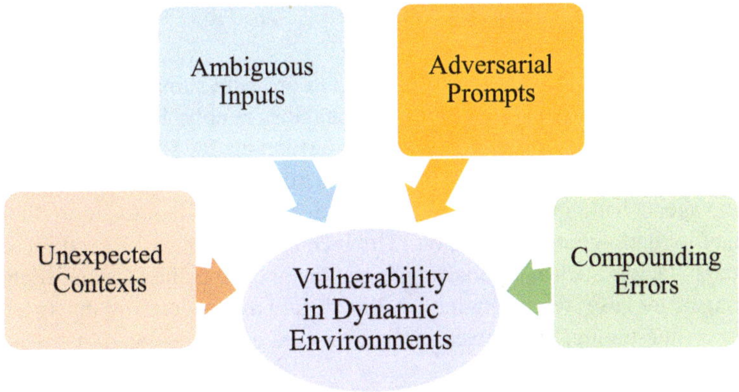

Fig. 3 Challenges in dynamic environments

in an early step can cascade or accumulate, leading to compounding failures even without the user's awareness.

To address the above challenges of agentic AI systems in the face of novelty, ambiguity, and maliciousness, potential research opportunities in causal modeling and reasoning, intent modeling, explainability and transparency, and robustness to shifts and adversarial inputs could be explored, contributing to more robust, transparent, and adaptable decision-making in dynamic environments.

More specifically, it should be pivotal to equip an agentic AI system with the capability of reasoning about cause-effect relationships, rather than just pattern recognition and matching. With such capability, agents could predict the outcomes of their actions and adapt policies to changing environments to prevent or mitigate unintended consequences. To achieve these, future research may explore causal discovery (Pearl 2011; Spirtes et al. 2001) in unstructured and high-dimensional data like language and perception and integrate causal models into the decision-making pipeline. Moreover, it is important to enable AI agents to reason about human preferences and beliefs even when they are under-specified; pragmatic reasoning and social reasoning could be explored to infer human intent, avoiding misaligned or harmful behaviors. Another research path could involve making the reasoning and decision processes of AI agents transparent and understandable to human users through, for example, explicit chain-of-thought traces in multi-step reasoning, dynamic dialogue between users and agents, and introspection where an agents can explain its plans and goals.

To ensure that an agentic AI system remains reliable under changing or new conditions, research questions such as out-of-distribution detection and incorporating epistemic uncertainty into action selection and planning must be addressed. To build the system's resilience to adversarial inputs, it should be promising to explore automated red-teaming to discover edge-case vulnerabilities and develop context-sensitive sandboxing strategies for managing high-risk AI behaviors.

4 Inconsistent Task Execution

Agentic AI systems are expected to plan and coordinate subtasks (possibly across multiple agents) to produce goal-aligned behavior reliably. However, in practice, they often display variability in their outputs (Atil et al. 2025; Fournier et al. 2025; Ouyang et al. 2025). Specifically, outputs vary across repeated runs for the same task or among agents handling similar subtasks, undermining reliability, reproducibility, predictability and agent coordination. The inconsistency in task execution can be attributed to factors such as stochastic generation, underspecified task prompts, non-deterministic tool use (search engines, APIs), drift in model behavior due to adaptive learning or fine-tuning, and unclear policies for executing repeated or overlapping tasks.

To ensure consistent task execution in agentic AI systems, a promising research direction is the development of integrated methods that encompass grounded reasoning, explicit action-validation cycles, and stabilized memory (Fig. 4). Such

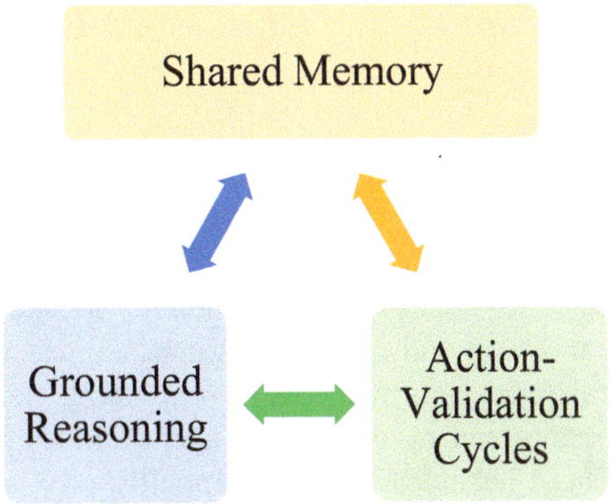

Fig. 4 Integrated methodology for addressing inconsistency

a combined solution method could lead to more reliable, reproducible, and trustworthy AI agents, promoting consistency in agent behaviors, actions and decisions. Its three interrelated components are elaborated below.

Grounded Reasoning: Through integrating external knowledge retrieval (documents, policies, prior actions or decisions) into the reasoning pipeline of AI agents, outputs can be grounded in static and verifiable sources rather than relying solely on stochastic generation or internal memory. More specifically, the grounded reasoning can be achieved by retrieving relevant context from indexed knowledge documents or sources, injecting the retrieved context directly into agent prompts or planning modules, and integrating it with memory recall to ensure coherence across task executions.

Action-Validation Cycles: Instead of generating a one-shot response, AI agents conduct reasoning and actions in iterative and reflective loops, where checkpoints are introduced to allow an agent to review its state and output, or make alignment with past decisions or outputs that can be retrieved at each step. In the case of inconsistencies being detected, the step may be retried. Such self-validation and dynamic plan adjustments could lead to improved state consistencies across iterations.

Stabilized and Shared Memory: To enable ground reasoning and action-validation loops, a stabilized memory mechanism is essential, which supports the retrieval of past assumptions, plans or decisions within each loop and the examination of causal relationship between past and current reasoning steps. The mechanism should also facilitate coordination among multiple agents via context sharing; the agents should be able to access common retrieval indexes and shared memory, ensuring they execute tasks from the same base of knowledge and thereby enabling task reproducibility and cross-agent consistency.

5 Unpredictable Emergent Behavior

In agentic AI systems, multiple autonomous agents often need to interact in complex and open-ended environments, potentially leading to unexpected or misaligned system-level outcomes, even when each individual agent behaves as intended (Garg 2025; Murugesan 2025). More specifically, different agents are autonomous, pursuing their goals independently in shared environments. However, they are also interactive and adaptive during task executions, leading to changes to the shared environments. Such interaction effects can create behaviors that are not intended or foreseen by any single agent and may not emerge until real-time executions. These behaviors are not explicitly programmed, are often resistant to traditional alignment or debugging tools, cannot be easily predicted based on local rules, and are often difficult to reverse once in motion. Consequently, they can be harmful and misaligned with human intent or values. Examples include destructive competition over shared resources, system-level goal mis-generalization, feedback loops amplifying small effects into catastrophic failures, unintended cooperation or collusion undermining fairness and trust, and cross-agent misalignment causing goal divergence and non-traceable harmful behavior.

Addressing emergent behaviors needs to overcome several difficulties in nonlinearity, lack of global perspective, and scale effects. To elaborate, the global behavior of the system cannot be fully understood by examining its individual agents in isolation as it is more than the sum of its parts. Small, insignificant changes in one module can cascade, leading to nonlinear and often unpredictable shifts in overall system behavior. Also, each agent acts based on its local goals, observations, and rules; no single agent has a global view of the entire system, making cross-agent coordination fragile. Additionally, some emergent behaviors only manifest at scale during real-time simulation or deployment, making them difficult to be detected during the development stage.

To mitigate the risk of unpredictable emergent behaviors, there is a need to treat a multi-agent AI system as a complex ecosystem with system-level monitoring and causal tracing, shared ontologies for inter-agent communications, coordination protocols for preventing conflicting behaviors, and collective incentive design for ensuring system-level alignment. Along this research direction, it is also significant to implement redundancy, fallback strategies, and self-healing mechanisms to manage unexpected emergent risks. In particular, the development of such an ecosystem involves addressing key open research problems, as outlined in Fig. 5 and detailed below.

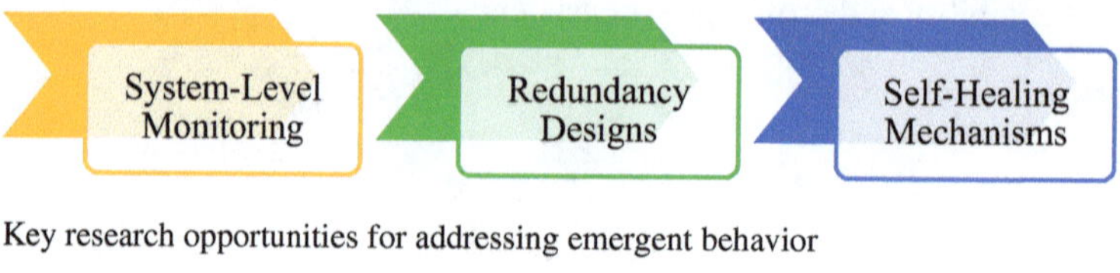

Fig. 5 Key research opportunities for addressing emergent behavior

System-Level Monitoring: The goal is to design real-time monitoring systems that can observe and evaluate agent behaviors both individually and collectively, and trigger interrupts to contain damage when anomalies are detected. Possible research problems include detecting emergent patterns or behavioral deviation from expected norms; summarizing multi-agent states through aggregating and interpreting metrics that reflect system-wide behaviors like resource usage, quality of coordination, and divergence from expected plans; designing safety–critical interrupt signals that can override local agent autonomy, automatically halt or adjust agent behaviors, or trigger system rollback or fallback strategies in the event of anomalies or failures being detected.

Redundancy Designs: The goal is to ensure that the system can continue to function even in the presence of agent or planning module failures. Potential research problems include designing passive and active redundancy techniques (Johnson 1989) that can restore system functionality in failure cases; and developing graceful degradation techniques that enable agents to continue their functionality but with degraded performance through, for example, shifting to more conservative or simpler policies when optimal performance cannot be maintained under unexpected stresses.

Self-Healing Mechanisms: The goal is to design mechanisms for post-failure correction or repair. Related research problems include designing self-correction mechanisms that can reverse recent actions causing the harmful or error states; developing rollback strategies that can revert to a previous safe state or decision path; and investigating distributed recovery agents capable of watching and repairing coordination failures.

6 Resource-Intensive Reliability Mechanisms

To ensure reliability and mitigate risks such as cascading failures and emergent behaviors discussed in prior sections, reliability mechanisms are essential components of AI systems where redundancies in the form of hardware, software, information, and time (Johnson 1989) are built in to prevent failures or harmful outcomes. However, these mechanisms are resource-intensive, particularly involving high computational overhead, energy consumption and infrastructure cost. Extra computational overhead arises from simultaneously running multiple redundant processes or models that consumes more GPU/CPU cycles and performing frequent consistency checks and validation steps. The excessive energy consumption results from the operation of additional hardware resources as well as continuous monitoring and fail-safe mechanisms, which drives up operational costs. Moreover, the extra resources (servers, data centers) increase the demand for cooling and maintenance, thereby raising overall infrastructure costs as well. All these factors can make AI systems expensive and less accessible, especially in resource-limited settings. Figure 6 outlines several potential mitigation strategies to address the resource-intensive reliability challenge in agentic AI systems, which are explained below.

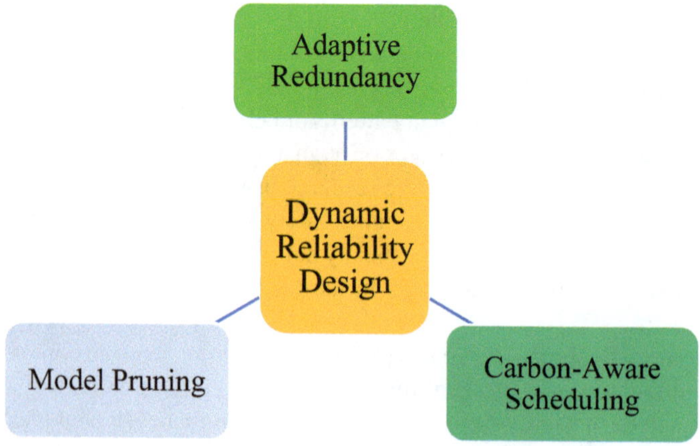

Fig. 6 Key research components of resource-aware reliability design

Adaptive Redundancy: Instead of maintaining uniform redundancy and reliability-related checks, the redundancy mechanism can be dynamically scaled and adjusted based on real-time risk levels and energy availability. In the case of low risk (e.g., low-stakes task, stable environment) or limited energy, the level of redundancy or frequency of checks can be reduced to save energy consumption and operational cost. In the case of high risk (e.g., uncertain conditions, critical and complex operations) or energy-rich environment, reliability measures can be ramped up.

Model Pruning: By maintaining robustness in critical components, redundant or non-essential parts of the AI model may be removed or simplified (e.g., pruning less important neurons in neural networks) (Jia et al. 2025; Mohanty et al. 2025; Yeom et al. 2021) to lower computational overhead with minimal impact on overall system reliability.

Carbon-Aware Scheduling Strategies: To reduce environmental impact without significantly impacting overall reliability, energy-heavy workloads could be scheduled to run when renewable, clean energy (e.g., solar, wind) is abundant or available (Hewage et al. 2025). By factoring in carbon intensity in task scheduling, the AI system may also operate more sustainably.

Each component explained above represents a promising research direction, contributing to energy-aware reliability designs that balance reliability and efficiency and making agentic AIs more feasible for energy-constrained environments and more sustainable in general.

Table 1 Differences between traditional and agentic AI systems

Criterion	Traditional systems	Agentic AI
Determinism	High (same output for the same input)	Low (stochastic behaviors; output may vary on the same prompt)
Action space	Explicitly defined and bounded	Large and dynamic; unexpected emergent behaviors
Reproducibility	High	Low (different plans or paths across runs)
Specification	Clearly defined function requirements	May be ambiguous, underspecified, or emergent
Failure modes	Often discrete and classifiable	Often context-dependent and can be cascading
Evaluation metrics	Objective criteria (accuracy, efficiency)	Ambiguous, may involve human-in-the-loop and multi-dimensional evaluation (task completion, safety compliance, user satisfaction)
Logging and tracing	Easy	Often difficult due to opaque reasoning and interactions with external tools

7 Reliability Testing and Evaluation

While traditional software systems benefit from mature and standardized reliability testing methodologies (e.g., ISO/IEC/IEEE 29119) (Pham 1995; Homès 2024), these techniques fall short when applied to agentic AI systems. The inherent non-determinism and dynamic action space along with other complex aspects of agentic AI systems (summarized in Table 1) demand new testing and evaluation paradigms.

To establish a robust testing and evaluation paradigm for agentic AI systems, several key research directions can be pursued, including.

- Developing a behavioral testing framework where conflicting goals, noisy or adversarial inputs, or unexpected tool failures can be injected to observe and evaluate the system's resilience and adaptability under uncertainty.
- Developing logging and tracing tools that record not only memory reads/writes and tool calls/return values but also reasoning chains and plan/action updates.
- Designing and automating a multi-dimensional reliability evaluation framework that assesses failure resilience, explanation quality, consistency, plan efficiency, and safety compliance.
- Developing and applying formal methods (e.g., model checking) to verify agent plans and behaviors for AI systems in high-assurance application domains like aviation and healthcare.

Acknowledgements This chapter partially draws upon insights presented in Chaps. 1 and 10. We would like to acknowledge the valuable work of these chapter authors whose findings and perspectives have contributed to challenges and future directions in the field offered in this chapter. This work was partially supported by the U.S. National Science Foundation under Grant No. 2302094.

References

Allam H, AlOmar B, Dempere J (2025) Agentic AI for IT and beyond: a qualitative analysis of capabilities, challenges, and governance. Artif Intell Bus Rev 1. Available online: https://theaibr.com/index.php/aibr/article/view/3

Atil B, Aykent S, Chittams A, Fu L, Passonneau RJ, Radcliffe E, Rajagopal GR, Sloan A, Tudrej T, Ture F, Wu Z, Xu L, Baldwin B (2025) Non-determinism of "deterministic" LLM settings, https://arxiv.org/abs/2408.04667. arXiv:2408.04667

Bandi A, Kongari B, Naguru R, Pasnoor S, Vilipala SV (2025) The rise of agentic AI: a review of definitions, frameworks, architectures, applications, evaluation metrics, and challenges. Future Internet 17(9):404. https://doi.org/10.3390/fi17090404

Delipetrev B, Tsinaraki C, Kostic U (2020) Historical Evolution of Artificial Intelligence. Technical Report. Publications Office of the European Union

Fournier F, Limonad L, David Y (2025) Agentic AI process observability: discovering behavioral variability. https://doi.org/10.48550/arXiv.2505.20127

Fu X, Zheng D, Liu X, Xing L, Peng R (2025) Systematic review and future perspectives on cascading failures in Internet of Things: modeling and optimization. Reliab Eng Syst Safety 254(Part A) 110582

Garg V (2025) Designing the mind: how agentic frameworks are shaping the future of AI behavior. J Comput Sci Technol Stud 7(5):182–193. https://doi.org/10.32996/jcsts.2025.7.5.24

Hewage TB, Ilager S, Rodriguez MA, Buyya R (2025) A Framework for carbon-aware real-time workload management in clouds using renewables-driven cores. IEEE Trans Comput 74(8):2757–2771

Homès B (2024) Fundamentals of software testing, 2nd edn. Wiley

Hu P, Mei T, Fan W (2017) Cascading failure forecast of complex power grids: a review. Scientia Sinica Technol 47(4):355–363

Jia Y, Liu B, Zhang X, Dai F, Khan A, Qi L, Dou W (2025) Model pruning-enabled federated split learning for resource-constrained devices in artificial intelligence empowered edge computing environment. ACM Trans Sensor Netw in Press,. https://doi.org/10.1145/3687478

Johnson BW (1989) Design and analysis of fault tolerant digital systems. Addison-Wesley

Joshi S (2025) Model risk management in the era of generative AI: challenges, opportunities, and future directions. Int J Sci Res Publ 15(5):299–309. https://doi.org/10.29322/ijsrp.15.05.2025.p16133

Liao W, Salinas S, Li M, Li P, Loparo K (2017) Cascading failure attacks in the power system: a stochastic game perspective. IEEE Internet Things J 4(6):2247–2259

Liu D, Tse CK (2019) Cascading failure of cyber-coupled power systems considering interactions between attack and defense. IEEE Trans Circuits Syst I Regul Pap 66(11):4323–4336. https://doi.org/10.1109/TCSI.2019.2922371

Mohanty L, Kumar A, Mehta V et al (2025) Pruning techniques for artificial intelligence networks: a deeper look at their engineering design and bias: the first review of its kind. Multimed Tools Appl 84:9591–9665. https://doi.org/10.1007/s11042-024-19192-x

Murugesan S (2025) The rise of agentic AI: implications, concerns, and the path forward. IEEE Intell Syst 40(2):8–14. https://doi.org/10.1109/MIS.2025.3544940

Ouyang S, Zhang JM, Harman M, Wang M (2025) An empirical study of the non-determinism of chatgpt in code generation. ACM Trans Softw Eng Methodol 34. https://doi.org/10.1145/3697010

Pearl J (2011) Causality: models, reasoning, and inference, second edition. https://doi.org/10.1017/CBO9780511803161

Pham H (1995) Software reliability and testing. IEEE Computer Society Press, Washington, DC, United States

Spirtes P, Glymour C, Scheines R (2001) Causation, prediction, and search. The MIT Press. https://direct.mit.edu/books/book/2057/causation-prediction-and-search. https://doi.org/10.7551/mitpress/1754.001.0001

Turing AM (1950) Computing machinery and intelligence. Mind 49:433–460
Xing L (2021) Cascading failures in internet of things: review and perspectives on reliability and resilience. IEEE Internet Things J 8(1):44–64
Xing L (2024) Reliability and resilience in the internet of things. Elsevier
Yeom S, Seegerer P, Lapuschkin S, Binder A, Wiedemann S, Müller K, Samek W (2021) Pruning by explaining: a novel criterion for deep neural network pruning. Pattern Recogn 115:107899. https://doi.org/10.1016/j.patcog.2021.107899
Zheng D, Fu X, Liu X, Xing L, Peng R (2025) Modeling and analysis of cascading failures in industrial internet of things considering sensing-control flow and service community. IEEE Trans Reliab 74(2):2723–2737

FSC
www.fsc.org
MIX
Papier | Fördert gute Waldnutzung
FSC® C083411

Kolibri 360 GmbH
Ferdinand-Jühlke-Straße 7
99095 Erfurt, Deutschland
produktsicherheit@kolibri360.de